I0482781

NRC Sensitivity and Uncertainty Analyses for a Proposed HLW Repository at Yucca Mountain, Nevada, Using TPA 3.1

Conceptual Models and Data

U.S. Nuclear Regulatory Commission
Office of Nuclear Material Safety and Safeguards
Washington, DC 20555-0001

Center for Nuclear Waste Regulatory Analyses
Southwest Research Institute

NRC Sensitivity and Uncertainty Analyses for a Proposed HLW Repository at Yucca Mountain, Nevada, Using TPA 3.1

Conceptual Models and Data

Manuscript Completed: Janaury 2001
Date Published: February 2001

Prepared by
S. Mohanty, CNWRA
T.J. McCartin, NRC

Center for Nuclear Waste Regulatory Analyses
Southwest Research Institute
6220 Culebra Road
San Antonio, TX 78238-5166

Division of Waste Management
Office of Nuclear Material Safety and Safeguards
U.S. Nuclear Regulatory Commission
Washington, DC 20555-0001

ABSTRACT

Total system performance assessment (TSPA) is playing an increasingly vital role in regulatory decision-making. Within the U.S. Nuclear Regulatory Commission (NRC) high-level waste program, TSPA studies are being performed to focus the activities of the NRC key technical issues (KTIs); evaluate the hypotheses of the U.S. Department of Energy (DOE) Repository Safety Strategy; and prepare for the NRC review of the DOE viability assessment (VA) for the Yucca Mountain (YM).

Conducting a TSPA for the proposed repository involves application of a total-system model that simulates the processes affecting repository performance, including propagation of the uncertainties associated with model parameters, conceptual models, and future system states. The simulation process, which implements a probabilistic framework, integrates a broad spectrum of site-specific data and information and produces estimates for a set of performance measures. Building on the previously developed NRC assessment methodology, a new Total-system Performance Assessment (TPA) code, designated TPA Version 3.1.4 code, was developed by NRC and the Center for Nuclear Waste Regulatory Analyses to evaluate the relative significance of the NRC-identified KTIs/subissues and to assess the assumptions and models in forthcoming DOE TSPAs, such as the DOE TSPA VA for the YM site.

The TPA Version 3.1.4 code is designed to estimate total-system performance measures of annual individual dose or risk. The TPA Version 3.1.4 code consists of an executive driver, a set of consequence modules, and a library of utility modules. The executive driver controls the probabilistic sampling of input parameters, the calculational sequence and data transfers among consequence modules, and the generation of output files. The various output files are used in parameter sensitivity analyses, post-processing of time-dependent risk curves, and synthesis of statistical distributions (e.g., cumulative distribution functions and complementary cumulative distribution functions) for appropriate performance measures. Consequence models simulate physical processes and events such as unsaturated zone infiltration; evolution of the near-field thermal-hydrologic environment; failure of waste packages; dissolution and release of waste; transport of waste in the ground-water system; extraction of ground water; and consumption of ground water which, if contaminated, would expose future populations to radiation. In addition to considering climate change, the TPA Version 3.1.4 code is designed to calculate the effects of disruptive events such as faulting, seismicity, and volcanism. Utility modules ensure the consistency of algorithms and data sets used repeatedly by various consequence modules.

This volume has been prepared to describe the methods and assumptions used in the TPA Version 3.1.4 code. It contains descriptions of

- Overall TSPA methodology

- Consequence modules that simulate physical processes and events that affect the release and transport of radionuclides

It should be noted that the TPA Version 3.1.4 code is only an interim version of the code. An improved version will soon supersede it.

CONTENTS

Page

ABSTRACT iii
FIGURES xi
TABLES xii
FOREWORD xiii
ACKNOWLEDGMENTS xv
ACRONYMS xvii
TABLE SHOWING ENGLISH/METRIC SYSTEM CONVERSION FACTORS xix
QUALITY OF DATA AND CODE DEVELOPMENT xxi

1   INTRODUCTION 1-1
    R. Baca, T. McCartin, M. Lee
    1.1   TOTAL-SYSTEM PERFORMANCE ASSESSMENT CODE 1-1
    1.2   DESCRIPTION OF THE PROPOSED REPOSITORY SITE 1-3
    1.3   REPORT CONTENT 1-6

2   OVERVIEW OF THE TOTAL-SYSTEM PERFORMANCE ASSESSMENT
    CONCEPTUAL MODELS 2-1
    T. McCartin, S. Mohanty, R. Baca
    2.1   CONCEPTUALIZATION OF REPOSITORY AND GEOLOGIC SETTING 2-1
    2.2   CONCEPTUAL MODELS IMPLEMENTED IN THE TOTAL-SYSTEM
          PERFORMANCE ASSESSMENT COMPUTER CODE 2-3
          2.2.1   Infiltration and Deep Percolation 2-5
          2.2.2   Near-Field Environment 2-6
          2.2.3   Radionuclide Releases from the Engineered Barrier System 2-6
          2.2.4   Treatment of Aqueous-Phase Transport in the Unsaturated
                  and Saturated Zones 2-8
          2.2.5   Airborne Transport for Direct Releases 2-9
          2.2.6   Exposure Pathways and Reference Biosphere 2-9
    2.3   RADIONUCLIDE INVENTORIES 2-10

3   CONCEPTUAL AND MATHEMATICAL MODELS 3-1
    3.1   UZFLOW MODULE DESCRIPTION 3-2
          S. Stothoff, C. Scherer, R. Janetzke
          3.1.1   Information Flow Within TPA 3-2
          3.1.1.1 Information Supplied to UZFLOW 3-2
          3.1.1.2 Information Provided by UZFLOW 3-2
          3.1.2   Conceptual Model 3-3
          3.1.2.1 Climate Model 3-4
          3.1.2.2 Variation in Climate Change Cycle 3-6
          3.1.2.3 Shallow Infiltration Model 3-7
          3.1.2.4 Deep Percolation Model 3-9
          3.1.3   Assumptions and Conservatism of the UZFLOW Approach 3-11
    3.2   NFENV MODULE DESCRIPTION 3-12
          S. Mohanty, R. Green, G. Rice, S. Stothoff, B. Leslie
          3.2.1   Information Flow Within TPA 3-12

CONTENTS (cont'd)

	3.2.1.1	Information Supplied to NFENV	3-12
	3.2.1.2	Information Provided by NFENV	3-13
	3.2.2	Conceptual Model	3-13
	3.2.2.1	Heat Transfer, Temperature, and Relative Humidity Calculation	3-13
	3.2.2.2	Near-Field Chemical Composition	3-23
	3.2.2.3	Near-Field Ground-water Percolation	3-26
	3.2.3	Assumptions and Conservatism of the NFENV Approach	3-32
3.3		EBSFAIL MODULE DESCRIPTION	3-34
		G. Cragnolino, S. Mohanty, T. Ahn, K. Gruss	
	3.3.1	Information Flow Within TPA	3-35
	3.3.1.1	Information Supplied to EBSFAIL	3-35
	3.3.1.2	Information Provided by EBSFAIL	3-35
	3.3.2	Conceptual Model	3-35
	3.3.2.1	Corrosion and Mechanical Failures	3-35
	3.3.2.2	Dry-Air Oxidation	3-36
	3.3.2.3	Humid-Air Corrosion	3-37
	3.3.2.4	Aqueous Corrosion	3-37
	3.3.2.5	Mechanical Failure	3-40
	3.3.3	Assumptions and Conservatism of the EBSFAIL Approach	3-41
3.4		SEISMO MODULE DESCRIPTION	3-44
		S. Hsiung, J. Stamatakos, S. Mohanty, A.-B. K. Ibrahim	
	3.4.1	Information Flow Within TPA	3-44
	3.4.1.1	Information Supplied to SEISMO	3-44
	3.4.1.2	Information Provided by SEISMO	3-44
	3.4.2	Conceptual Model	3-44
	3.4.2.1	Impact Load and Stress Calculation	3-44
	3.4.2.2	Joint Spacing and Rock Conditions in the Topopah Spring Welded Tuff Unit	3-48
	3.4.3	Assumptions and Conservatism of the SEISMO Approach	3-49
3.5		EBSREL MODULE DESCRIPTION	3-50
		R. Codell, S. Mohanty, T. Ahn	
	3.5.1	Information Flow Within TPA	3-50
	3.5.1.1	Information Supplied to EBSREL	3-50
	3.5.1.2	Information Provided by EBSREL	3-51
	3.5.2	Conceptual Model	3-51
	3.5.2.1	Radionuclide Inventory and Mass Transfer	3-51
	3.5.2.2	Advective Mass Transfer	3-53
	3.5.2.3	Diffusive Mass Transfer	3-54
	3.5.2.4	Waste Package Inventory	3-56
	3.5.2.5	Spent-Fuel Dissolution Rate	3-57
	3.5.2.6	Spent-Fuel Surface Area	3-58
	3.5.2.7	Cladding	3-59
	3.5.2.8	Water Dripping Abstraction	3-59
	3.5.2.9	Computational Approach	3-61
	3.5.3	Assumptions and Conservatism of the EBSREL Approach	3-62

CONTENTS (cont'd)

3.6   UZFT MODULE DESCRIPTION 3-65

T. McCartin, R. Janetzke, N. Coleman

     3.6.1   Information Flow Within TPA 3-65

     3.6.1.1   Information Supplied to UZFT 3-65

     3.6.1.2   Information Provided by UZFT 3-65

     3.6.2   Conceptual Model 3-65

     3.6.2.1   Unsaturated Zone Flow Model 3-66

     3.6.2.2   Unsaturated Transport 3-68

     3.6.2.3   Efficiency of Simulating Flow and Transport 3-70

     3.6.3   Assumptions and Conservatism of the UZFT Approach 3-70

3.7   SZFT MODULE DESCRIPTION 3-73

G. Wittmeyer, R. Rice, R. Janetzke, T. McCartin, J. Winterle

     3.7.1   Information Flow Within TPA 3-73

     3.7.1.1   Information Supplied to SZFT 3-73

     3.7.1.2   Information Provided by SZFT 3-75

     3.7.2   Conceptual Model 3-75

     3.7.2.1   Saturated Zone Flow Model 3-75

     3.7.2.2   Saturated Zone Transport Model 3-76

     3.7.3   Assumptions and Conservatism of the SZFT Approach 3-77

3.8   DCAGW MODULE DESCRIPTION—FARMING RECEPTOR GROUP 3-79

P. LaPlante, G. Wittmeyer, M. Jarzemba, S. Mohanty, R. Rice, R. Fedors,
C. McKenney, L. Deere

     3.8.1   Information Flow Within TPA 3-79

     3.8.1.1   Information Supplied to DCAGW 3-79

     3.8.1.2   Information Provided by DCAGW 3-79

     3.8.2   Conceptual Model 3-80

     3.8.2.1   Development of Radionuclide Concentrations in
                Water for a Farming Receptor Group 3-80

     3.8.2.2   Development of Dose Conversion Factors 3-81

     3.8.3   Assumptions and Conservatism of the DCAGW Approach 3-82

3.9   DCAGW MODULE DESCRIPTION—RESIDENTIAL RECEPTOR GROUP 3-86

P. LaPlante, G. Wittmeyer, M. Jarzemba, S. Mohanty, R. Rice, R. Fedors,
C. McKenney, L. Deere

     3.9.1   Information Flow Within TPA 3-86

     3.9.1.1   Information Supplied to DCAGW 3-86

     3.9.1.2   Information Provided by DCAGW 3-87

     3.9.2   Conceptual Model 3-87

     3.9.2.1   Development of Radionuclide Concentrations in
                Water for a Residential Receptor Group 3-87

     3.9.2.2   Development of Dose Conversion Factors 3-90

     3.9.3   Assumptions and Conservatism of the DCAGW Approach 3-90

3.10   FAULTO MODULE DESCRIPTION 3-91

J. Stamatakos, A. Ghosh, S. Mohanty, A.-B. K. Ibrahim

     3.10.1   Information Flow Within TPA 3-91

     3.10.1.1   Information Supplied to FAULTO 3-91

CONTENTS (cont'd)

		3.10.1.2	Information Provided by FAULTO	3-91
		3.10.2	Conceptual Model	3-92
		3.10.3	Assumptions and Conservatism of the FAULTO Approach	3-93
	3.11		VOLCANO MODULE DESCRIPTION	3-96

B. Hill, C. Connor, J. Trapp, S. Mohanty

		3.11.1	Information Flow Within TPA	3-96
		3.11.1.1	Information Supplied to VOLCANO	3-96
		3.11.1.2	Information Provided by VOLCANO	3-97
		3.11.2	Conceptual Model	3-97
		3.11.3	Assumptions and Conservatism of the VOLCANO Approach	3-99
	3.12		ASHPLUMO MODULE DESCRIPTION	3-100

M. Jarzemba, B. Hill, C. Connor

		3.12.1	Information Flow Within TPA	3-101
		3.12.1.1	Information Supplied to ASHPLUMO	3-101
		3.12.1.2	Information Provided by ASHPLUMO	3-101
		3.12.2	Conceptual Model	3-101
		3.12.3	Assumptions and Conservatism of the ASHPLUMO Approach	3-105
	3.13		ASHRMOVO MODULE DESCRIPTION	3-107

M. Jarzemba

		3.13.1	Information Flow Within TPA	3-107
		3.13.1.1	Information Supplied to ASHRMOVO	3-107
		3.13.1.2	Information Provided by ASHRMOVO	3-107
		3.13.2	Conceptual Model	3-107
		3.13.3	Assumptions and Conservatism of the ASHRMOVO Approach	3-111
	3.14		DCAGS MODULE DESCRIPTION—FARMING RECEPTOR GROUP	3-112

P. LaPlante, M. Jarzemba, C. McKenney, R. Abu-Eid

		3.14.1	Information Flow Within TPA	3-112
		3.14.1.1	Information Supplied to DCAGS	3-112
		3.14.1.2	Information Provided by DCAGS	3-112
		3.14.2	Conceptual Model	3-112
		3.14.3	Assumptions and Conservatism of the DCAGS Approach	3-116
	3.15		DCAGS MODULE DESCRIPTION—RESIDENTIAL RECEPTOR GROUP	3-119

P. LaPlante, M. Jarzemba, C. McKenney, R. Abu-Eid

		3.15.1	Information Flow Within TPA	3-119
		3.15.1.1	Information Supplied to DCAGS	3-119
		3.15.1.2	Information Provided by DCAGS	3-119
		3.15.2	Conceptual Model	3-119
		3.15.3	Assumptions and Conservatism of the DCAGS Approach	3-120
4			FUTURE IMPROVEMENTS TO THE TPA CODE	4-1
5			REFERENCES	5-1

CONTENTS (cont'd)

APPENDICES

APPENDIX A REFERENCE DATA SET
M.R. Byrne; J. Weldy; S. Mohanty; T. McCartin; R. Codell; J. Stamatakos; S. Stothoff; B. Leslie; R. Pabalan; D. Turner; G. Wittmeyer; S. Hsiung; P. LaPlante; G. Cragnolino; B.J. Davis; T. Ahn; R. Green; M. Jarzemba; C. McKenney; W. Murphy; H.L. McKague; B. Hill; C. Connor; N. Coleman; J. Winterle; J. Bradbury; R. Fedors; and A. Ghosh

APPENDIX B EBSREL FLOW FACTOR DERIVATION
R. Codell

FIGURES

Figures		Page
1-1	Proposed repository site	1-5
2-1	Repository system	2-2
2-2	Saturated zone conceptual model representation showing the 20-km (12.4-mi) receptor location boundary and the center lines (dashed) of the four saturated zone streamtubes	2-4
2-3	Cm-243, Cm-244, Cm-245, and Cm-246 decay chains	2-13
3-1	UZFLOW conceptual model	3-3
3-2	Boundary of proposed Yucca Mountain repository	3-5
3-3	Response of *MAI* to changes in *MAP* during a climate cycle: (a) mean climatic signal only and (b) with perturbation	3-10
3-4	Discretization of repository into nine rectangular regions for the NFENV module	3-15
3-5	Mountain-scale heat transfer model with heated rectangular regions	3-16
3-6	Plan view of repository showing emplacement drifts and waste packages	3-17
3-7	Waste package on an emplacement cart in a drift	3-18
3-8	Internals of a large waste package	3-21
3-9	Time snapshots of the vertical distribution of temperature caused by repository heating	3-25
3-10	Conceptualization of drift-scale thermal hydrologic model	3-27
3-11	(a) Waste package equivalent beam and (b) waste package vertical supports for SEISMO	3-45
3-12	Schematic of bathtub model with incoming and outgoing water conduits	3-52
3-13	Schematic drawing for advective and diffusive mass transfer from the waste package to the host rock	3-54
3-14	UZFT contaminant transport legs	3-71
3-15	Saturated zone streamtube model showing four streamtubes extending from just upgradient of the repository footprint southward to the major ground-water withdrawal area of Amargosa farms.	3-74
3-16	Farmer exposure pathways	3-83
3-17	Simulated faults within the repository boundary	3-94
3-18	Schematic of subsurface disruption area associated with a volcanic conduit using waste package dimension [5.68 m (18.6 ft) long, 1.8 m (5.9 ft) wide], waste package spacing [19 m (62 ft) center-to-center], and emplacement drift spacing [22 m (72 ft)]	3-98
3-19	ASHPLUMO contaminant release and deposition	3-102
3-20	A plot of the areal radionuclide density as a function of time for radionuclides in the curium-245 decay chain	3-111
3-21	Examples where fraction of resuspended mass that emanated from the contaminated volcanic ash layer (f_R) (a) is unity and (b) is less than unity	3-115

TABLES

Tables   Page

2-1   List of 43 Nuclides in the TPA Version 3.1.4 Code Database 2-11

2-2   Radionuclide Chains For Ground-Water Release 2-12

3-1   Initial Fluid Composition and pH Corresponding to J-13 Well Water 3-24

3-2   Dripping Parameter Correlation Matrix 3-61

3-3   Stratigraphic Thicknesses (m) for Each of the Seven Repository Subareas 3-66

3-4   A Listing of the Nonradionuclide-Specific Data Used in Estimating the Surficial Radionuclide Areal Density for Members of the Cm-245 Decay Chain 3-110

3-5   A Listing of the Radionuclide-Specific Data Used in Estimating the Surficial Radionuclide Areal Density for Members of the Cm-245 Decay Chain 3-110

FOREWORD

In accordance with the provisions of the Nuclear Waste Policy Act of 1982, the U.S. Nuclear Regulatory Commission (NRC) has the responsibility to evaluate a license application for geological repositories constructed for emplacement of high-level nuclear waste (HLW) (i.e., commercial spent fuel and vitrified HLW). This act was amended in 1987 to designate one site for detailed characterization in the unsaturated region of the tuffaceous rocks of Yucca Mountain (YM), in part located on the Nevada Test Site in southern Nevada. To execute its mandated prelicensing and licensing function, the NRC staff will review the critical aspects of the U.S. Department of Energy (DOE) Viability Assessment (VA), any future draft environmental impact statement for the proposed repository site, and any potential license application. In addition to these reviews, NRC will also develop preliminary comments on the sufficiency of the at-depth site characterization program and waste form proposal. Although the DOE VA is not a licensing document, the NRC staff will review the document as a basis for future reviews of the DOE draft license application. A major component of the approach the staff will use to review these products is the Total-system Performance Assessment (TPA).

In support of these review activities, the NRC staff is focusing on detailed technical assessments to understand and quantify the isolation characteristics and capabilities of the proposed repository system at the YM site. Concurrent with these assessment studies, the Center for Nuclear Waste Regulatory Analyses (CNWRA), is enhancing the TPA computer code. The TPA code is designed to simulate the behavior of the geologic repository, taking into account the essential characteristics of the natural barriers and engineered barrier system, as well as the evolutionary changes in the geologic setting (i.e., climate, seismicity, faulting, and volcanism). This document presents the latest version of the TPA code, designated as 3.1.4. This is an interim version of the code. An improved and personal computer-based version is expected to supersede it in the near future.

It is important to point out that the TPA Version 3.1.4 code was developed to allow the NRC and CNWRA staffs to perform interim evaluations of the DOE Total System Performance Assessment approaches and parameter values used to estimate the performance of the proposed HLW repository. Because investigations at the YM site are ongoing and the analyses are iterative, the TPA Version 3.1.4 code was developed with flexibility to analyze a variety of designs and site characterization factors. It is also important to note that the particular conceptual models and assignment of initial model parameter values (and distributions) in this description of the TPA code do not constitute regulatory acceptance. As DOE's ongoing repository investigations develop more information, it can be expected that different conceptual models and parameter values and distributions will be used in performance assessments. Thus, estimates of the performance using the TPA Version 3.1.4 code are to be considered preliminary and to not represent a regulatory determination of total system performance for the YM site.

ACKNOWLEDGMENTS

This NUREG was prepared to document work performed by the U.S. Nuclear Regulatory Commission (NRC) and the Center for Nuclear Waste Regulatory Analyses (CNWRA), for NRC, under Contract No. NRC-02-97-009. The activities reported were performed on behalf of the NRC Office of Nuclear Material Safety and Safeguards, Division of Waste Management. The report is a joint NRC/CNWRA product and does not necessarily reflect the NRC's views or regulatory position.

The authors wish to thank G. Wittmeyer; M.P. Miklas, Jr.; E.C. Pearcy; and N. Sridhar, of the CNWRA, for their thorough technical reviews and numerous helpful technical comments. The authors also thank W. Patrick, B. Sagar, and K. McConnell for the programmatic review and the valuable suggestions for improving the quality of the report. The authors appreciate the thorough editorial review conducted by the staff of the Publications Services Department at Southwest Research Institute—especially the efforts of B. Ford; C. Gray; S. Harley; B. Long; and A. Woods. The authors are grateful for the secretarial support provided by C. Garcia and A. Burne. Thanks are also expressed to K. Poor, of Portage Environmental, Inc.; J. Bogan (consultant); and C. Scherer (Southwest Research Institute), for their assistance in reviewing the document, testing the code, and conducting quality assurance-related tests and preparing related documentation. The authors acknowledge the tests conducted on the initial version of the code by many CNWRA and NRC staff members.

In preparing this document and in developing the TPA Version 3.1.4 code, numerous individuals made strong technical contributions. Assistance from R. Rice (consultant), L. Deere, and P. LaPlante, in preparation of this volume, is gratefully acknowledged. Insofar as was possible, those who contributed to writing the report, developing the conceptual models, and programming and testing the code have been recognized by having their names listed after the appropriate chapter and section headings in the table of contents. The efforts of C. Lui, the Program Element Manager for the Total System Performance Assessment and Integration key technical issue, in coordinating the preparation of this document, are also recognized.

ACRONYMS AND INITIALISMS

1-D	one-dimensional
2-D	two-dimensional
AML	areal mass loading
BWR	boiling-water reactor
CCDF	complementary cumulative distribution function
CDF	cumulative distribution function
CM	consequence module
CNWRA	Center for Nuclear Waste Regulatory Analyses
CP	compliance point
CSR	critical simulation region
DCF	dose conversion factor
DEM	digital elevation model
DOE	Department of Energy
DVM	distributed velocity method
EBS	engineered barrier system
ENFE	evolution of the near-field environment
EPA	U.S. Environmental Protection Agency
EPRI	Electric Power Research Institute
GPD	gallons per day
GWd	gigawatt-day
GWTT	ground-water travel time
HLW	high-level waste
IA	igneous activity
ICRP	International Commission on Radiological Protection
IPA	Iterative Performance Assessment
JS	joint spacing
KTI	key technical issue
LA	license application
LHS	Latin Hypercube Sampling
MAI	mean annual infiltration
MAP	mean annual precipitation
MAT	mean annual temperature
MAV	mean annual atmospheric vapor density
MCS	Monte Carlo Sampling
MTU	metric tons of uranium
NFENV	Near-field environment
NRC	Nuclear Regulatory Commission
PA	performance assessment
PDF	probability density function
PNNL	Pacific Northwest National Laboratory

ACRONYMS AND INITIALISMS (cont'd)

ppm	parts per million
PWR	pressurized-water reactor
QA	quality assurance
RH	relative humidity
RT	radionuclide transport
SF	spent fuel
SZ	saturated zone
TEDE	total effective dose equivalent
TPA	Total-system Performance Assessment
TSPA	total system performance assessment
UDEC	Universal Distinct Element Code
UZ	unsaturated zone
WP	waste package
YM	Yucca Mountain
YMR	Yucca Mountain region

TABLE SHOWING ENGLISH/METRIC SYSTEM CONVERSION FACTORS

The preferred of measurement today is the "System Internationale" or the metric system. However, for some physical quantities, many scientists and engineers prefer the familiar and continue to use the English system (foot-pound units). With few exceptions, all units of measure cited in this report are usually in the metric system. The following table provides the appropriate conversion factors to allow the user to switch between these two systems of measure. Not all units nor methods of conversion are shown. Unit abbreviations are shown in parentheses. All conversion factors are approximate.

QUANTITY	TO INCH-POUND UNITS	FROM METRIC UNITS[1]	CONVERSION FACTOR[2]
SPACE AND TIME			
length	statute mile (mi) foot (ft) inch (in)	kilometer (km) meter (m) centimeter (cm)	0.6214 3.2808 0.3937
area	square mile (mi^2) acre square foot (ft^2) square inch (in^2)	square kilometer (km^2) square kilometer square meter (m^2) square centimeter (cm^2)	0.3861 247.1 10.7639 0.1550
volume	cubic yard (yd^3) cubic foot (ft^3) cubic inch (m^3)	cubic meter (m^3) cubic meter liter (l) centimeter (cm^3)	1.3080 35.3147 0.0353 0.0610
velocity	feet/second (ft/sec)	meters/second (m/sec)	3.2808
acceleration	feet/square second (ft/sec^2)	meters/square second (m/sec^2)	3.2808
MECHANICS			
mass (weight)	pounds (lb) short ton	kilogram (kg) metric ton (t)	2.2046 1.1023
density	pounds/cubic foot (lbs/ft^3)	kilograms/cubic meter (kg/m^3)	0.0624
force	pound-force (lbf)	Newton (N) dyne (dyn)	0.2248 $.2248 \times 10^5$
pressure	atmosphere (atm) pound-force/square foot (lb/ft^2)	kilopascal (kPa) dyne/square centimeter (dyn/cm^2)	0.0099 0.0021
power	horsepower (hp)	kilowatts (kW)	1.3405
work	foot-pound-force (ft-lbf)	joule (J)	0.7376
HEAT			
temperature	degrees Fahrenheit ($^\circ$F)	degrees Celsius ($^\circ$C) Kelvin (K)	$^\circ F = 1.8 ^\circ C + 32$ $^\circ F = 1.8 K - 459.67$
IONIZING RADIATION			
activity (of a radionuclide)	curie (Ci)	megabecquerel (MBq)	2.7027×10^5
absorbed dose	rad	gray (Gy)	100
dose equivalent	rem	sievert (Sv)	100

[1]Not all metric units are shown. Most metric units can be arrived at by multiplying the value by 10^n.

[2]Multiply quantity in metric units by the appropriate conversion factor to obtain inch-pound units. For additional unit conversions. refer to C.J. Pennycuick, *Conversion Factors: SI Units and Many Others*, Chicago. The University of Chicago Press. 1998.

QUALITY OF DATA AND CODE DEVELOPMENT

DATA: CNWRA-generated data contained in this report meet quality assurance (QA) requirements described in the CNWRA QA Manual. Data from other sources, however, are freely used. The respective sources of non-CNWRA data should be consulted for determining levels of QA.

CODE: The TPA Version 3.1.4 code has been developed following the procedures described in the CNWRA Technical Operating Procedure (TOP), TOP-018, which implements the QA guidance contained in the CNWRA QA Manual.

1 INTRODUCTION

To more effectively fulfill its statutory responsibilities for the prelicensing and licensing actions for the proposed high-level waste (HLW) repository at the Yucca Mountain (YM) site, the U.S. Nuclear Regulatory Commission (NRC) has focused its repository program on resolving issues most significant to overall system performance. This has led to a major restructuring of the NRC HLW program, to encompass ten key technical issues (KTIs) (Sagar, 1997). Limiting the program scope to these KTIs was necessary to focus and streamline the regulatory review and decision-making process. In addition, the use of the KTI framework provides a vehicle for ensuring early, clear, and sound technical feedback to the U.S. Department of Energy (DOE) regarding possible licensing concerns in the DOE safety case for the proposed YM repository.

Some of the core objectives of the restructured NRC HLW repository program are to

- Evaluate the DOE repository program [including field studies, laboratory experiments, and performance assessments (PAs)] to provide early feedback to DOE
- Evaluate the attributes of the DOE Waste Containment and Isolation Strategy (now called the Repository Safety Strategy); Total System Performance Assessment (TSPA) — Viability Assessment (VA); and TSPA — License Application (LA)
- Provide input for the NRC management decision-making on prioritization of the KTIs and related subissues
- Provide support for the NRC rulemaking for the HLW repository

A prerequisite to pursuing the aforementioned objectives is development of a generalized computer code, specifically tailored for evaluating the total system performance for the proposed repository at the YM site. Although NRC previously developed computer codes (Codell, et al., 1992; Wescott, et al., 1995) to analyze repository performance, a more general and versatile TSPA code was necessary to: (i) accommodate the most current DOE repository design [e.g., repository layout, waste package (WP), and emplacement design]; (ii) quantify total system performance in terms of the new compliance performance measure (i.e., annual individual dose or risk) expected in the forthcoming U.S. Environmental Protection Agency (EPA) standard and new NRC HLW regulation; and (iii) include the latest site data (e.g., tracer testing at C-well complex), knowledge base, and improved models. In addition, the computer code was specifically designed to allow a wide spectrum of individuals with diverse technical backgrounds to easily analyze repository performance. This report contains descriptions of the conceptual and mathematical models included in the code.

1.1   TOTAL-SYSTEM PERFORMANCE ASSESSMENT CODE

One of the basic purposes for applying the Total-system Performance Assessment (TPA) Version 3.1.4 code to the proposed repository is to secure a detailed and quantitative understanding of (i) the key factors controlling the degradation of the engineered barrier system (EBS); (ii) release of the waste from the repository; (iii) subsequent transport of the waste through various environmental pathways; and (iv) possible human exposure at the location(s) of the designated receptor group. To achieve this understanding, the total repository system is modeled in a probabilistic manner (Thompson and Sagar, 1993) that considers significant physical and chemical processes, phenomenological interactions and couplings, and potentially disruptive events and processes. This probabilistic approach, although computationally intensive, yields a range of potential future evolutions of the repository system. In addition, this approach is widely favored because it avoids many of the technical shortcomings associated with completely deterministic scenario-based assessments such as inability to account for uncertainty in model parameter and site characterization data (Thompson, 1988).

To date, two TSPAs for the proposed YM repository have been conducted by NRC and the CNWRA using the probabilistic approach. The first TSPA, referred to as Iterative Performance Assessment (IPA) Phase I (Codell, et al., 1992), was conducted to assemble and demonstrate the NRC assessment methodology. The second TSPA, designated as IPA Phase 2 (Wescott, et al., 1995), improved on the previous effort by developing a more comprehensive and integrated computer code (TPA Version 2.0 code). Those analyses yielded many valuable insights into the features, events, and processes influencing isolation performance of the proposed YM repository; these insights were put to use in the NRC and CNWRA reviews of early DOE TSPAs for YM. In the current effort, the IPA activity emphasizes updating and advancing the NRC TSPA capability, in preparation for the forthcoming review of the DOE TSPA-VA (TRW Environmental Safety Systems, Inc., 1996) as well as for assisting in focusing the NRC KTI activities on technical issues significant to repository performance.

Consistent with the recommendations of IPA Phase 2 (Wescott, et al., 1995), the new TPA Version 3.1.4 code was developed to include

- Most recent site data and knowledge base for conceptual models
- Updated abstractions for features, events, and processes
- Streamlined methodology for data transfers between executive and consequence modules
- More flexible design to accommodate future code modifications
- Improved computational algorithms permitting more detailed analyses
- Improved capabilities for parameter importance analysis and ranking

Federal regulations for HLW disposal in geologic repositories are being developed by EPA and NRC. Site data continue to be acquired—thus repository and WP designs continue to evolve. Therefore, it was deemed important to enhance the TPA Version 3.1.4 code by increasing the flexibility to evaluate alternative repository and design features; analyze the effect of different areal mass loadings (AMLs); assess the significance of various disruptive scenario classes; assess the impact of radionuclide dilution in the saturated zone (SZ); and compute the peak dose for a 10,000-yr compliance period and longer time periods that may be of interest.

TPA Version 3.1.4 code is composed of an executive driver program with a set of consequence modules and a library of utility modules. The executive driver controls the sampling of stochastic input parameters, the calculational sequence and data transfers between modules, and the production of output files. Various intermediate output files are generated for later use in representations of repository performance including performance of individual components (e.g., WP, SZ), parameter importance analyses, post-processing of time-dependent risk curves, and synthesis of statistical distributions [e.g., cumulative distribution functions (CDFs) and complementary cumulative distribution functions (CCDFs)] for appropriate performance measures. Utility modules ensure that consistent data sets are used by all consequence modules and facilitate the discretization of the repository system and surrounding geologic media.

Technical contributions from the NRC KTI activities resulted in the development of several TPA modules. For example, new conceptual and mathematical models were developed for infiltration, deep percolation, and unsaturated zone (UZ) flow. Other noteworthy KTI contributions include

- "Unsaturated and Saturated Flow under Isothermal Conditions KTI"—Improved information on climate and the distribution of infiltration over the repository area;

- "Container Lifetime and Source Term KTI"—New WP failure and source term models and parameters;

- "Evolution of the Near-Field Environment KTI"—Estimates of the chemical composition of the aqueous environment contacting the WP;

- "Radionuclide Transport KTI"—Updated estimates of near-field elemental solubilities and far-field sorption coefficients;

- "Thermal Effects on Flow KTI"—Improved capability to predict time-dependent temperature and relative humidity for use in the source term module;

- "Structural Deformation and Seismicity KTI"—Improved hydrostratigraphic model and a new faulting consequence module;

- "Igneous Activity KTI"—Improved technical basis for the probability and consequence characteristics for volcanic activity;

- "Repository Design and Thermal-Mechanical Effects KTI"—Improved modeling capability for a new seismic consequence module; and

- "Activities Related to Development of the NRC High-Level Waste Regulations KTI"—Realistic representation of plume dilution induced by receptor group well pumping.

In addition to contributing to the development of the new TPA Version 3.1.4 code, KTI members participated in the testing and documentation of the code, as well as in preparation of the input data set for the reference case.

1.2   DESCRIPTION OF THE PROPOSED REPOSITORY SITE

Although the TSPA methodology implemented in the TPA Version 3.1.4 code is somewhat general, the conceptual and mathematical models included in the code were largely selected to be specific to the characteristics and conditions of the YM site. For completeness, a brief description of the YM site geology, hydrology, climatology, and current repository design are presented in this section.

Located in Nye County, the YM site (see Figure 1-1) is approximately 120 km (75 mi) northwest of Las Vegas, in southern Nevada. The site is entirely on Federal land controlled by DOE and the Bureau of Land Management. The extreme northern portions of the site abut the Nellis Air Force Range. YM itself is located in the southern part of the Great Basin, the northernmost subprovince of the Basin and Range Physiographic Province. Climate at the site is generally arid to semiarid, with an average precipitation of about 180 mm/yr (7 in./yr) in the form of winter snow and summer thunderstorms. The YM area exhibits sparse vegetation and has a relatively low population density.

YM is an irregularly shaped cuesta 40-km (25 mi) long by 6- to 10-km (3.7- to 6.2-mi) wide. The crest of the mountain ranges between altitudes of 1500 m (4920 ft) and 1930 m (6330 ft) or about 650 m (2132 ft) higher than the floor of Crater Flat to the west of the site. The mountain is dominated by a subparallel series of north-trending ridges and valleys controlled by steeply dipping faults. YM is bounded by Crater Flat on the west, by Jackass Flat-Fortymile Wash on the east and southeast, the Amargosa Desert to the south, and by the Timber Mountain Caldera complex to the north. The stratigraphy at YM is composed of a gently dipping sequence of Miocene ash-flow tuffs, lavas, and volcanic breccias more than 1800-m (5904-ft) thick. The rock unit being considered for the repository facility is a densely welded ash-flow tuff of the Topopah Springs Member of the Paintbrush Tuff. There are data for the hydraulic and transport properties of the UZ in Flint and Flint (1990); Wittwer, et al. (1995); Rautman, et al. (1995); Schenker, et al. (1995); and Flint, et al. (1996).

The SZ in the YM region (YMR) can be divided into two major aquifer systems, a fractured tuff aquifer below the repository location extending more than 10 km (6.2 m) downgradient, and an alluvial aquifer extending southward from the terminus of the tuff aquifer to the Amargosa Desert. Patterns of ground-water flow in the tuff aquifer appear to be determined by the combined effects of hydraulic boundary conditions, dipping layers, and the presence of fault zones and laterally extensive shear fracture zones (Geldon, 1993). In contrast, ground-water flow in the alluvial aquifer is likely controlled by the interrelationship of recharge and discharge areas, interbasin transfers, and water well pumping in the Amargosa Desert area. Modeling studies of regional ground-water flow have been published by the U.S. Geological Survey (e.g., Czarnecki and Waddell, 1984).

Depth to ground water varies from over 350 m (1148 ft) directly below the repository block at YM to less than 10 m (33 ft) along a short reach of the Amargosa River in the south central portion of the Amargosa Desert. Water quality through the YM and Amargosa Desert regions is generally adequate to drink, water stock, and irrigate crops. However, in far southern Amargosa Desert, near Franklin Lake Playa, evaporites from lacustrine deposits cause the ground water to have total dissolved solids concentrations in excess of 10,000 parts per million (ppm) (Winograd and Thordarson, 1975). Near the communities of Amargosa Valley (Lathrop Wells) and Death Valley Junction, there are small clusters of domestic water wells. Agricultural uses of ground water are currently confined to the portion of the southern Amargosa Desert that lies 30 km north-northwest of Franklin Lake Playa, where depths to water range from 10 to 40 m (33 to 131 ft) and the topography is suitable for flood and center pivot irrigation.

The repository proposed for the YM site consists of an underground facility designed to accommodate 70,000 metric tons of HLW. The specific layout of the underground facility will depend on the thermal loading strategy that, as yet, has not been finalized. The waste disposed of in the repository is, at this time, expected to consist of both commercial spent fuel (SF), DOE SF, and defense HLW. For simplicity, the TPA Version 3.1.4 code assumes the inventory is SF. The current WP design for SF disposal consists of a large cylinder [i.e., approximately 1.8-m (5.9-ft) diameter by 5.6-m (18.6-ft) length] made up of a 20-mm (0.8-in.)

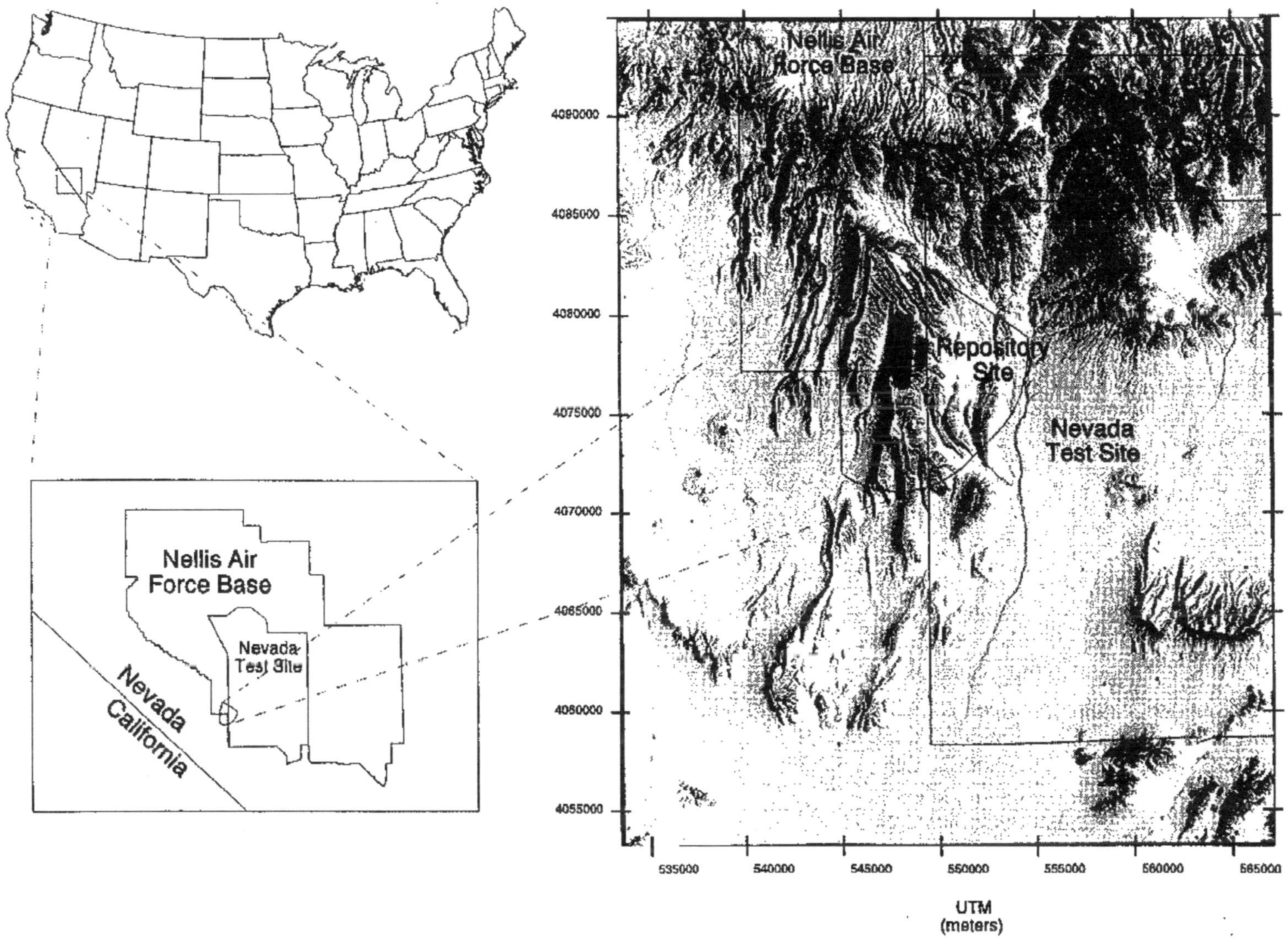

Figure 1-1. Proposed repository site.

wall-thickness inner overpack, and outer overpack with a 100-mm (4-in.) wall thickness. The container would be emplaced, in-drift, on v-shaped supports. If licensed, the underground facility would remain open 100 yr or more before final closure. Design decisions regarding backfilling or leaving the drifts open have not been made at this juncture (TRW Environmental Safety Systems, Inc., 1995).

1.3   REPORT CONTENT

This document was prepared to provide a single reference for the TPA Version 3.1.4 code. This report provides information on the system parameters, conceptual models, and code outputs essential to interpretation of calculational results. Additionally, the documentation communicates the developers' perspectives on the degree of conservatism adopted in the modeling approaches, major assumptions, and input parameters.

An overview to the modeling approach is presented, in Chapter 2, that describes the primary conceptual models and their interrelationships. Chapter 3 illustrates the conceptual and mathematical models implemented in 13 consequence modules; this chapter also delineates the major assumptions and conservatism of each module. Chapter 4 outlines future TPA Version 3.1.4 code developments, and Chapter 5 lists the technical references for this report.

Appendix A provides a roadmap to the primary input file (*tpa.inp*) and is expected to be helpful in understanding the TPA Version 3.1.4 code reference data set. The reference data set is subdivided by modules and provides a description of the parameters and the basis for values selected for the reference case. The parameter description follows the same order as in the primary data file and Chapter 3. Appendix B presents the derivation of empirical flow factors used in the source term module.

2 OVERVIEW OF THE TOTAL-SYSTEM PERFORMANCE ASSESSMENT CONCEPTUAL MODELS

Analysis of repository performance is anticipated to be complex, with substantial uncertainties because of the first-of-a-kind nature of the repository, extended period of performance (at least 10,000 yrs), and reliance on engineered and natural barriers. Detailed simulation models that include all the couplings, heterogeneities, and complexities cannot be incorporated into PA models and still maintain reasonable computer execution times and meet hardware limitations. Therefore, a key part in developing the TPA computer code is determining the level of detail in the processes, design, and attributes of the site necessary to produce a credible analysis that provides meaningful insights on performance without an unreasonable computational burden. The TPA Version 3.1.4 code includes the repository system description (e.g., representation of the site and repository design within the TPA Version 3.1.4 code), conceptual models, and parameter values. A discussion of the repository system description and conceptual models is presented in this section to provide a general overview of the TPA Version 3.1.4 code. Chapter 3 contains more detailed information on the conceptual and mathematical models. The TPA input parameter values and the basis for their selection are presented in Appendix A.

2.1   CONCEPTUALIZATION OF REPOSITORY AND GEOLOGIC SETTING

For ease of use and computational efficiency, the TPA Version 3.1.4 code replaces the intricate repository layout and the complex geologic setting with relatively simple conceptual representations. The repository layout, for example, is represented by an idealized planar feature discretized into a set of subareas, whereas the geology is replaced by a sequence of layers, each of which is homogenous. Properties and conditions for each subarea are assumed to be uniform. Except for the influence of the thermal load, flow and transport processes in and below a given subarea are independent of those processes in other subareas. Thus, flow is entirely vertical, with no lateral diversion in the UZ.

As illustrated in Figure 2-1, quadrilateral subareas of uniform thickness are used to represent individual subregions of the repository. In the current application, the repository is divided into seven subareas; however, the TPA Version 3.1.4 code has the capability to use much finer discretizations of both the repository and the geologic setting beneath it. The number of WPs in each subarea is assumed to be proportional to the fraction of total repository area represented. Radionuclide releases from the EBS are calculated by modeling a single prototypical WP for each subarea and for each failure type. Performance characteristics of the WP in each subarea are calculated by considering the evolution of such characteristics as water flux, thermal and chemical conditions, and geologic processes [e.g., seismicity, fault displacement, and igneous activity (IA)].

The geologic setting is composed of the UZ (i.e., geologic media between the land surface and the water table) and the SZ (i.e., ground-water aquifer beneath the repository and extending to the location of the receptor group). For simplicity, the stratigraphy is assumed to be laterally continuous and uniform within a subarea, to represent the UZ in a separate hydrostratigraphic sequence for each subarea. This simplification is consistent with the assumption that, in general, flow in the UZ is primarily vertical, that is, little or no lateral diversion of flow along hydrostratigraphic units. The geologic setting also includes features, events, and processes, such as seismicity, tectonism, and volcanism, that may adversely affect the performance of the repository.

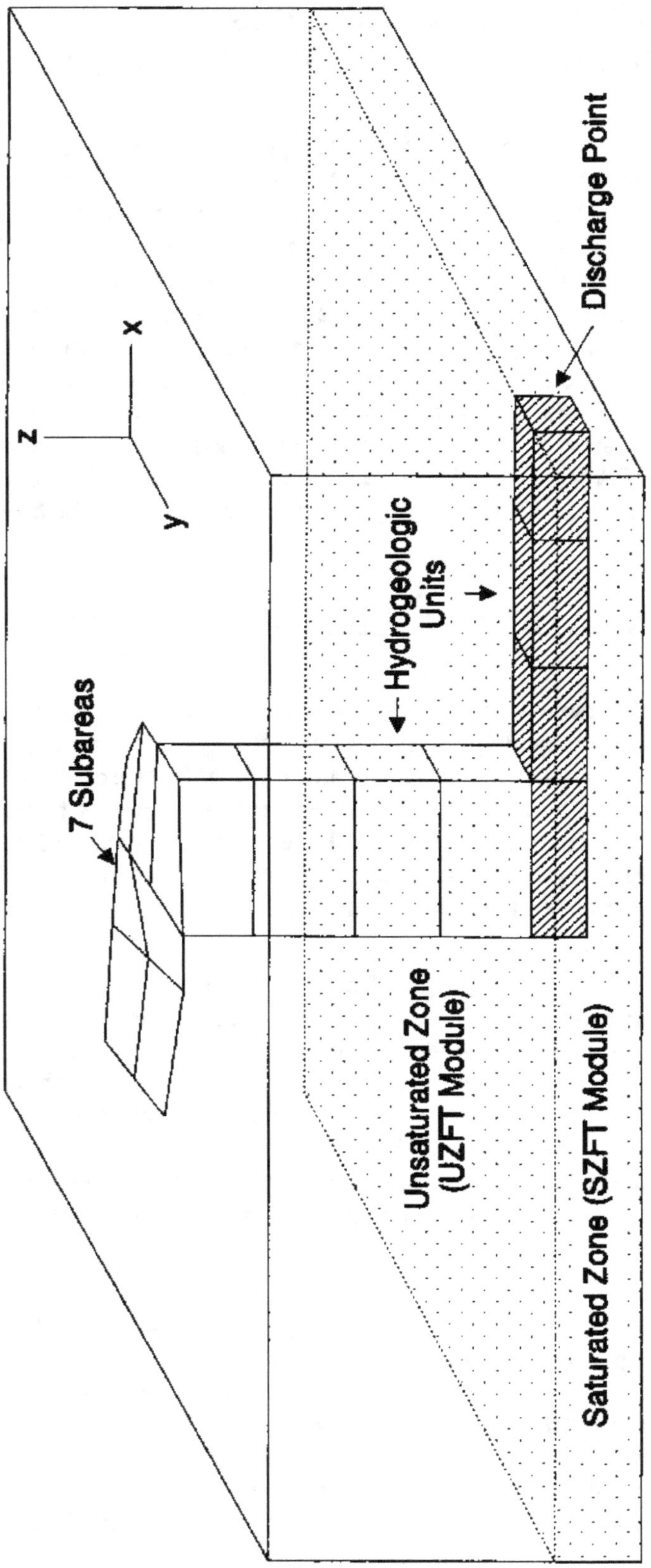

Figure 2-1. Repository system.

Direct release of radionuclides to the accessible environment because of an extrusive igneous event is also modeled in the TPA Version 3.1.4 code. The physical characteristics of the extrusion and the assumption of a uniform distribution of WPs in the repository are used to determine the number of WPs affected by the event. Radionuclides are transported to the receptor location, based on characteristics of the eruption and meteorological conditions, where the concentration of radionuclides in soil, resulting from the deposition of volcanic ash containing SF particles, is calculated. This soil concentration is used in calculating the annual dose to the average member of the receptor group.

2.2 CONCEPTUAL MODELS IMPLEMENTED IN THE TOTAL-SYSTEM PERFORMANCE ASSESSMENT COMPUTER CODE

In developing the TPA computer code, several conceptual models were formulated, integrated, and implemented in various abstracted mathematical models. These basic conceptual models, which describe the interactions/couplings of physical and chemical processes believed to be present in a proposed geologic repository (at YM), can be grouped into the following generic categories:

- Infiltration and deep percolation
- Near-field environment
- . Radionuclide releases from the EBS
- Aqueous-phase radionuclide transport (RT) through the UZ and SZ

To model flow and transport in the SZ, the TPA conceptual model consists of four distinct streamtubes over the width of the repository footprint normal to UZ flow (Figure 2-2). Each of the seven streamtubes in the UZ is connected to one of the four streamtubes in the SZ. Radionuclide releases from each of the UZ streamtubes provide the source term to the SZ streamtubes. The SZ streamtubes are treated as separate conduits and have flow velocities that vary along the individual flow paths. The mass flow rate of radionuclides exiting all SZ streamtubes is used to compute the average concentration at the well head. This, in turn, is used in calculating annual dose to the average member of the receptor group. The average concentration accounts for all releases, from the ground-water pathway to the location of the receptor group; spatial extent of the releases in the SZ at the location of the receptor group; extent of the production zone containing the radionuclides (all radionuclides are assumed to be released in one production zone); and the influence of the pumping rate attributed to water use by the receptor group.

- Airborne transport from direct radionuclide releases
- Exposure scenario and reference biosphere

These conceptual models are designed to apply to the current DOE repository design and specific site characteristics of the YM area and provide flexibility for examining alternative designs and uncertainties in site and engineered material performance. In some of these generic categories, alternative conceptual models have also been incorporated into the code.

These conceptual models represent a range of system states, including disruptive events. The consequences of disruptive events (e.g., seismicity, fault displacement, and igneous activity) are evaluated with the TPA Version 3.1.4 code by assessing the effects on failure (producing releases to ground water) and

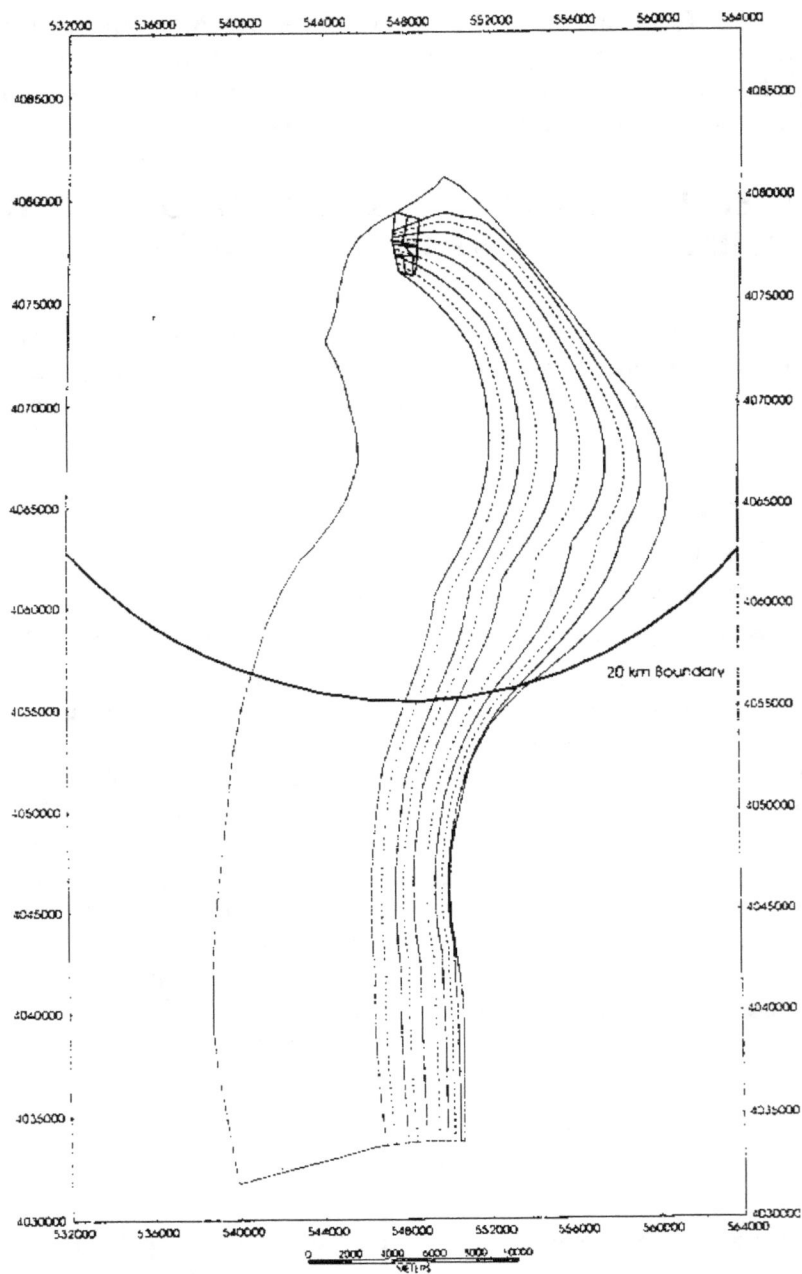

Figure 2-2. Saturated zone conceptual model representation showing the 20-km (12.4-mi) receptor location boundary and the center lines (dashed) of the four saturated zone streamtubes.

direct releases of radionuclides (airborne releases to the biosphere). In the TPA computer code, the probability of a disruptive event affecting the repository is used to calculate a risk curve. (A risk curve can be determined by a calculation, separate from the TPA Version 3.1.4 code execution, that combines the consequences calculated with the TPA Version 3.1.4 code and the appropriate probabilities.)

The following discussion provides a general overview of the key aspects of the major conceptual models implemented in the TPA Version 3.1.4 code. More detailed descriptions of these models, including the mathematical basis, assumptions, and calculational methodologies, are presented in Chapter 3.

2.2.1 Infiltration and Deep Percolation

A one-dimensional (1-D) modeling approach is used in the TPA Version 3.1.4 code (Section 3.1 -- "UZFLOW Module") to describe how meteoric water at the land surface moves vertically downward (i.e., without lateral flow) through the UZ, to the repository horizon, and ultimately to the water table. In the 1-D conceptual model, the deep percolation flux (q_{perc}) is constrained to be equal to the shallow infiltration rate (q_{infil}). The annual average q_{infil} is estimated based on:

- Present-day q_{infil}
- Change in climate with time
- Elevation and soil depth over the repository subarea

Uncertainty in the present day infiltration rate estimate is accounted for in the TPA Version 3.1.4 code by treating it as a statistically sampled input parameter. Temporal variations are incorporated by varying the present day q_{infil} over the 100,000-yr period assumed for long-term climatic changes. The effects of site-specific soil-cover thickness and elevation are used to reflect the spatial variation over each of the subareas.

The variation of q_{infil} because of changes in climate was developed through consideration of paleo-climatic information and a process-level auxiliary analysis (Stothoff, et al., 1997). The q_{infil} response function depends on two independent variables--present-day mean annual precipitation (*MAP*) and temperature (*MAT*), as well as the present-day shallow infiltration rate. After computing q_{infil}, the water flux at the repository horizon is then partitioned into:

- Water flux diverted around the failed WP
- Water flux entering the failed WP

Thus, for the purposes of the TPA computer code, the net water flux carrying dissolved radionuclides is a fraction of the total water flux arriving at the repository. It is this net water flux that is used in the TPA Version 3.1.4 code (Section 3.5 -- "EBSREL Module") to calculate the radionuclide source term for each subarea.

2.2.2 Near-Field Environment

Physical and chemical processes in the near field of the repository, such as heat transfer, water-rock geochemical interactions, and refluxing of condensate water, are expected to affect WP performance. In the TPA Version 3.1.4 code, a range of near-field characteristics is depicted in the abstracted mathematical models for heat and water flow and table look-ups for chemical parameters (Section 3.2 -- "NFENV Module"). For the purpose of estimating WP failure times and radionuclide release rates, the near-field environment is characterized by:

- Drift wall rock and WP surface temperatures
- Relative humidity (RH) (i.e., ratio of vapor pressure at the WP surface to the vapor pressure at the drift rock wall)
- Water chemistry (e.g., pH, chloride concentration, and carbonate ion concentration)
- Water reflux during the thermal phase

The average rock temperature in the repository horizon is calculated assuming a conduction-only model (i.e., the time history of temperature for each subarea is calculated, accounting for the amount of emplaced waste). The WP surface temperature is calculated using a multimode heat transfer (i.e., conduction, convection, and radiation) model. Vapor pressure is computed using the standard thermodynamic equation relating vapor pressure to temperature.

Estimates of the pH and chloride concentration histories of water films on the WP surface were developed in a separate process-level auxiliary analysis using the multi-component geochemical module of the MULTIFLO code (Lichtner and Seth, 1996). MULTIFLO was applied to calculate the pH and chloride concentration for water percolating through the matrix of the tuffaceous rock. Because the chloride concentration in the water film is likely to be higher than that in the rock mass, the chloride concentration history is scaled by a statistically sampled parameter. The TPA Version 3.1.4 code provides the option of either using a look-up table that uses the temperature-dependent pH (not currently used) and chloride concentration generated with the MULTIFLO code or constant values specified in the input file (Appendix A).

The amount of water percolating through the drifts varies over time primarily because of the coupled processes of heat transfer and fluid flow (e.g., vaporization, condensation, and refluxing). Water refluxing produced by these thermohydrologic effects is important over the first few thousand years, after which natural percolation determines the rate of water flow into the repository. Two lumped-parameter water reflux models are included in the TPA Version 3.1.4 code, both based on bulk flow balances. The first model considers episodic reflux associated with time-dependent perching above the repository. The second model assumes that refluxing water can be sufficient to depress the boiling isotherm in fractures and reach the WP during times when the surface temperature exceeds the boiling point of water. Each reflux model produces estimates of the total water flux into the repository during the thermal period.

2.2.3 Radionuclide Releases from the Engineered Barrier System

In the TPA Version 3.1.4 code, the performance of a prototypical WP is modeled for each repository subarea, considering the failure time and radionuclide release rates for each of the WP failure categories (Section 3.3 -- "EBSFAIL Module"). When this prototypical WP fails, all WPs in that subarea under a specified failure category are assumed to have failed. The estimation of both WP failure times and liquid releases depends on the nature and extent of corrosion, near-field environment, percolation flux in the drift, and external processes that may impose static loads, dynamic loads, or both. WP failures are grouped into three basic categories: (i) corrosion and mechanical failure; (ii) disruptive event; and (iii) initially defective WP failures. After determining the WP failure time, the TPA Version 3.1.4 code calculates the aqueous-phase radionuclide releases from the WP by considering the dissolution of radionuclides from the SF matrix and advective diffusive transport from the WP directly to the UZ beneath the repository.

Corrosion failure of the WP is defined to occur at the time when the inner overpack is fully penetrated by a single pit and the waste form is therefore accessible to water. The abstracted corrosion model uses a conceptual framework that assumes the formation of a water film containing a salt solution but does not explicitly consider water dripping on the container. The corrosion processes considered in the model abstraction consist of:

- Dry air oxidation
- Humid air corrosion
- Aqueous corrosion

WP surface temperature and the chloride concentration in the water film influence the mode, and hence, the rate of corrosion. The predominant mode of corrosion however, depends on the critical RH, as well as the container material. Mechanical failure of the WP is considered to be the result of fracture of the outer steel overpack, because thermal embrittlement arising from prolonged exposure at temperatures sufficiently elevated to cause substantial degradation of mechanical properties (it is assumed conservatively that the inner overpack is fractured when the outer overpack is fractured).

Disruptive event failures are taken into account by modeling the effects of events such as seismicity (Section 3.4 -- "SEISMO Module"); fault displacement (Section 3.10 -- "FAULTO Module"); and igneous activity (Section 3.11 -- "VOLCANO Module"). In the case of seismicity, WP failures are caused by rock falls that mechanically load and deform the WP (the drift is assumed not to be backfilled for rockfall to damage the WP). Displacements along yet undetected faults or new faults (because DOE will not emplace the WPs within a setback distance from known and well-characterized faults) that exceed a preestablished threshold are assumed to fail WPs within the fault zone. For IA, simulated magmatic intrusions intersecting the repository are assumed to cause WP failure; WPs within a dike but outside the vent hole are assumed to fail and expose the SF to water, whereas those within the vent hole are assumed to be entrained in the magma and released directly to the biosphere. For both IA and fault displacement, failures are modeled by superimposing the physical dimensions of the perturbation (i.e., length, width, and orientation of the fault and the igneous intrusion) on the repository footprint, to determine the total number of WPs potentially affected in each repository subarea. Separate failure times are calculated for seismicity, fault displacement, and igneous activity.

In most applications of the TPA Version 3.1.4 code, it is assumed that a small number of WPs are failed at the time of repository closure. These initially failed WPs are attributed to fabrication defects or damage to the WP as a result of improper emplacement. For conservatism, the number of initially defective WPs is typically assumed to be 0.1 percent[1] of the total number of containers.

Radionuclide releases from the WP are calculated by considering the alteration rate of SF (i.e., rate at which radionuclides in fuel become available for release); radionuclide solubility limits; and transport mechanisms out of the WP. The TPA Version 3.1.4 code incorporates numerous parameters (e.g., fraction of SF that is wet, particle size of the SF, alteration rate of uranium oxides (UO_{2+x}), and credit for cladding) that control the release of radionuclides from the SF matrix. After radionuclides are leached from the SF waste form, the calculated releases are adjusted to ensure consistency with the radioelement solubility limits.

A parameter value is used to specify the fraction of failed WPs in the subarea that is wetted -- the number of failed WPs available to contribute to the source term (auxiliary calculations are typically done to justify parameter selection). To compute the time-dependent source term, the TPA provides two alternative conceptual models: (i) a bathtub model -- the WP must fill with water before the radionuclides are released; and (ii) a flow-through model -- radionuclides are released by water dripping on the waste form. For the bathtub model, the WP is treated as a stirred tank, with the tank capacity dependent on the statistically sampled water outlet height. Water will fill the WP until the capacity (height) is reached and, thereafter, the amount of water entering the WP will equal the amount of water flowing out. Water leaving the WP transports dissolved radionuclides into the UZ below the repository. Additionally, the TPA Version 3.1.4 code has the option to account for diffusive transport of radionuclides remaining in the fluid of the bathtub or WP through an assumed backfill. The flow-through model is a variant of the bathtub model except water does not have to first fill the bathtub before release and the fraction of fuel wetted is independent of the water level.

2.2.4 Treatment of Aqueous-Phase Transport in the Unsaturated and Saturated Zones

Movement of aqueous-phase radionuclides from the repository horizon, through the UZ and SZ and ultimately to the receptor group, is modeled in the TPA Version 3.1.4 code (Section 3.6 -- "UZFT Module" and Section 3.7 -- "SZFT Module") using the previously described streamtube approach. Each streamtube encompasses one or more repository subareas and is composed of a vertical section from the repository to the water table and horizontal sections in the SZ. The transport module NEFTRAN II (Olague, et al., 1991) simulates the spectrum of processes (e.g., advection, dispersion, matrix diffusion, sorption, and decay) occurring within individual streamtubes. For a set of about 20 radionuclides, this module simulates vertical transport through the UZ and horizontal transport through the SZ.

Time-dependent flow velocities in the UZ are calculated using the hydraulic properties of each major hydrostratigraphic unit. The transport module simulates the transport of radionuclides through either the porous rock matrix or fractures.[2] Radionuclide retardation by chemical sorption in the rock matrix can significantly reduce the transport rates and is therefore included in the model. Retardation on fracture surfaces, however, is neglected for conservatism, because the significance of this mechanism has yet to be demonstrated.

---

[1]Tschoepe, et al. (1994) suggest fabricated metallic component reliabilities of 99.9 to 99.99 percent.

[2]Transport though rock matrix takes place if the deep percolation (q_{perc}) is less than the hydraulic conductivity of the rock matrix (K_{matrix}) or through fractures when q_{perc} exceeds K_{matrix}.

Although ground-water flow in the SZ is assumed to be at steady state, RT within individual streamtubes is time-dependent because the source term varies with time. Streamtubes in the SZ exhibit variable cross-sections along the flow path; this variable streamtube geometry was determined from a separate two-dimensional (2-D) modeling study of the subregional flow (Baca, et al., 1996). The conceptual model of the SZ assumes that flow in the tuff aquifer is in localized conductive zones (i.e., permeable fracture zones) whereas flow in the alluvium is presumed uniformly distributed in the alluvial aquifer. Although the streamtube approach neglects dilution effects arising from lateral dispersion, credit is taken for sorption in the alluvium, which is likely to retard aqueous phase transport of many radionuclides. Additionally, matrix diffusion from flowing pores and fractures into the more-or-less stagnant matrix pore water within the rock is included in the SZ transport model.

2.2.5 Airborne Transport for Direct Releases

Radiologic risks associated with the extrusive component of IA are calculated in the TPA Version 3.1.4 code by modeling airborne releases of radionuclides for simulated extrusive events. The volcanism module assumes that the magma intercepts WPs, moves upward to the land surface, and then ejects the ash and SF mixture into the atmosphere. The physical characteristics of each simulated extrusion (e.g., vent size, event energetics, and duration) and atmospheric conditions are treated as statistical parameters in calculations of ash dispersal and deposition patterns, ash blanket thickness, and radionuclide soil concentrations. Three primary factors determining the ash plume geometry and transport rates include:

- Power and duration of the eruption
- Wind speed and direction
- SF particle sizes

The ash transport model developed by Suzuki (1983) was modified by Jarzemba, et al. (1997) and incorporated into the TPA Version 3.1.4 code, to calculate distribution of the released radionuclides. The time-dependent radionuclide areal densities are calculated taking into account the thickness of the ash blanket, leaching and erosion rates, and radionuclide decay rates. The calculated doses attributed to direct releases are strongly influenced by the time of the event (early events result in larger doses in part, because of, the contribution to the estimated doses from short-lived fission products present in the SF).

2.2.6 Exposure Pathways and Reference Biosphere

Dose calculations are performed in the TPA Version 3.1.4 code for exposure pathways that consider an average person of a designated receptor group. These calculations are expressed by the total effective dose equivalent (TEDE). Alternative receptor groups are currently included in the exposure scenario. One receptor group is a farming community 20 km (12.4 mi) from the repository location, whereas the second is a residential community at a specified distance typically less than 20 km (12.4 mi). The average member of the designated receptor group is assumed to be exposed to radionuclides transported through the ground-water pathway, air pathway, (as a result of direct releases arising from the extrusive component of IA), or both.

Geographic location and lifestyle characteristics assigned to each receptor group are two primary aspects defining the receptor group and are specified in the TPA Version 3.1.4 code by selection of appropriate input options (Appendix A). In addition, the farming community receptor group is assumed to include persons that use the contaminated water for:

- Drinking [i.e., 2 L/day (0.53 gal./day)]
- Agriculture, typical of Amargosa Valley area practices (e.g., growing alfalfa and gardening)

The farming community receptor group is assumed to be exposed to surface contamination through:

- Consumption of contaminated farm products (i.e., ingestion)
- Breathing air with ash-SF particles (i.e., inhalation)
- External exposure

In contrast, the residential receptor group is assumed to be composed of persons who use contaminated ground water only for drinking, but are also exposed to surface contamination (created by ash-SF particle deposition from the extrusive component of IA) through inhalation and direct exposure.

Site-specific dose conversion factors (DCFs) for each radionuclide and pathway are contained in TPA data files. These DCFs are used to convert radionuclide concentrations in the ground water and soil to TEDE values. The individual DCFs are mean values generated through separate pathway calculations using the GENII-S code (Leigh, et al., 1993). In the ground-water pathway, for example, the DCFs are applied to the concentrations at the well head. Two separate sets of DCFs are included in the TPA Version 3.1.4 code to represent two distinct reference biospheres associated with the present arid climate and the projected future pluvial climate. In addition to computing the TEDE history for each stochastic simulation, the TPA Version 3.1.4 code scans these dose calculations to identify the magnitude and timing of the peak dose.

2.3 RADIONUCLIDE INVENTORIES

In the TPA Version 3.1.4 code, waste inventories and associated thermal output are based on the assumptions of an average of 65 percent pressurized water reactor (PWR) waste with a 42 GWd/MTU burnup and 35 percent boiling-water reactor (BWR) waste, with a 32-GWd/MTU burnup. This provides an average waste burnup of 38.5 GWd/MTU. The average age of the waste at the time of emplacement is set at 26 yrs.

After reviewing the literature (Barnard, et al., 1992; Wilson, et al., 1994; TRW Environmental Safety Systems, Inc., 1995), a list of 43 radionuclides (Table 2-1) was compiled for determining HLW inventories. All these nuclides are considered for the direct release calculations; however, in ground-water release calculations, the computational burden imposed by short-lived nuclides is avoided by specifying only a selected list of nuclides. Short lived radionuclides, not expected to arrive at the receptor location because of radioactive decay, were not considered in the ground-water pathway. Radionuclide chains considered for ground water and direct releases are presented in Table 2-2 and Figure 2-3, respectively. The direct-release calculations consider chains with multiple parents, whereas the ground-water release calculations consider chains with a single parent.

Table 2-1. List of 43 Nuclides in the TPA Version 3.1.4 Code Database

Nuclide Number	Nuclide Name	Nuclide Number	Nuclide Name
1	U-238	23	Pu-240
2	Cm-246	24	U-236
3	Pu-242	25	U-232
4	Am-242m	26	Sm-151
5	Pu-238	27	Cs-137
6	U-234	28	Cs-135
7	Th-230	29	I-129
8	Ra-226	30	Sn-126
9	Pb-210	31	Sn-121m
10	Cm-243	32	Ag-108m
11	Am-243	33	Pd-107
12	Pu-239	34	Tc-99
13	U-235	35	Mo-93
14	Pa-231	36	Nb-94
15	Ac-227	37	Zr-93
16	Cm-245	38	Sr-90
17	Pu-241	39	Se-79
18	Am-241	40	Ni-63
19	Np-237	41	Ni-59
20	U-233	42	Cl-36
21	Th-229	43	C-14
22	Cm-244		

Note: U-uranium; Cm-curium; Am-americium; Pu-plutonium; Th-thorium; Ra-radium; Pb-lead; Pa-protactinium; Ac-actinium; Np-neptunium; Th-thorium; Sm-samarium; Cs-cesium; I-iodine; Sn-tin; Ag-silver; Pd-palladium; Tc-technetium; Mo-molybdenum; Nb-niobium; Zr-zirconium; Sr-strontium; Se-selenium; Ni-nickel; Cl-chlorine; and C-carbon.

Table 2-2. Radionuclide Chains For Ground-Water Release

Chain No.	Chain	Chain No.	Chain
1	Cm-246 → U-238	8	Tc-99
2	Cm-245 → Am-241 → Np-237	9	Ni-59
3	Am-243 → Pu-239	10	Cl-36
4	Pu-240	11	C-14
5	U-234 → Th-230 → Ra-226 → Pb-210	12	Se-79
6	Cs-135	13	Nb-94
7	I-129		

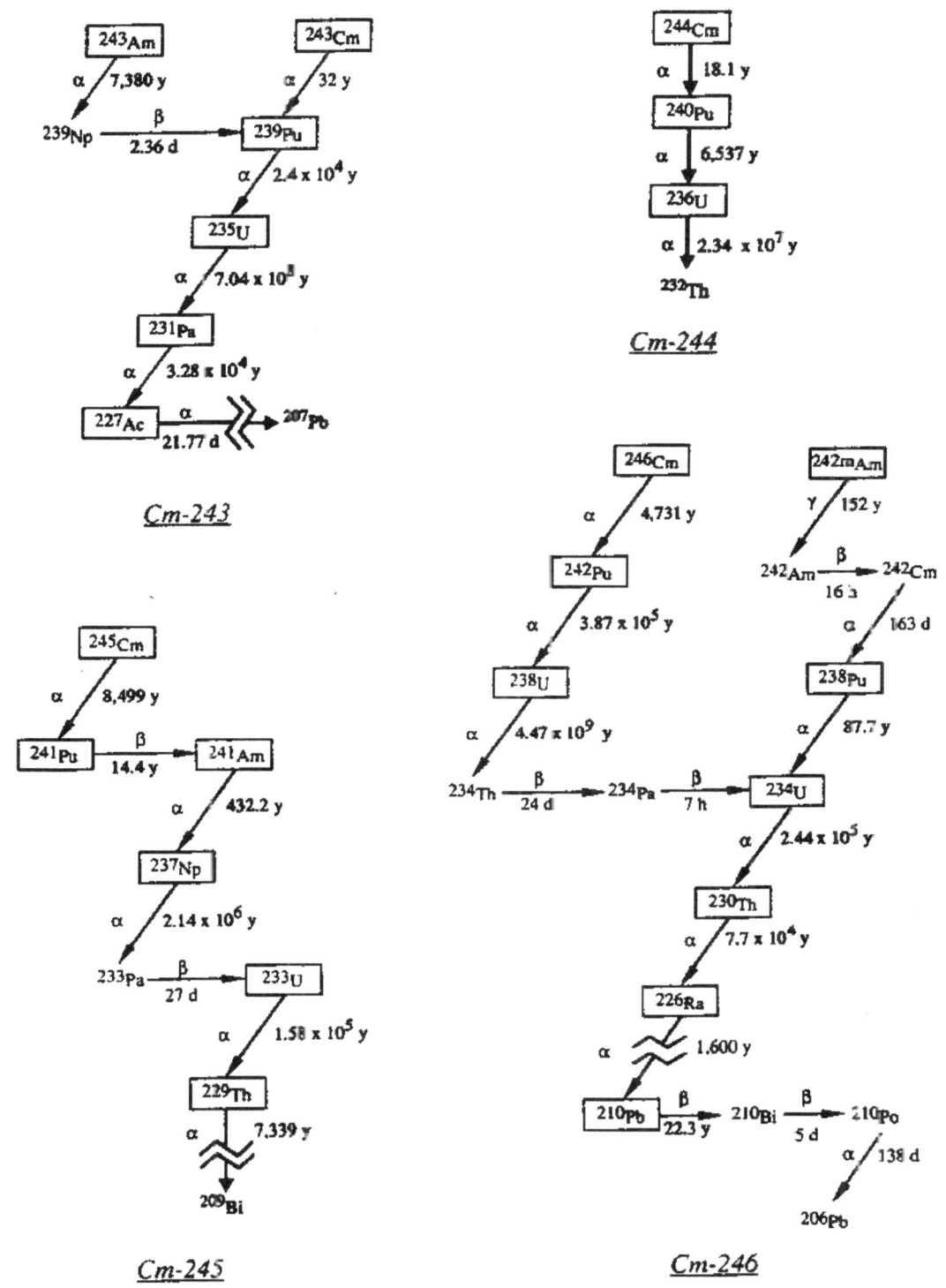

Figure 2-3. Cm-243, Cm-244, Cm-245, and Cm-246 decay chains.

3 CONCEPTUAL AND MATHEMATICAL MODELS

This chapter describes conceptual and mathematical models used in the TPA Version 3.1.4 code. For each module, a description of the conceptual model and its implementation in TPA Version 3.1.4 code is presented via the following sections

    (i)   Information Flow—general identification of the input supplied by the user and information supplied by other modules, and the information that this module provides to other modules;

    (ii)   Conceptual Model—description of the technical approach and implementation within the code; and

    (iii)   Assumptions—listing of some of the important assumptions and, where appropriate, their effect on conservatism in the final results.

The information in this chapter is intended to provide an understanding of the relationship of the input parameters to the calculational approach in the code, and to describe the types of results available for analyzing performance.

Numerous types of input parameters are used in the TPA Version 3.1.4 code. A primary input file controls the attributes of the simulation (e.g., inclusion of disruptive events, number of realizations to perform when conducting probabilistic analyses, and length of simulation) and specification of parameters. It is anticipated that the TPA Version 3.1.4 code will typically be used in a probabilistic mode (i.e., performing a large number of trials based on Monte Carlo Simulation (MCS) or Latin Hypercube Sampling (LHS) of the input parameters) to account for uncertainty and variability in describing the repository system (e.g., porosity, retardation, solubility limits). The input parameters included in the primary input file can be sampled. Other input data, not considered appropriate for sampling, are used to specify constants of the analysis (e.g., elevation of the subarea, radionuclide half-life) or abstractions of detailed models (e.g., definition of streamtube characteristics based on a 2-D analysis). A variety of data files is used to supply the nonvariable information. For completeness, this section identifies all input relevant to a particular module.

Two general types of consequence modules are incorporated in the TPA Version 3.1.4 code: (i) base case modules primarily associated with the ground-water pathway—UZFLOW, NFENV, EBSFAIL, EBSREL, UZFT, SZFT, and DCAGW; and (ii) disruptive event modules associated either with damage to the WP or direct release of radionuclides in the air pathway from the extrusive component of IA—SEISMO, FAULTO, VOLCANO, ASHPLUMO, ASHRMOVO, and DCAGS. Overall, the consequence modules estimate and provide intermediate output information for:

    •   Water infiltration from the land surface to the subsurface and, subsequently, into the emplacement drifts onto WPs;

    •   Environmental conditions around the WPs (e.g., temperature, humidity, pH, chloride ion concentration, and carbonate ion concentration) that affect WP degradation or radionuclide release;

    •   WP failure times caused by corrosion and mechanical processes, seismicity-induced rockfall, faulting, IA, and WPs assumed to be initially defective;

- Release of radionuclides from failed WPs to the ground-water transport pathway;

- Release of radionuclides from the extrusive component of IA to the air pathway;

- Transport of radionuclides in ground water through the UZ and SZ;

- Transport of radionuclides in volcanic ash;

- Doses to receptors through the radiological contamination in ground water; and

- Doses to receptors through the radiological contamination of the ground surface (released from the repository by an extrusive volcanic event).

3.1 UZFLOW MODULE DESCRIPTION

The UZFLOW module calculates the amount of water infiltrating from the ground surface into the UZ above the repository. Water that has infiltrated the subsurface affects the repository near-field environment and is available for WP corrosion, dissolution of material in the WP, and transport of material released from WPs. Specifically, UZFLOW determines the temporal and spatial variation of percolation water flux at the repository horizon in the absence of thermal effects from the repository. Although thermal effects are not accounted for in UZFLOW, the NFENV module, which simulates the near-field environment, explicitly evaluates how repository heating affects percolation through the repository. The calculation of percolation flux includes changes in precipitation and temperature that may occur at the ground surface throughout the life of the repository. Additional information on modeling of climate change and its effect on percolation flux can be found in Stothoff, et al. (1997).

3.1.1 Information Flow Within TPA

3.1.1.1 Information Supplied to UZFLOW

Input provided to the UZFLOW module consists of detailed information on ground surface elevation, soil thickness, future climatic conditions, and parameters that specify the present infiltration and the variation in climate (provided in the primary input file). Ground surface elevation is supplied as a digital elevation model (DEM) in the static file entitled *elevdem.dat*. A DEM with the same discretization, *soildem.dat*, provides soil thicknesses. The input files *climato1.dat* and *climato2.dat* provide data for representing time-dependent future climatic conditions. Data file *climato1.dat* provides noise data (normally distributed random numbers) used in computing time-varying *MAP* and mean annual temperature (*MAT*). Data file *climato2.dat* provides a time history of climate change expressed as a fraction of the full glacial maximum in temperature and precipitation. All other input parameters to UZFLOW, specifying present infiltration and the magnitude of climate variation, are in the UZFLOW section of the primary input file.

3.1.1.2 Information Provided by UZFLOW

UZFLOW passes to EXEC the time-varying volumetric flow rate of water infiltrating toward the repository, which then provides these values to NFENV module for use in estimating ground-water reflux during the repository thermal phase—and the UZFT module in determining transport velocities.

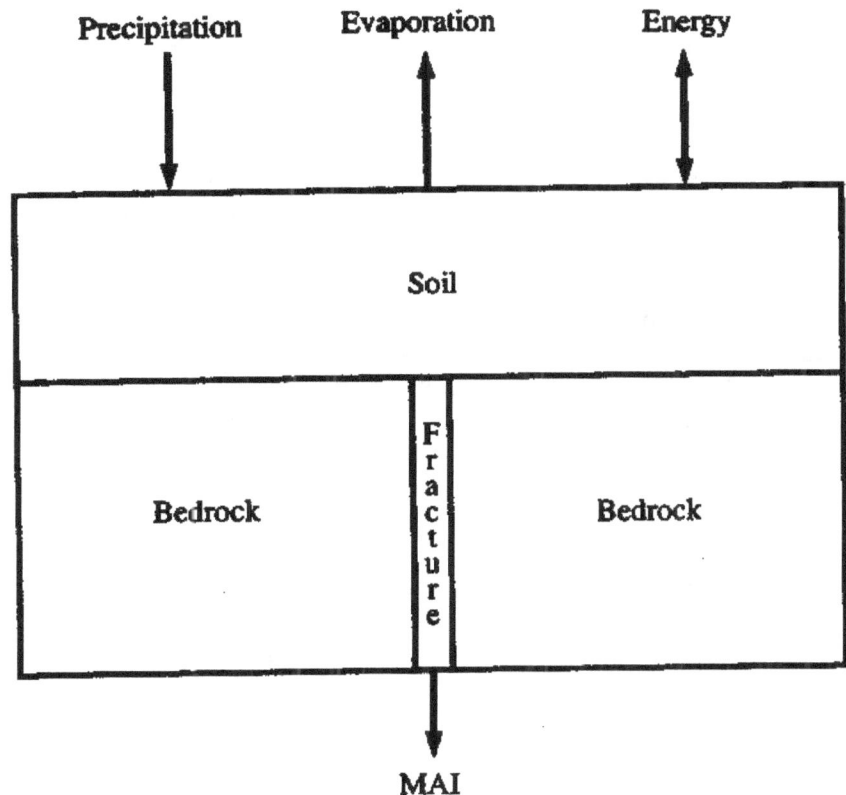

Figure 3-1. UZFLOW conceptual model.

3.1.2 Conceptual Model

The UZFLOW module is conceptually divided into three components: (i) a climate model for specification of the time history of climatic change; (ii) a model for calculation of shallow infiltration rate with climate change; and (iii) a deep percolation model for calculation of areal-average percolation flux at the repository level.

The climate model assumes that climate variables change in response to a glacial cycle, with a period of roughly 100,000 yrs, with shorter-term perturbations superimposed on the cycle. Two climate variables, precipitation and temperature, have the dominant impact on infiltration (Stothoff, et al., 1997). The time history of the dominant variables is calculated in the TPA Version 3.1.4 code using input specifications from the *climato1.dat* and *climato2.dat* data files and the UZFLOW section of the primary input file. The elevation-dependent spatial variability of these climatic variables, as well as vapor density, is superimposed on the glacial perturbations using the elevations in the *elevdem.dat* data file.

Figure 3-1 illustrates the shallow infiltration conceptual model, including water and energy balances, for a system of shallow surficial soil above a fractured impermeable bedrock. Water and heat enter and exit the soil at the ground surface. A portion of the water infiltrates after precipitation does not evaporate and moves into the deep subsurface (below the root zone) via fracture flow. Thus water escaping evaporation becomes deep percolation. Offline simulations of shallow infiltration using this conceptual model were performed for a simulated time period of at least a decade using hourly time steps (Stothoff, et al., 1997). The resulting time-varying infiltration, accounting for short-term variability in climatic parameters, is abstracted into relationships between mean annual infiltration (*MAI*) and long-term-average climate (i.e., *MAP* and *MAT*). The time history of *MAI* is obtained from the time history of *MAP* and *MAT*, with individual *MAI* values calculated for each grid block lying within the repository footprint in the *elevdem.dat* data file.

The gridblock values of *MAI* are used to determine the total deep percolation for each repository subarea (Figure 3-2). The time step for deep percolation is specified in the primary input file but is typically on the scale of centuries. The deep-percolation time scale is much longer than is considered in shallow infiltration simulations, justified by the assumption of strong temporal smoothing of fluxes as waters pass from the ground surface to the repository horizon. The smoothing is conceived of as a local process, so that fluxes generally pass vertically to the repository without systematic lateral diversion.

3.1.2.1 Climate Model

Infiltration rates are strongly affected by precipitation and evapotranspiration (which in turn is strongly affected by air temperature). Over the period of repository performance, it is anticipated that average precipitation and air temperature will change. At the last full glacial maximum, available evidence suggests that *MAP* at YM may have been 1.5–3 times larger than under current climatic conditions, whereas *MAT* may have been cooler by 5–10 °C (9–18 °F) (Nuclear Regulatory Commission, 1997). The effect of such large climatic changes can be examined using the UZFLOW module.

The shallow infiltration model uses the time history of *MAP* and *MAT* to calculate *MAI*. In arid environments such as exist at YM, interannual variability in precipitation can vary by a factor of 10, so that the concepts of *MAP* and *MAT* implicitly require averaging over several decades or more. The shallow infiltration model incorporates climatic variability at time scales shorter than a decade.

The climate model assumes that the bulk effects of climatic change can be represented by a glacial-change cycle, with shorter time-scale climate changes (century-to-century variability in precipitation and temperature) superimposed on the overall change in climate with time (bulk signal). The timing of bulk climatic change is felt to be more certain than the magnitude of climatic changes, with a full cycle of climatic changes occurring roughly over 100,000 yrs. For purposes of this analysis, *MAP* and *MAT* are assumed to be known at the start of the simulation (i.e., the present day), whereas changes in *MAP* and *MAT* under full glacial conditions are considered quite uncertain.

The UZFLOW model incorporates the effects of climate change by sampling only those parameters that have the greatest uncertainty for each TPA realization. For example, climate change can be characterized by changes in *MAP* as a multiplier of current conditions (e.g., 1.5–3 times current values) and changes in *MAT* as an effect of current conditions [e.g., 5–10 °C (9–18 °F) cooler]. Accordingly, the MAP multiplier at full glacial maximum and the *MAT* offset at full glacial maximum are specified in the primary input file. A function describing the temporal variation of bulk climatic change is obtained from file *climato2.dat*. As the simulation proceeds from the start through the climate change cycle, *MAP* and *MAT* cycle from current values through full glacial maximum conditions and back to current conditions. *MAP* and *MAT* have the same formula for bulk climatic change, so they vary at the same rate over the glacial cycle.

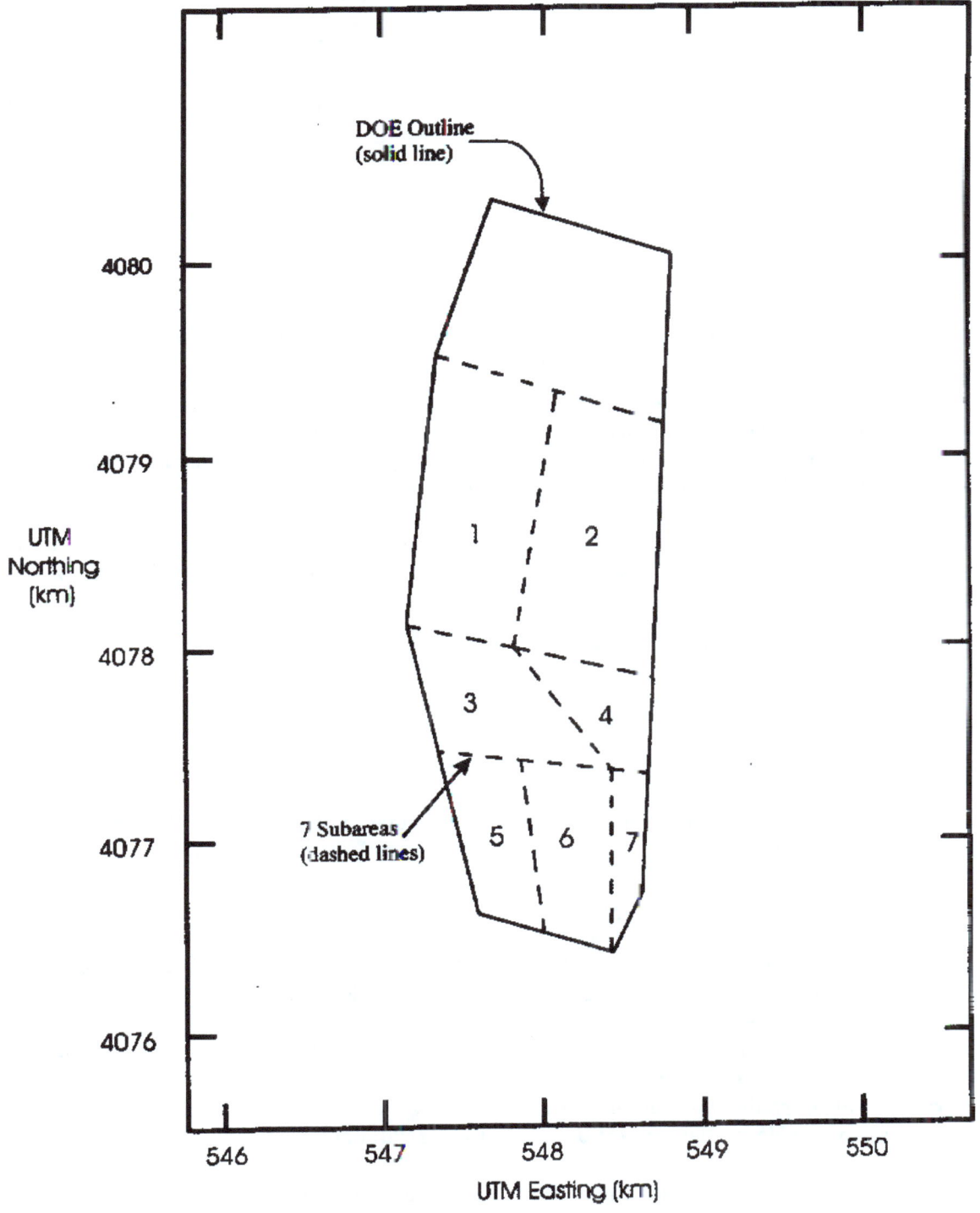

Figure 3-2. Boundary of proposed Yucca Mountain repository.

The bulk climatic change file (*climato2.dat*) supplies the code with data generated using the following formula, determined by matching the last several Milankovich cycles:

$$p = 1 - \frac{1}{4}\left[1 - \sin\left(\frac{\tau}{14.33 - 1.22}\right)\right]^2 \tag{3-1}$$

$$f = \frac{p - p_0}{1 - p_0} \tag{3-2}$$

where τ is time (ky) before present (in the Milankovich cycles used to develop the formula), f is the relative change between the current climate and the full glacial maximum, p is a periodicity function, and p_0 represents the present-day value of p (i.e., $\tau = 0$). The input file has values at each 1000 yrs that are linearly interpolated to obtain intermediate values. Note that Eqs. (3-1) and (3-2) predict that conditions are more than halfway to full glacial maximum (i.e., $f > 0.5$) for roughly two-thirds of the total climatic cycle, although conditions may be slightly drier and hotter than present for about 10 kys.

The climate history generated from the data file is evaluated for uniform time intervals, with length specified in the primary input file as TimeStepForClimate[yr]. At the midpoint for each interval, *MAT* and *MAP* are evaluated using:

$$MAP(t) = MAP_0 + f(t)\ (MAP_{fgm} - MAP_0) \tag{3-3}$$

$$MAT(t) = MAT_0 + f(t)\ (MAT_{fgm} - MAT_0) \tag{3-4}$$

where t is time from start of simulation; MAT_0 and MAP_0 represent *MAT* and *MAP* at the start of the simulation; and MAT_{fgm} and MAP_{fgm} represent *MAT* and *MAP* at full glacial maximum.

3.1.2.2 Variation in Climate Change Cycle

Relatively short-term climatic changes can significantly change the climate from the bulk predictions (e.g., mini-glacial periods), thereby causing periods of higher or lower *MAI*. As *MAI* is exponentially dependent on climate (discussed in Section 3.1.2.3 -- "Shallow Infiltration Model"), the expected long-term average infiltration is larger if these mini-glacial periods are considered, even if the climatic changes are evenly distributed above and below the mean cycle. If the variability is small, it may be adequate to use the bulk climate formula to predict infiltration. During relatively short simulations, these departures from the bulk conditions can provide substantial portions of the total infiltration.

The UZFLOW module provides a mechanism to examine the importance for repository performance of short-term climatic perturbations. A perturbation to the bulk value of both *MAT* and *MAP* (developed as described previously) is calculated for each of the uniform time intervals. It is assumed that the *MAT* and *MAP* perturbations are correlated with each other within a time interval, but perturbations are not correlated between successive intervals (i.e., *MAT* and *MAP* variations in one time interval are not affected by variations in previous intervals and do not affect variations in future time intervals). Within a time period, k, perturbations are generated using:

$$\chi_{kj} = \rho_{ij}\varepsilon_i \tag{3-5}$$

where

i, j	=	MAT or MAP
χ_{kj}	=	a vector of correlated perturbations
ε_i	=	a vector of independent standard normal perturbations
ρ_{ij}	=	the lower triangle of the Cholesky decomposition of a correlation matrix

The correlation between MAT and MAP perturbations is CorrelationBetweenMAPAndMAT in the primary input file. The correlation must lie between -1 and 1, and a reasonable value is about -0.8 (i.e., wetter conditions are generally associated with cooler conditions).

The actual values for MAT and MAP within time step k are calculated using:

$$v_{kj} = \chi_{kj}\sigma_j + m_j \tag{3-6}$$

where j represents MAT or MAP, v_{kj} is the calculated value of variable j, σ_j the standard deviation of variable j, and m_j the value of variable j, using the bulk climatic interpolation. For MAP, standard deviation values are called StandardDeviationOfMAPAboutMeanInOneTimePeriod[mm/yr] and for MAT, they are termed StandardDeviationOfMATAboutMeanInOneTimePeriod[degC] in the primary input file. Both standard deviation values are assumed constant over the length of the TPA simulation.

It is important to note that the time variability in the MAI formula is for time periods of less than a decade. Therefore, the standard deviations for MAT and MAP used in Eq. (3-6) should capture variability of 10-yr average climate. Reasonable standard deviation values might be 20–30 mm/yr (0.8–1.2 in./yr) (MAP) and 0.4–0.5 °C (0.7–0.9 °F) (MAT), based on a 50-yr sequence of daily meteorologic observations at Beatty, Nevada.

3.1.2.3 Shallow Infiltration Model

The shallow infiltration model in UZFLOW calculates the mean annual UZ net-infiltration flux leaving the root zone for the deeper subsurface. The net-infiltration flux is strongly linked to the MAT and MAP climatic parameters, calculated as described earlier.

Infiltration in arid zones, such as exists at YM, tends to occur during wet years, possibly separated by years, to decades of dry years. Simulations of bare-soil infiltration conducted by Stothoff, et al. (1997) indicate that MAI is strongly dependent on soil thickness, when the soil thickness is less than 50 cm (20 in.) and that different soil thicknesses exhibit different sensitivities to changes in MAP and MAT. Generally, thin soil layers allow infiltration to quickly reach fractures and thus percolate below the root zone into the deeper subsurface. At YM, most of the repository footprint is overlain by thin soils [i.e., < 50 cm (20 in.)]. Accordingly, to calculate MAI over the repository footprint, the shallow infiltration model uses soil thickness and infiltration values on a fine scale grid, compared with a typical subarea. Then, the fine-scale calculations are averaged to the larger subarea scale.

Simulations conducted external to the TPA Version 3.1.4 code (Stothoff, et al., 1997) looked at effects of bedrock fracturing, soil thickness, and meteorologic conditions on shallow infiltration. These simulation results indicated that *MAP*, *MAT*, soil thickness, and soil hydraulic properties strongly affect *MAI*, whereas mean annual atmospheric vapor density (*MAV*) and solar loading have secondary impacts on *MAI*. Both *MAT* and *MAP* vary in such a way that *MAI* increases with elevation, whereas the variation of *MAV* with elevation decreases *MAI* with elevation. Consideration of solar loading will tend to locally extend *MAI* on north-facing slopes and locally diminish *MAI* on south-facing slopes, but does not significantly change overall *MAI* estimates. The systematic impact of *MAV* changes overall estimates of *MAI* and therefore is incorporated in UZFLOW, but the local effect of solar loading is not considered in UZFLOW.

The relationship among *MAI*, *MAP*, *MAT*, *MAV*, and soil thickness (*b*) used in UZFLOW was abstracted from bare-soil infiltration simulations (Stothoff, et al., 1997) as:

$$\ln (MAI) = [a_0 - a_1 B^{b_1} + a_2 \ln (P) - a_3 T - a_4 V^{b_2}][1 - a_5 \ln (MAI)] \quad (3\text{-}7)$$

where:

B	=	normalized soil thickness = $(b/0.02)$ [*b* in m]
P	=	normalized *MAP* = $(MAP/162.8)$ [*MAP* in mm/yr]
T	=	normalized *MAT* = $(MAT - 17.38)/273.15$ [*MAT* in °C]
V	=	normalized *MAV* = $(MAV/4.522 \times 10^{-6})$ [*MAV* in gm/cm³]
a_i	=	regression constant *i*
b_i	=	regression exponent *i*

The normalization constants for *MAP*, *MAT*, and *MAV* are the corresponding 10-yr average values of the hourly meteorologic data for Desert Rock, Nevada, from 1984 through 1993 (National Climatic Data Center, 1984–1994). The 10-yr data set was used for the nominal meteorologic boundary condition for all 1-D simulations.

The a_0 constant is used as an adjustable parameter accounting for all uncertainties in characterizing *MAI*. A sampled parameter within the UZFLOW section of the primary input file specifies the areal-average *MAI* within the repository footprint under current climatic conditions. Internally to the code, the a_0 constant in Eq. (3-7) is adjusted at the start of every realization until the predicted areal-average *MAI* matches the sampled value.

Simple relationships between elevation and *MAP*, *MAT*, and *MAV* were developed for current climatic conditions (Stothoff, et al., 1997).

$$MAP(Z) = \exp(4.26 + 0.000197\ Z) \quad (3\text{-}8)$$

$$MAT(Z) = 25.83 - 0.0084\ Z \quad (3\text{-}9)$$

$$MAV(Z) = \exp(-11.96 - 0.000341\ Z) \quad (3\text{-}10)$$

where Z is the ground surface elevation above sea level in meters. MAP has units of mm/yr, MAT is in units of °C, and MAV has units of gm/cm³.

To characterize the fine-scale variation in soil thickness, MAP, MAT, and MAV, a digital elevation model (DEM) is used. Both elevation (in meters) and soil depth (in meters) are required for every grid cell. The DEM provided with the TPA Version 3.1.4 code uses the finest scale available for YM (i.e., 30×30-m grid cells).

The elevation data are used to calculate MAV for each grid cell under initial climatic conditions, after which the values are assumed to remain constant over the length of the simulation. The calculated values of MAP and MAT in each grid cell i are adjusted according to the time history provided by the climate model, by adding the climate-induced change to the elevation-dependent component:

$$MAP_i(t) = MAP_i(Z) + MAP(t) - MAP_0 \qquad (3\text{-}11)$$

$$MAT_i(t) = MAT_i(Z) + MAT(t) - MAT_0 \qquad (3\text{-}12)$$

For example, if MAT is calculated to be 1 °C (1.8 °F) cooler than at the start of the simulation, the elevation-dependent values of MAT are uniformly cooled by 1 °C (1.8 °F). For each climate time step, hundreds of estimates of MAI are obtained, one for each grid cell in each subarea. The estimates for all grid cells within a subarea are averaged to obtain the subarea value. MAP and the resulting repository-average MAI history are shown in Figure 3-3, with and without climatic perturbations imposed (assuming MAP at full glacial maximum is double the MAP at the start of the simulation). The extreme sensitivity of MAI to climate under bare-soil conditions is evident.

All else being equal, MAI in areas with vegetation will not exceed MAI in areas without vegetation. Under current climatic conditions at YM, potential evaporation is so large that neglecting transpiration from vegetation introduces only a relatively small error in infiltration calculations. Under glacial conditions, however, the ratio of potential evaporation to precipitation is smaller, desert shrubs become larger and more numerous, and the error in MAI predictions introduced by not considering transpiration may become significant. By not considering transpiration, the approximation used in the TPA Version 3.1.4 code almost certainly results in MAI that is larger than would have been observed in previous glacial maxima.

The constants and exponents in Eq. (3-7) depend on the fracture-filling material, although the general form of the relationship is the same for all filling materials. The effects of soil and fracture hydraulic properties on MAI have also been examined, but are not included in Eq. (3-7).

3.1.2.4 Deep Percolation Model

Deep percolation at the repository horizon within each subarea is assumed to be the average value of MAI within the subarea, with fluxes throughout the repository equilibrating rapidly relative to climatic change. The deep percolation model assumes that redistribution of percolating water from the root zone to the repository level occurs in each subarea. Because of gravity, percolating water will tend to move vertically

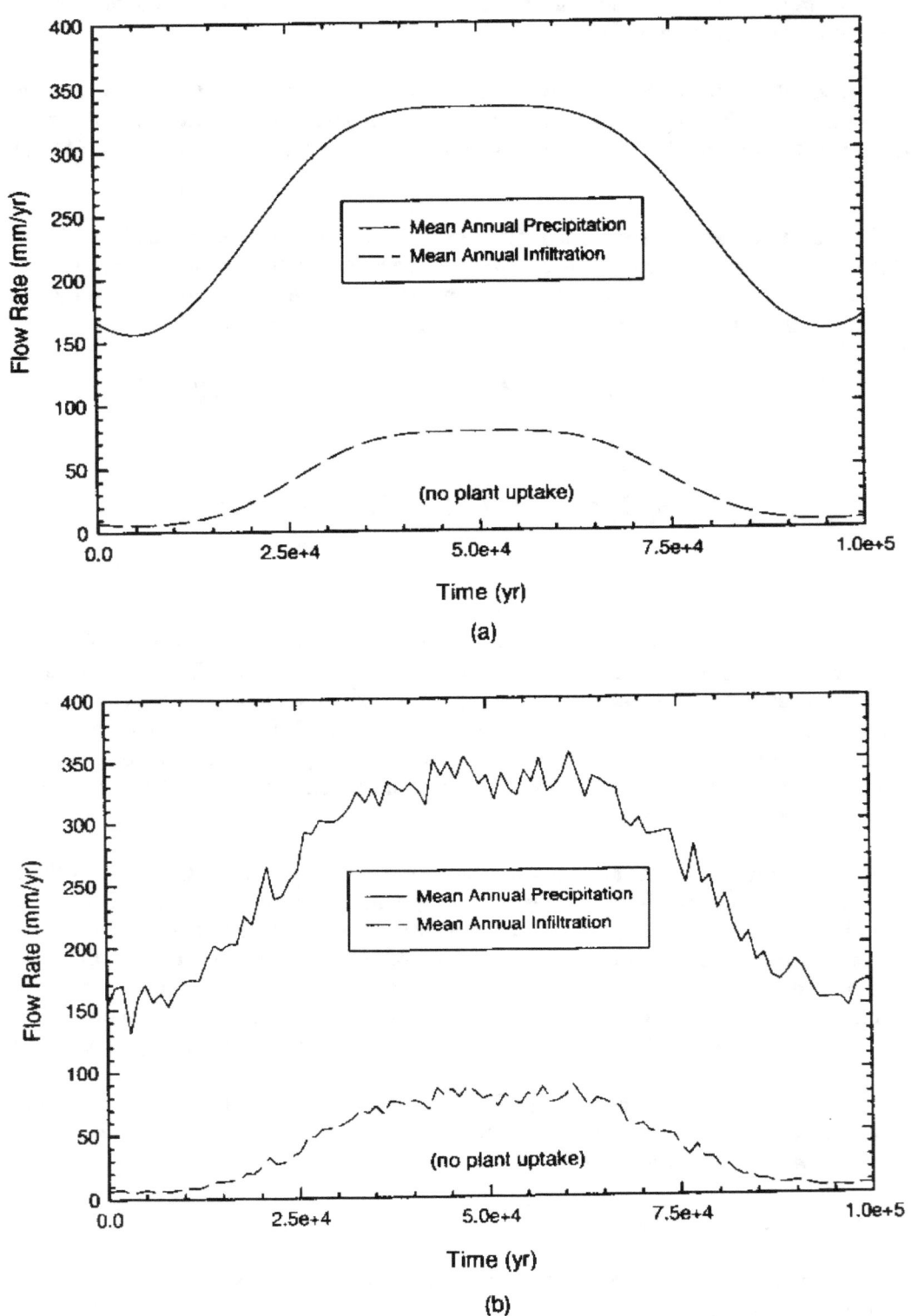

Figure 3-3. Response of *MAI* to changes in *MAP* during a climate cycle: (a) mean climatic signal only; and (b) with perturbation.

in the UZ unless capillary or permeability barriers force lateral redistribution. The resulting model conservatively neglects lateral diversion away from the repository and assumes the subarea-average shallow-infiltration flux moves vertically down to the repository level.

3.1.3 Assumptions and Conservatism of the UZFLOW Approach

The UZFLOW module is an abstraction of the processes controlling MAI and redistribution of infiltrating waters above the repository horizon. The main processes that control infiltration are: (i) climatic conditions or variation of precipitation and temperature; (ii) near-surface processes such as evapotranspiration and runoff; and (iii) lateral diversion of subsurface flow.

Annual infiltration is expected to vary over the long term (i.e., thousands of years) because of variations in temperature and precipitation. Based on historical data and paleoclimatic markers, a full cycle of climatic changes is assumed to occur roughly every 100,000 yrs and conditions may be slightly drier and hotter than present for about the next 10,000 yrs, after which conditions can be expected to become cooler and wetter. This approach is considered a reasonable representation of the long-term evolution of climate. Although short-term perturbations of climate will occur (i.e., some years wetter and cooler and some years drier and hotter than expected), the net effect on estimated doses at a receptor location will be significantly dampened by the anticipated long transport times from the repository to the receptor location (travel times may be on the order of hundreds to thousands of years or longer).

Estimates of infiltration are sensitive to a number of processes affecting near-surface conditions (e.g., temperature, precipitation, soil depth, evaporation, incident solar radiation, plant transpiration, and surface water runoff). Results of detailed 1-D simulations of mass and energy fluxes under a range of conditions representative of the surface of YM were used to develop abstracted, predictive equations for MAI as a function of MAT, MAP, MAV, and soil depth. The detailed model simulations do not account for plant transpiration and runoff that tends to reduce net infiltration. Plant transpiration is not considered to have a significant effect on infiltration over the next 10,000 yrs if, as assumed, climate remains similar to present-day conditions. Beyond 10,000 yrs, neglecting plant transpiration will cause infiltration estimates to become increasingly overestimated because the climate is expected to become wetter and cooler with resultant increases in vegetation and plant transpiration. Also, the abstractions do not account for temporal soil depth variation and composition; both influence infiltration that might occur as the climate changes to wetter and cooler conditions.

After penetrating the subsurface, infiltrating water is assumed to pass in pulses vertically through the fracture system of the Tiva Canyon welded unit to the Paintbrush Tuff nonwelded unit. Here the pulses are assumed to spread and dissipate such that water passes uniformly through to the underlying Topopah Springs welded unit and continues vertically to the repository horizon. It is assumed that contacts between the welded units and the nonwelded unit do not cause systematic lateral diversion, because of the existence of numerous small faults in the nonwelded unit; diversion would tend to reduce fluxes on the western portion of the repository and might increase fluxes on the eastern portion of the repository. The current approach, which neglects lateral diversion of infiltrating water, is considered conservative because lateral diversion is anticipated to reduce the number of containers that encounter dripping water.

There are several specific assumptions related to the UZFLOW module:

- It is assumed that the future climate at YM will follow the general form of the paleoclimatic record, with shorter variations imposed on the general trend.

- Both MAP and MAT are assumed to vary according to the same formula derived from Milankovich cycle data.

- Infiltration is assumed directly related to climatic influences, increasing as precipitation increases and temperature decreases. The linkage between climate and infiltration is calculated assuming that infiltration occurs under bare-soil conditions. (Bare-soil conditions conservatively overpredict infiltration and are most appropriate for present-day climate. As conditions become cooler and wetter, predictions become increasingly conservative.)

- For the computation of shallow infiltration, it is assumed that hydrologic properties and soil genesis will not change during the simulation.

- It is assumed that *MAT* decreases linearly with increasing elevation, whereas *MAP* and *MAV* density vary exponentially with elevation, and the rates of change remain the same over time.

3.2   NFENV MODULE DESCRIPTION

The NFENV module calculates the time-dependent hydrothermal environment of the WP such as

- Average repository-horizon rock temperature
- WP surface and SF temperatures
- RH at the WP surface
- Flow rate of ground water into the near field
- pH and chloride concentrations of ground water flowing onto the WP

The near-field environment includes areas inside the drift and those portions of the geologic setting that the repository modifies and that may affect repository performance. Detailed information on calculation of the pH and chemistry of ground water flowing onto WPs can be found in Mohanty, et al. (1997).

3.2.1 Information Flow Within TPA

3.2.1.1 Information Supplied to NFENV

The NFENV module uses input data specified in the primary input file and other data files (*rectedge.dat* and *multiflo.dat*) for computing temperature, ground-water reflux, and ground-water chemistry. In addition, for each subarea and realization, EXEC passes to NFENV the time-varying volumetric flow rate of water infiltrating toward the repository calculated by UZFLOW for the computation of ground-water reflux. The *rectedge.dat* file provides to NFENV the coordinates for the discretization used exclusively for the computation of repository-horizon average rock temperature. Other input parameters to NFENV for computing temperature and RH, such as thermal diffusivities and effective thermal conductivities, are specified in the NFENV section of the primary input file. Temperature and RH values computed external to the TPA Version 3.1.4 code can be used via the *tefkti.dat* file when the appropriate flag is set in the primary input file. Chloride concentrations are read from the data file *multiflo.dat*.

3.2.1.2 Information Provided by NFENV

The NFENV module passes to EXEC the values for temperature, RH, pH, and the chloride concentration of the representative WP in a subarea for use in EBSFAIL. In addition to the temperature history, NFENV also provides the time history of percolation flux at the repository level to EBSREL through EXEC.

3.2.2 Conceptual Model

The NFENV conceptual model description is divided into three portions: (i) heat transfer, temperature, and RH calculation on a mountain and drift scale; (ii) chemical composition of ground-water flow in the near field; and (iii) calculation of near-field thermally driven reflux and percolation of ground-water.

3.2.2.1 Heat Transfer, Temperature, and Relative Humidity Calculation

Mountain-Scale Heat Transfer

The repository-horizon average rock temperature is computed using an analytic conduction-only model for mountain-scale heat transfer. The model is based on a heated rectangular region residing in a semi-infinite medium. The modeled repository region was divided into nine rectangular regions (Figure 3-4) to cover the repository area corresponding to the seven subareas shown in Figure 3-2. The discretization was based on an AML of 80 MTU/acre or greater and thus did not cover the whole area within the repository boundaries. Each heated rectangular region is defined by a depth of H below the ground surface width, of $2B$ and length of $2L$ (Figure 3-5). Because more than one rectangular region exists, the temperature increase in the semi-infinite medium is the sum of contributions from each heated region. The general solution for the temperature increase at any point in space and time is given by [Claesson and Proberts (1996); Carslaw and Jaeger (1959)]:

$$\Delta T(x,y,z,t) = \int_0^t \frac{\alpha q_{rep}''(t')}{4k\sqrt{\pi}} \frac{1}{\sqrt{4\alpha(t-t')}} \left[\text{erf}\left(\frac{L-x}{\sqrt{4\alpha(t-t')}} \right) + \text{erf}\left(\frac{L+x}{\sqrt{4\alpha(t-t')}} \right) \right]$$

$$\left[\text{erf}\left(\frac{B-y}{\sqrt{4\alpha(t-t')}} \right) + \text{erf}\left(\frac{B+y}{\sqrt{4\alpha(t-t')}} \right) \right] \left[\exp\left(\frac{-z^2}{4\alpha(t-t')} \right) - \exp\left(\frac{-(z-2H)^2}{4\alpha(t-t')} \right) \right] dt' \quad (3-13)$$

where:

$\Delta T(x,y,z,t)$	=	increase in temperature at any time at any point in space and time in the semi-infinite medium from one heated rectangular region [°C]
$q_{rep}''(t)$	=	time-dependent repository heat flux [W/m²]
α	=	thermal diffusivity of the semi-infinite medium [m²/s]

k	=	thermal conductivity of the semi-infinite medium [W/(m °C)]
L	=	half-length of the heated rectangular region [m]
B	=	half-width of the heated rectangular region [m]
H	=	depth of the heated region below the ground surface [m]
t	=	actual time after activation of heat flux [s]
t'	=	time of integration [s]
x,y,z	=	location of interest [m]

The ground surface is assumed to be exposed to atmospheric conditions and has a constant temperature (currently not affected by climate change). The analytic equation is valid below the ground surface, $z < H$. The repository heat flux is related to the AML and heat output per MTU of waste:

$$q_{rep}''(t) = AML \; Q_{mtu}(t) \tag{3-14}$$

Likewise, the thermal output for a single WP is related to the WP inventory:

$$Q_{wp}(t) = MTU_{wp} \; Q_{mtu}(t) \tag{3-15}$$

where:

AML	=	areal mass loading [MTU/m^2]
MTU_{wp}	=	MTU in a representative WP
$Q_{per \; mtu}(t)$	=	time-dependent heat output per MTU of waste [W/MTU]

Waste is assumed to be emplaced uniformly throughout the nine rectangular regions, so there is no spatial variation in the waste heat output. Figure 3-6 shows a plan view of the repository with parallel emplacement drifts with WPs periodically spaced. The temperature increase at any point is caused by the contribution from all subregions. The average rock temperature is computed at an elevation of half the drift diameter at the center of each of the subareas as specified in the primary input file and used throughout the TPA Version 3.1.4 code as a common basis for transferring and calculating information. The subareas should not be confused with the nine rectangular regions in NFENV. The nine rectangular regions are used by the NFENV module to predict only the temperatures in the subareas. The analytic mountain-scale conduction model predicts the rock-wall temperature (T_{rock}) as a function of time. Having computed T_{rock} for the subarea, the WP surface temperature can be calculated.

Drift-Scale Heat Transfer

A multimode (i.e., conduction, convection, and radiation) heat transfer model is used for modeling drift-scale heat transfer. Figure 3-7 shows a WP resting on an emplacement cart within a drift. A simplified thermal network is used to predict the WP surface temperature and the maximum SF temperature given T_{rock} (from the mountain-scale model described previously) and $Q_{wp}(t)$, and is described in the following sections.

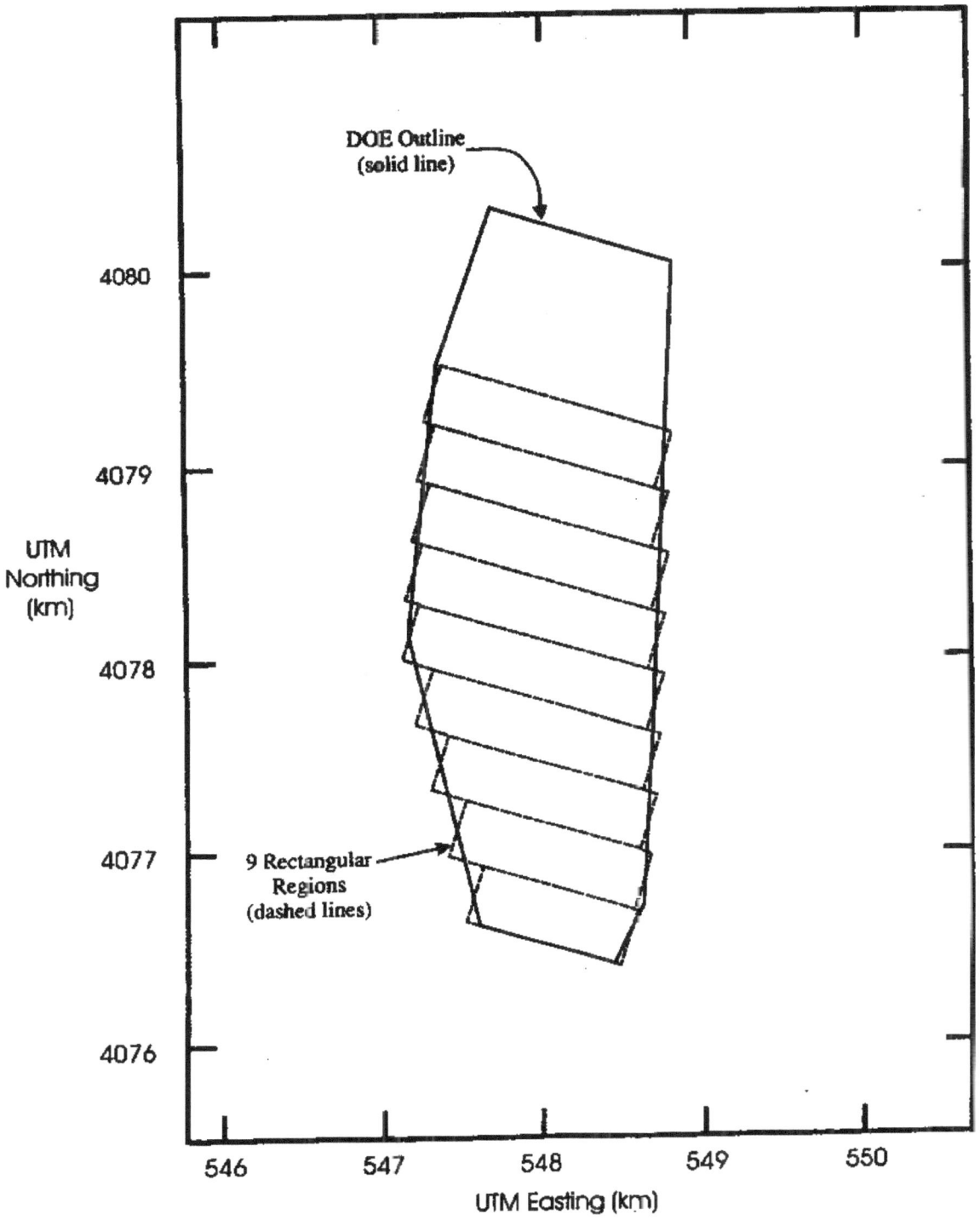

Figure 3-4. Discretization of repository into nine rectangular regions for the NFENV module.

NUREG–1668

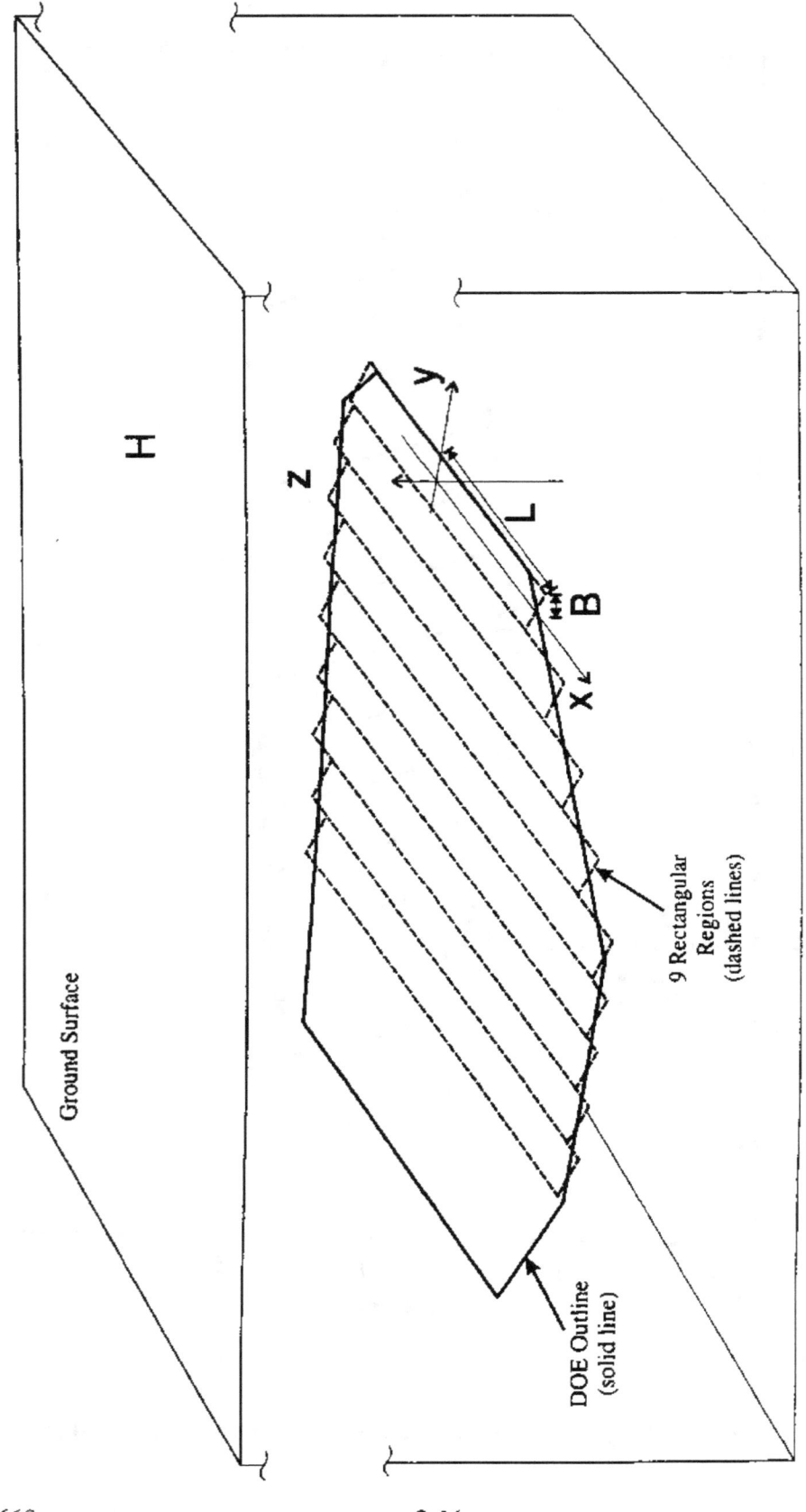

Figure 3-5. Mountain-scale heat transfer model with heated rectangular regions.

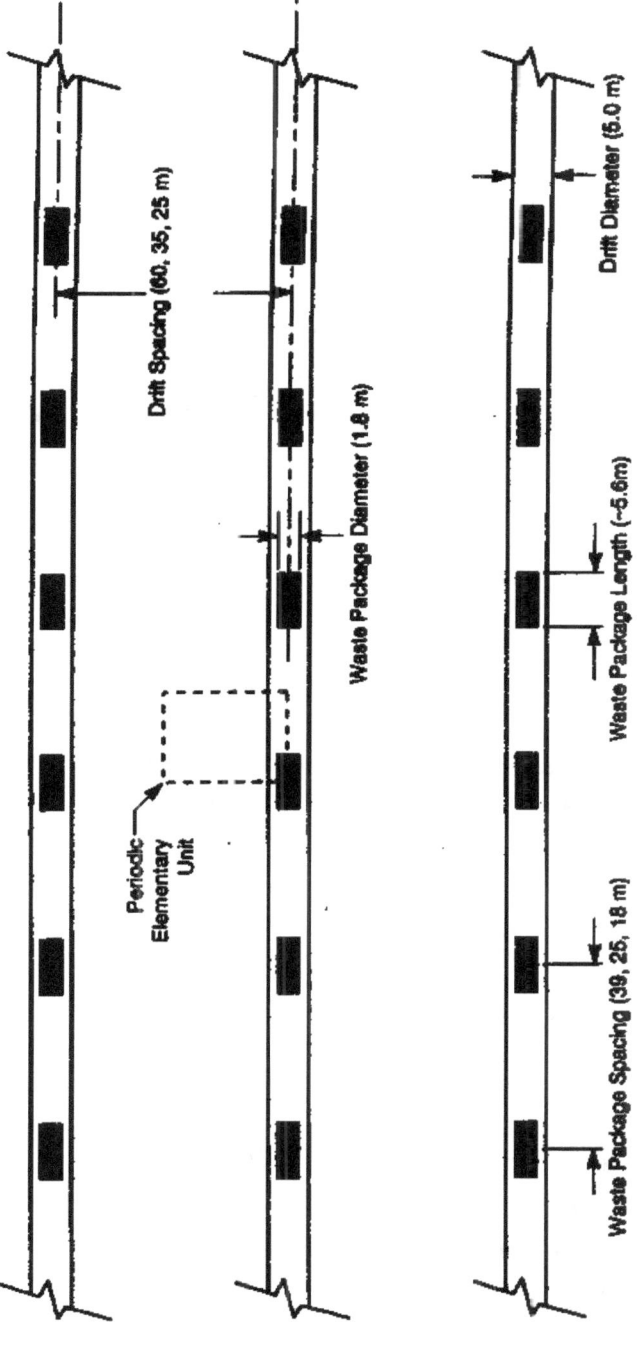

Figure 3-6. Plan view of repository showing emplacement drifts and waste packages.

NUREG–1668

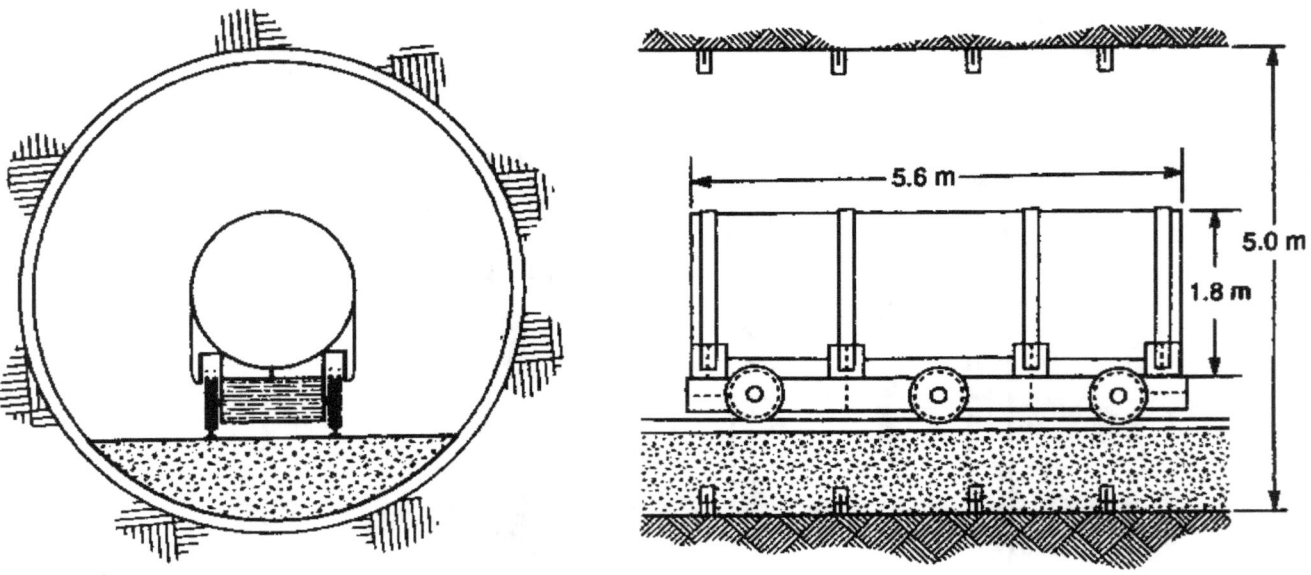

Figure 3-7. Waste package on an emplacement cart in a drift.

Waste Package Surface Temperature

Equation (3-16) is used to solve for WP surface temperature, given the rock temperature, thermal output of the WP, and thermal conductances. Heat is transferred by both thermal radiation and natural convection in the unbackfilled region above a WP, and by conduction through the package support and floor material as:

$$Q_{wp} = \left(G_{rad} + G_{conv} + G_{cond} \right) \left(T_{wp.surf} - T_{rock} \right) \qquad (3\text{-}16)$$

where:

Q_{wp}	=	time-dependent heat output for a WP [W]
G_{rad}	=	effective thermal conductance for radiative heat transfer [W/°C]
G_{conv}	=	effective thermal conductance for convective heat transfer [W/°C]
G_{cond}	=	effective thermal conductance for conduction [W/°C]
$T_{wp.surf}$	=	WP surface temperature [°C]
T_{rock}	=	near-field rock temperature [°C]

The thermal conductance for radiative heat transfer above the WP is based on a linearization of the Stefan-Boltzmann law and accounts for the emissivity of the WP and drift rock wall (Incropera and DeWitt, 1990):

$$G_{rad} = \left(\frac{3}{4}\right) \frac{4\sigma(273.15 + T_{rock})^3}{\dfrac{1 - \varepsilon_{wp}}{\varepsilon_{wp}\pi D_{wp}L_{wp}} + \dfrac{1}{F_{wp\text{-}rw}\pi D_{wp}L_{wp}} + \dfrac{1 - \varepsilon_{rw}}{\varepsilon_{rw}\pi D_{rw}L_{rw}}} \qquad (3\text{-}17)$$

where:

σ = Stefan-Boltzmann constant [5.67×10^{-8} W/(m^{2}K^4)]

ε_{wp} = emissivity of the WP surface [unitless]

D_{wp} = diameter of the WP [m]

L_{wp} = length of the WP [m]

$F_{wp\text{-}rw}$ = radiative view factor from the WP to the rock wall (=1) [unitless]

ε_{rw} = emissivity of the drift rock wall surface [unitless]

D_{rw} = diameter of the rock wall [m]

L_{rw} = length of drift wall per WP drift [m] (estimated to be ~18 m for 80 MTU/acre)

The top three-quarters of the WP are available for the radiative/convective heat transfer and the bottom quarter of the package participates in conduction through the pedestal/floor. Thermal conductances for convective transfer above the WP and conductive transfer below the package are computed from:

$$G_{conv} = \left(\frac{3}{4}\right) \frac{2\pi k_{eff,air}(2L_{wp})}{\ln \dfrac{D_{rw}}{D_{wp}}} \qquad (3\text{-}18)$$

$$G_{cond} = \left(\frac{1}{4}\right) \frac{2\pi k_{floor}(2L_{wp})}{\ln \dfrac{D_{rw}}{D_{wp}}} \qquad (3\text{-}19)$$

where:

$k_{eff,air}$ = effective thermal conductivity of buoyant air estimated to be 30 times stagnant air conductivity [W/(m°C)] (Manteufel, 1997)

k_{floor} = thermal conductivity of the concrete pedestal/floor material [W/(m°C)]

The effective axial length for conductive and convective transfer from the WP to the drift wall should be larger than the length of the WP and smaller than the package spacing length within a drift. A reasonable value for this length is 2 times the WP length.

For the option of backfilled drifts, heat transfer through the top three-quarters can be predicted, using an effective conductivity for the backfill material of:

$$G_{bf} = \left(\frac{3}{4}\right) \frac{2\pi k_{eff.bf}\left(2L_{wp}\right)}{\ln \dfrac{D_{rw}}{D_{wp}}} \tag{3-20}$$

where:

G_{bf} =   effective thermal conductance for backfill region [W/°C]

$k_{eff.bf}$ =   effective thermal conductivity of backfill material [W/(m°C)]

Waste Package Inner Wall Temperature

After computing the outer WP surface temperature, the inner surface temperature of the wall is calculated. The wall of the WP consists of two cylindrical layers for the inner and outer overpacks. The thicknesses and properties of the walls are parameters specified in the primary input file. The inner wall temperature is related to the WP heat, according to:

$$Q_{wp} = G_{shell}\left(T_{in,surf} - T_{wp.surf}\right) \tag{3-21}$$

where:

G_{shell} =   thermal conductance for WP shell [W/°C]

$T_{in.surf}$ =   inner surface temperature of the WP wall [°C]

A schematic of the internals of a WP is shown in Figure 3-8. The shell conductance consists of a contribution from the outer carbon steel layer and inner nickel-based alloy layers and is calculated according to

$$G_{shell} = \frac{L_{wp}}{\dfrac{t_{ss}}{\pi D_{ss} k_{ss}} + \dfrac{t_{cs}}{\pi D_{cs} k_{cs}}} \tag{3-22}$$

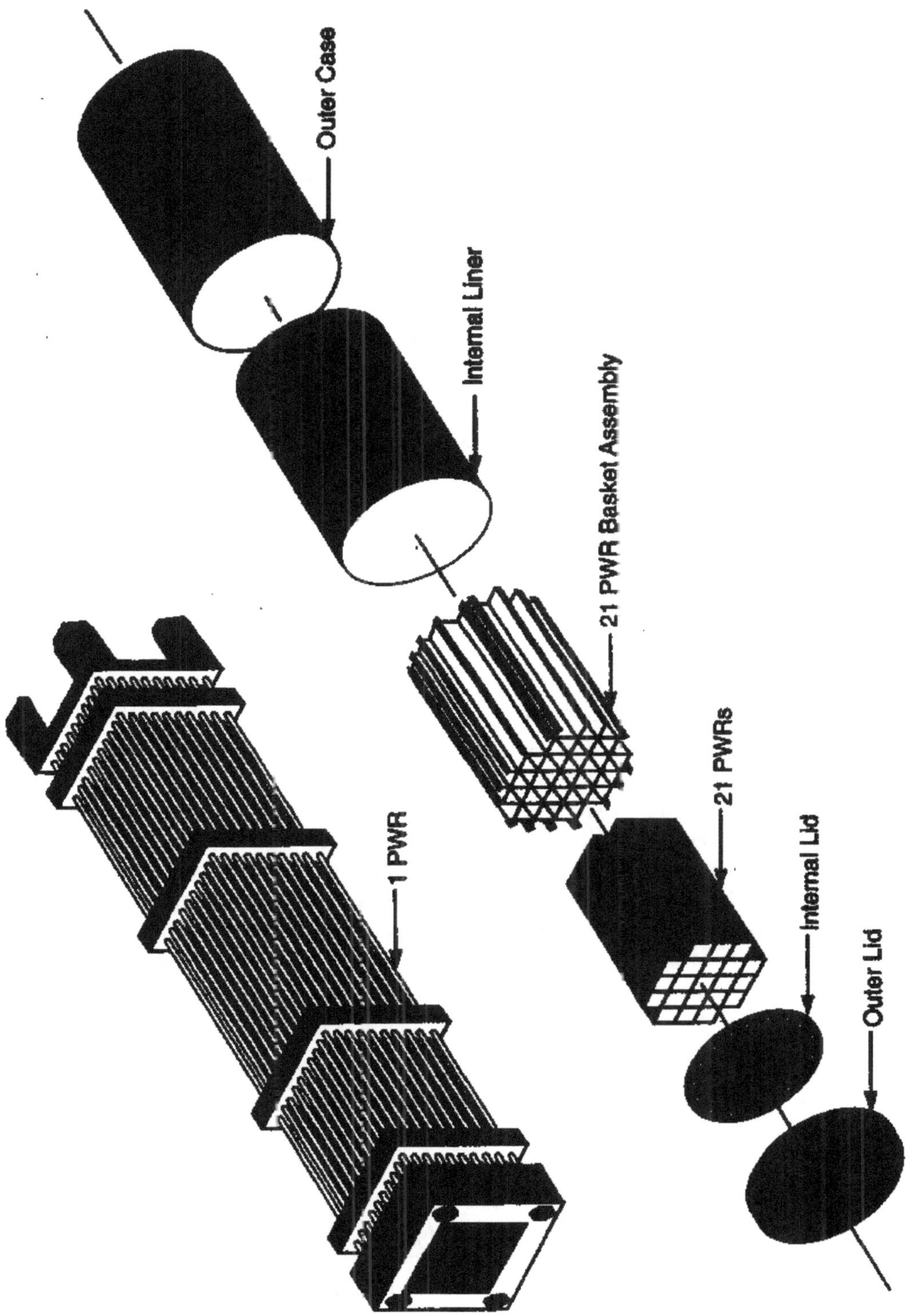

Figure 3-8. Internals of a large waste package

Labels in figure: Outer Case, Internal Liner, 21 PWR Basket Assembly, 21 PWRs, 1 PWR, Internal Lid, Outer Lid

where:

t_{ss}	=	thickness of the inner stainless steel layer [m]
D_{ss}	=	diameter of the inner Ni-based alloy layer [m]
k_{ss}	=	thermal conductivity of stainless steel [W/(m°C)]
t_{cs}	=	thickness of the outer carbon steel layer [m]
D_{cs}	=	diameter of the outer carbon steel layer [m]
k_{cs}	=	thermal conductivity of carbon steel [W/(m°C)]

Maximum Spent-Fuel Temperature

Using the WP inner surface temperature, the maximum SF temperature($T_{max, sf}$) is calculated using a conduction shape factor formula that accounts for the volumetric heat generation in the interior region of the package, which includes the SF assemblies and the basket assembly (Manteufel and Todreas, 1994):

$$Q_{wp} = G_{int} \left(T_{max,sf} - T_{in,surf} \right) \tag{3-23}$$

where the conductance of the cylindrical interior region is computed from:

$$G_{int} = k_{sf} \, S \, L_{wp} \tag{3-24}$$

where:

k_{sf}	=	effective thermal conductivity of the basket and SF in the WP [W/(m°C)]
S	=	conduction shape factor for a heated cylindrical region equal to 4π

The effective thermal conductivity of the SF accounts for the region between the inner wall and the basket material, the basket material, and the individual assemblies. There are multiple modes of heat transfer, including thermal radiation, buoyant convection primarily in the larger void regions, and conduction in the basket material, fuel rods, and regions with primarily stagnant gas. At high temperatures, the heat transfer, is dominated by radiative transfer. At lower temperatures, heat transfers can be dominated by conduction. Effective thermal conductivity is a function of temperature and is a parameter specified in the primary input file.

Relative Humidity

The WP temperature and RH are required inputs for the corrosion and release models. The RH is defined as the ratio of the actual vapor pressure to the vapor pressure at the WP surface:

$$RH = \frac{P_v \left[\min \left(T_b, \, T_w \right) \right]}{P_v \left(T_{wp} \right)} \tag{3-25}$$

where:

P_v	=	vapor pressure that is a function of temperature [Pa]
min (T_b, T_w)	=	minimum of T_b and T_w
T_b	=	boiling point temperature [~370 °K at repository]
T_w	=	drift wall temperature [°K]
T_{wp}	=	WP surface temperature [°K]

Below boiling conditions, the definition of RH used in Eq. (3-25) is equivalent to the mole fraction definition of RH frequently found in thermodynamic textbooks (e.g., Van Wylen and Sonntag, 1978; Moran and Shapiro, 1992). RH is generally defined as the actual mole fraction of water vapor in the air divided by the maximum or saturation mole fraction of water vapor in the air at the same temperature and pressure. Below boiling conditions, mole fractions are related to the vapor partial pressures so that this definition is equivalent to Eq. (3-25). Above boiling conditions, the vapor partial pressure cannot exceed the atmospheric pressure within the drift even in the presence of backfill because of its high porosity. When the WP surface temperature exceeds the boiling point, it is preferable to define RH as a ratio of the two vapor pressures specified previously, which is consistent with the technical literature (Hartman, 1991; Bejan, 1988; Fyfe, 1994).

As an alternative to the temperature and RH models described previously, the use of tabular input of temperature and RH into TPA Version 3.1.4 code can be selected in the primary input file, whereupon this information is obtained from an external data file, *tefkti.inp*. Results from detailed 2-D and 3-D modeling that can more accurately predict repository edge heat losses can be incorporated in the form of tabular data.

3.2.2.2 Near-Field Chemical Composition

The chemical composition of fluid able to come in contact with the WP is an important consideration in the modeling of WP integrity and RT in the TPA Version 3.1.4 code. NFENV is designed to provide the chemical composition of the environment as a function of time in the immediate neighborhood of the WP. Chemistry parameters, which include solution pH, oxygen fugacity, chloride and bicarbonate concentration, dissolved silica, and alkalinity, among other environmental factors, can have important consequences on the rate of corrosion of the WP, dissolution of SF, and formation of alteration products. Of special concern in the chemical composition of a partially saturated environment are evaporative effects produced by the heat released from the WP. Chemistry can also be affected by refluxing of evaporated water and by deep percolation at long time periods.

The maximum silica concentration estimated from equilibrium with quartz, chalcedony, and cristobalite gives, respectively, $a_{SiO_2} = 0.000835$, 0.00137, and 0.00218 molal. By contrast, J-13 well water has a silica concentration of 0.0011 molal (Table 3-1). Bounds on the pH are more difficult to obtain. Likewise, calcium and carbonate have no obvious upper bounds since their concentrations will be pH-dependent. Turner (1998) screened available ground-water chemistry data compiled by Perfect, et al. (1995), which may provide some limits for current ambient conditions in the vicinity of YM.

To provide a more detailed calculation of the near-field fluid composition, the computer code MULTIFLO (Lichtner and Seth, 1996; Seth and Lichtner, 1996) is used to provide quantitative data in tabular form to the TPA Version 3.1.4 code. MULTIFLO simulates the transport of reacting chemical constituents coupled to evaporation and condensation processes involving two-phase fluid flow. The code sequentially couples two-phase fluid flow of liquid water, water vapor, and air with reactive transport of aqueous and gaseous species. Homogeneous reactions in the aqueous phase and heterogeneous reactions between the aqueous and gaseous phases are assumed to be in local equilibrium. Mineral reactions are treated irreversibly through prescribed kinetic rate laws.

Table 3-1. Initial Fluid Composition and pH Corresponding to J-13 Well Water (Adapted From Harrar, et al., 1990)

Species	Molarity $\times 10^4$	Species	Molarity $\times 10^4$
Ca^{+2}	2.90 – 3.70	$SiO_2(aq)$	9.50 – 11.40
Na^+	18.30 – 21.70	Cl^-	1.78 – 2.37
K^+	1.00 – 1.40	pH	6.8 – 8.3
HCO_3^-	1.93 – 2.34	—	—

MULTIFLO computes temperature and the associated moisture redistribution as functions of time, using a repository-scale model, represented as a single disk-shaped uniform heat source, in contrast to the drift-scale model presented earlier in this chapter. Therefore, the detailed geometry of the WPs is lost. The fractured YM tuff host rock is described using the equivalent continuum model (ECM) in which fracture and matrix properties are averaged into a single continuum representation. Model predictions indicate the formation of zones of enhanced liquid saturation above and below the repository during the dry-out period after emplacement of the waste. As capillary suction draws liquid water toward the heat source, evaporation occurs, resulting in an increase in salinity and pH as carbon dioxide (CO_2) is degassed from solution. In the cooler condensate region, the salinity is expected to decrease as the dilute condensing solution is mixed with the ambient ground water.

The aim of the MULTIFLO calculations is to estimate the composition of the fluid at the boundary of the condensate zone that, as a first approximation, may be considered to be similar to the fluid that comes in contact with the surface of the WP. Within the confines of the repository-scale model and the ECM approximation of fracture-matrix interaction, however it is not possible to fully represent the eventual return of liquid water from the condensation front lying above the repository to the WP. This flow most likely would take place along open fractures in the form of gravity-driven flow manifested as dripping.

In the initial phase of work to determine the near-field environment and provide input to the TPA Version 3.1.4 code, a 1-D repository-scale calculation using MULTIFLO Version 1.0 code was performed with an AML of 80 MTU/acre corresponding to 26-yr-old SF with a mix of 65 percent PWR and 35 percent BWR assemblies. With this heat load, the calculations show that a liquid phase is always present and complete dryout does not occur after emplacement of the waste. The initial fluid composition corresponds to J-13 well water, as given in Table 3-1. The YM host rock was modeled in these preliminary calculations as pure quartz with 10 percent porosity. This assumption is considered reasonable for the purpose of estimating the change in salinity caused by evaporation and condensation, but may not be valid for the pH, if reactions with other silicate minerals prove to be important. Additional work (e.g., increased dimensionality) is need to determine the adequacy of this 1-D approach for representing the local conditions of the near field.

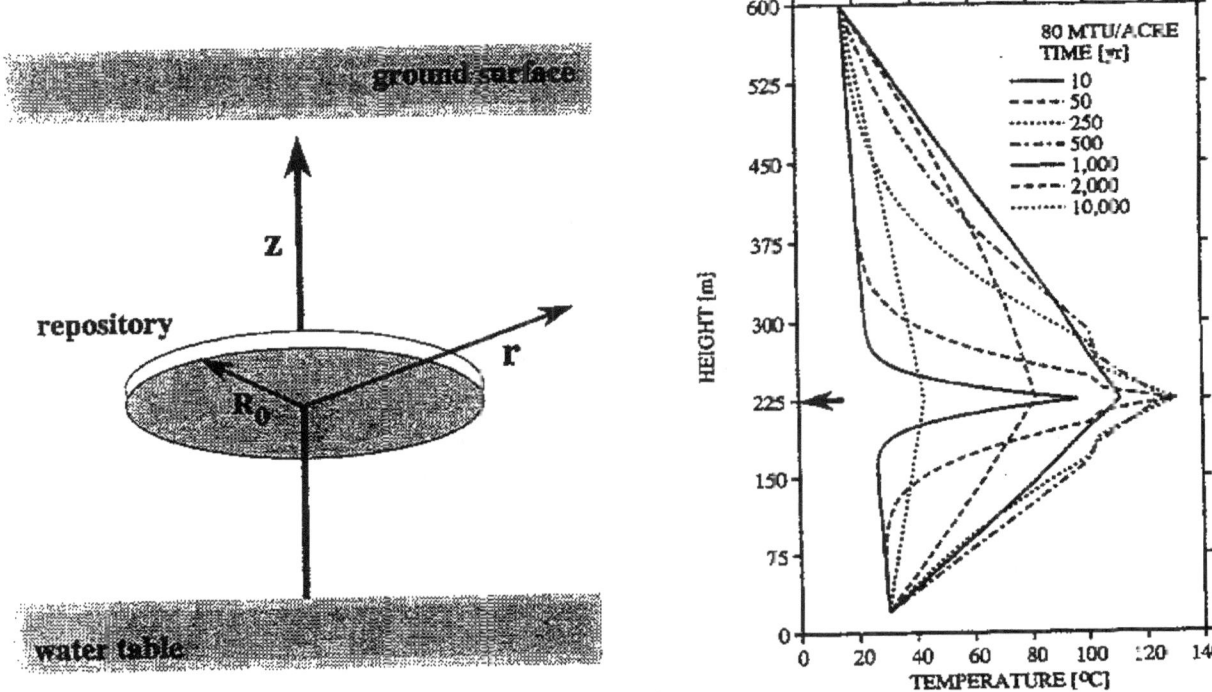

Figure 3-9. Time snapshots of the vertical distribution of temperature caused by repository heating.

The temperature profile shown in Figure 3-9 was computed at different times. The plateau of approximately 100 °C (212 °F) indicates the presence of a heat pipe. Above and below the repository horizon, zones of enhanced moisture form as vapor condenses. The pH increases in the vicinity of the repository to approximately 10 at 250 yrs after waste emplacement and then begins to decrease slowly as the repository cools. The chloride concentration increases by an order of magnitude at 250 yrs after waste emplacement, and then declines at much longer times.

Within the MULTIFLO code, a simple conceptual model was developed to determine chloride concentrations as a function of temperature, evaporation, and precipitation of sodium (Na)-bearing minerals. To estimate the expected changes in chloride concentrations caused by evaporation, a simple upper bound on the chloride concentration may be derived by assuming equilibrium with halite (NaCl), for example. As evaporation occurs, the sodium and chloride concentrations must remain approximately proportional to one another. Therefore, assuming that the activity coefficients of Na$^+$ and chloride (Cl$^-$) are similar at high ionic strength

$$a_{Na^+} = \alpha \, a_{Cl^-} \tag{3-26}$$

where α is the constant of proportionality and a_{Na^+} and a_{Cl^-} are sodium and chloride activities.

NUREG–1668

Equilibrium with halite implies

$$K = a_{Na^+} \cdot a_{Cl^-} = \alpha a_{Cl^-}^2 \qquad (3\text{-}27)$$

where K is the equilibrium constant. Therefore,

$$a_{Cl^-} = \sqrt{\frac{K}{\alpha}} \qquad (3\text{-}28)$$

The log for halite at 100 °C (212 °F) is equal to 1.578 (Wolery, 1992). Taking $\alpha = 11.11$, derived from J-13 well water (Table 3-1), it follows that $a_{Cl^-} = 1.85$ molal and $a_{Na^+} = 20.5$ molal. To estimate concentrations, activity coefficient corrections must be included. If the sodium concentration is lowered by precipitation of other Na-bearing solids, the chloride concentration in equilibrium with halite could be substantially increased and the concentration of sodium reduced.

Although MULTIFLO provides chloride concentration, pH, and oxygen fugacity as time-dependent functions, only chloride concentration is used in the TPA Version 3.1.4 code as a function of time. The pH value is currently set at a constant value of 9, the highest value obtained from the MULTIFLO simulations. Chloride concentrations, as provided by MULTIFLO, are multiplied by a parameter representing uncertainties and limitations in the MULTIFLO results (i.e., MULTIFLO results are representative of ground-water chemistry in the porous matrix) and chloride concentration on a WP could be higher than predicted by MULTIFLO because of evaporation on the WP.

3.2.2.3 Near-Field Ground-Water Percolation

NFENV also provides estimates of ground-water reflux and percolation rates. The amount of water percolating through the near field is different from the infiltration computed in UZFLOW because of thermohydrologic effects resulting from decay heat. WP temperatures may exceed the boiling point of water and vaporize water in the rock surrounding the WPs. The vapor will flow away from the WPs and condense where temperatures are below boiling. The condensate may then flow back toward the WPs. This return flow of condensate is called refluxing.

The area between the WPs and the condensate zone is called the boiling or dry-out zone. Although dry-out and condensate zones would form both above and below the repository, only the zones that form above the repository are considered in the reflux models. Water that fully penetrates the dry-out zone would be available to contact the WPs, possibly accelerating the corrosion of WP materials and facilitating transport of radionuclides released from failed WPs.

Two reflux models have been developed -- REFLUX1 and REFLUX2. Figure 3-10 is a schematic illustrating the movement of water above the repository. Although Figure 3-10 and the following discussion refer primarily to REFLUX1, in general, the concepts apply to both models. Differences are noted in the relevant sections.

Figure 3-10 illustrates the vaporization of ambient rock water in the area above the boiling point temperature, T_{boil}. The T_{boil} isotherm is at an elevation Z_{boil}, whose location depends on the rate at which heat is generated by the waste and the heat loss away from the near-field environment (calculated in the

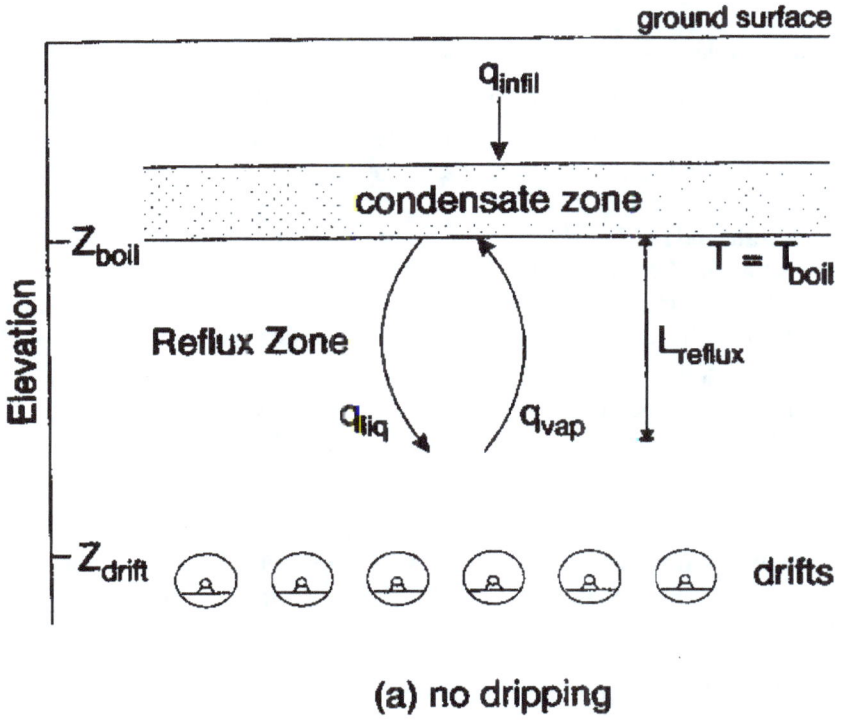

(a) no dripping

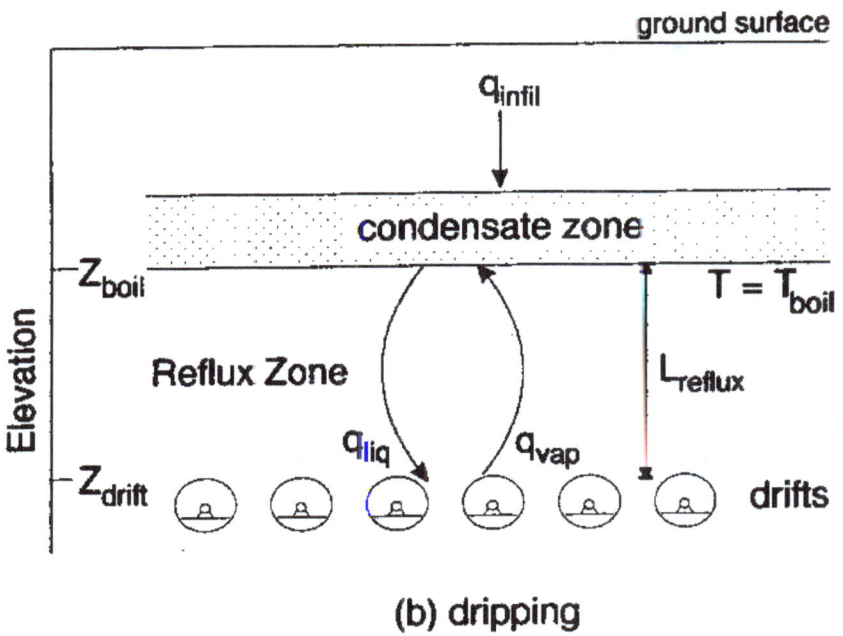

(b) dripping

Figure 3-10. Conceptualization of drift-scale thermal model.

mountain-scale heat-transfer model). Above the T_{boil} isotherm is the condensate zone where water vapor condenses to a liquid that is available to move down toward the repository. Below the T_{boil} isotherm is a boiling zone through which liquid water from the condensate zone flows until it is heated to the point of vaporization and subsequently rises back to the condensate zone.

Together the condensate and boiling zones constitute a reflux zone containing water driven from the rock surrounding the repository by heat generated in the repository. No reflux of ground water into the repository can occur when the bottom of the reflux zone is at or above the elevation of the top of the drift (which can be the case early in the heating phase) (Figure 3-10a). As heat generated by the WPs diminishes over time, the elevation of the T_{boil} isotherm and reflux zone moves closer to the elevation of the repository, and dripping in fractures or reflux can occur at the repository. The thickness of the reflux zone, L_R, depends on q_{infil} and the local heat flux (i.e., the temperature gradient). When $(Z_{boil} - L_R)$ is below the elevation of the top of the drift, Z_{drift}, water begins to drip into the drift (Figure 3-10b). Any liquid passing below the level of the repository is assumed to continue to the water table. The thermal load reflux is calculated in the NFENV module, using one of two models. The source of water in REFLUX1 is only the water infiltrating from the surface, whereas the sources of water for REFLUX2 are infiltrating water and water from the dry-out zone surrounding the repository.

NFENV uses the time-dependent temperature profiles generated by the heat transfer models previously described, along with time-dependent water flux (q_{infil}) from UZFLOW, to calculate time-dependent flux (q_{drip}) in fractures entering the near-field environment and available for dripping onto one WP. In the development of q_{drip}, NFENV considers the time-dependent amount of perching caused by repository heating and the time-dependent reflux of liquid and vapor. The REFLUX models are intended to determine q_{drip} for matrix flow and fracture flow. NFENV partitions percolation flux from the reflux zone, q_{perc}, from both models on a per WP cross-sectional area basis to obtain q_{drip}.

REFLUX1 Methodology

The REFLUX1 subroutine of NFENV is used to model redistribution of water in the near-field environment owing to heat from emplaced WPs. Vaporized water is assumed to move away from the repository until conditions are cool enough for condensation (i.e., conditions are cooler than boiling). Although a condensate zone would form on either side of the repository, only the zone above the repository is considered in REFLUX1. Above the boiling isotherm, a saturated condensate zone forms, bounded below by the boiling isotherm. Cool water from the condensate zone is assumed to percolate within fractures toward the repository within a reflux zone, with the liquid boiling and returning to the condensate zone as vapor, as shown in Figure 3-10.

A mass balance model is used to model the thickness of the condensate zone above the repository:

$$\frac{\partial V_c}{\partial t} = q_{infil} - q_{perc} \tag{3-29}$$

where:

V_c = volume of the condensate zone per unit area $[m^3/m^2]$

$$t \quad = \quad \text{time [yr]}$$

$$q_{\text{infil}} \quad = \quad \text{infiltration flux entering the condensate zone [m/yr]}$$

$$q_{\text{perc}} \quad = \quad \text{percolation flux leaving the reflux zone [m/yr]}$$

When $V_c = 0$ (no condensate zone exists) and when V_c is at a maximum value (set in the primary input file with the parameter PerchedBucketVolumePerSAarea[m3/m2]), q_{perc} is directly set to q_{infil}. A reasonable value for the maximum perched-water volume might be the fracture porosity times the thickness of the Welded Topopah Spring unit above the repository.

Although the condensate zone is intermediate between the two limiting cases, the position of the bottom of the reflux zone determines q_{perc}. If the reflux zone is completely above the drifts, all infiltration flux is stored in the condensate zone and $q_{\text{perc}} = 0$. Within the reflux zone, the liquid flux, q_{liq}, is everywhere balanced by the vapor flux, q_{vap}. If the reflux zone intercepts the repository horizon, q_{perc} is determined by evaluating q_{liq} at the repository horizon, assuming that q_{liq} decreases linearly from the boiling isotherm to the bottom of the reflux zone:

$$q_{\text{perc}} = q_{\text{reflux}} \left(1 - \frac{Z_{\text{boil}} - Z_{\text{drift}}}{L_{\text{reflux}}} \right) \tag{3-30}$$

where:

$$q_{\text{reflux}} \quad = \quad \text{downward liquid flux at the bottom of the condensate zone [m/s]}$$

$$L_{\text{reflux}} \quad = \quad \text{thickness of the reflux zone [m]}$$

$$Z_{\text{boil}} \quad = \quad \text{elevation of the boiling isotherm [m]}$$

$$Z_{\text{drift}} \quad = \quad \text{elevation of the drifts [m]}$$

The maximum thickness of the reflux zone, L_{reflux}, is a parameter in the primary input file called LengthOfRefluxZone[m], while the downward liquid flux at the boiling isotherm, q_{reflux}, is named MaximumFluxInRefluxZone[m/s]. Offline calculations are required to define these parameters.

The elevation of the boiling isotherm, Z_{boil}, which defines the bottom of the condensate zone and the top of the reflux zone, is determined by assuming that temperature varies linearly near the repository. Alteration of the temperature field, because of moisture redistribution and the effects of salt concentration and capillarity on T_{boil}, is neglected. The temperature gradient is calculated by assuming the heat generated by the repository is carried by conduction within the boiling isotherm; knowing the thermal conductance of the matrix and the thermal flux caused by the repository (obtained from NFENV calculations), the temperature gradient can be determined. The temperature at the repository (obtained from NFENV heat transfer calculations) is projected according to:

$$Z_{\text{boil}} = Z_{\text{drift}} + \left(T_{\text{boil}} - T_{\text{repos}} \right) \left(\frac{K_{\text{therm}}}{0.5 q_{\text{heat}}} \right) \tag{3-31}$$

where:

Z_{boil} = elevation of the boiling isotherm [m]

Z_{drift} = elevation of the drifts [m]

T_{repos} = temperature of the repository [°C]

K_{therm} = thermal conductivity [W K^{-1} m^{-1}]

q_{heat} = thermal flux due to the repository [W/m^2]

During much of the boiling phase for the repository, q_{infil} will exceed q_{perc} and liquid (condensate) will accumulate in the condensate zone. If the condensate zone grows beyond the user-specified limit (L_{reflux}), the condensate zone overflows and this is added to any fluxes in q_{perc} caused by the reflux zone contacting the drifts.

Once boiling around the repository stops, the reflux zone ceases to exist and any water contained in the condensate zone is released over a few time steps. During this post-boiling period, at each time step q_{perc} is set to q_{infil} plus one-half of the water remaining in the condensate zone. After a few time steps, q_{perc} approaches q_{infil}. Note that if q_{reflux} is significant, the condensate zone may empty as the boiling isotherm returns to the drift level.

Although dripping in a drift can occur $\left(q_{perc} > 0\right)$, the dripping flux, q_{drip}, is calculated on a per-WP basis (i.e., q_{perc} is multiplied by the area occupied by WP) and passed to EBSREL for determination of the amount of water that contacts and enters the WP.

REFLUX2 Methodology

REFLUX2 was created as an alternate to REFLUX1 to calculate the amount of water able to contact the WPs based on thermal and gravity driven refluxing. Conceptually, in REFLUX2 the quantity of refluxing water can be sufficient to penetrate the boiling isotherm in the fractures and reach the WP during times when the surface temperature of the WP exceeds boiling. The primary difference between REFLUX1 and REFLUX2 is that in REFLUX2 some condensate return flow may penetrate the boiling isotherm no matter how far the boiling isotherm is located above the WP. In addition, unlike REFLUX1, REFLUX2 considers rock (matrix) water as well as deep percolation of infiltration. As such, REFLUX2 is intended to determine flow toward the repository, considering both infiltration from the surface and water originating in the matrix of the dry-out zone surrounding the repository in the boiling phase.

Similar to REFLUX1, REFLUX2 is based on a reflux cycle located above the WPs. A particular parcel of water may participate in the reflux cycle many times. With every cycle, however, some portion of the refluxing water may escape as either vapor or liquid and flow away from the heat source, possibly toward the water table. Alternatively, the refluxing cycle can gain water from two sources: infiltration from ground surface (from UZFLOW) and water vaporized from the dry-out zone in rock surrounding the near field.

The amounts of water contributed to the refluxing cycle by infiltration and from vaporized ambient rock pore water are calculated separately. The total amount of refluxing water derived from both infiltration and rock pore water vaporization in year N is:

$$R_T = R_I + R_D \qquad (3\text{-}32)$$

where:

R_I = the amount of infiltration-derived water [m³/m²]

R_D = the amount of refluxing ambient rock pore water [m³/m²]

Reflux Derived from Infiltration

The amount of R_I that refluxes during year N is:

$$R_I = \sum_{j=0}^{N-1} \frac{q_{\text{infil}_{N-j}}}{A_{sa}} \left(1 - L_I\right)^j \qquad (3\text{-}33)$$

where:

$q_{\text{infil}_{N-j}}$ = infiltration for a given subarea from UZFLOW for the year $(N-j)$ [m³/yr]

A_{sa} = area for a given subarea [m²]

N = years after start of simulation [unitless]

L_I = fraction of infiltration that escapes the reflux cycle each year, user input [unitless]

The variable L_I (Reflux2LossI) is specified in the primary input file.

Reflux Derived from the Rock Pore Water

The total amount of water, D [m³/m²] vaporized from prewaste-emplacement ambient rock pore water and available for refluxing is defined by:

$$D = (T)(n)(S - S_r) \qquad (3\text{-}34)$$

where:

T = thickness of dry-out zone [m]

n = porosity of rock [unitless]

S = liquid saturation [unitless]

S_r = residual saturation [unitless]

The thickness of the dry-out zone (Reflux2Thickness), porosity (Reflux2Porosity), liquid saturation (Reflux2SatInit), and residual saturation (Reflux2SatResid) are specified in the primary input file.

A portion of refluxing water may escape each reflux cycle (L_D) lowering the amount of refluxing ambient rock pore water, R_D [m³/m²]. The time required to complete one water reflux cycle may vary from

1 yr to several hundreds to thousands of years. The amount of rock pore water that refluxes in a given year is

$$R_D = \frac{\left[D \left(P - L_D \right)^{N-1} \right]}{P^N}$$ (3-35)

where P is the number of one-year cycles required for water to complete one reflux cycle. The variables L_D (Reflux2LossD) and P (Reflux2Period) are specified in the primary input file.

The predominant source of refluxing water may change with time. At early times, refluxing water may be dominated by water derived from the rock pore water. At later times, water derived from infiltration may dominate if rock pore water near the WPs has been vaporized and removed from the refluxing cycles. In this case, the amount of refluxing water will converge to the value of R_I. If no losses from the refluxing cycle are experienced, the amount of refluxing water will grow at a rate equal to infiltration. In REFLUX2 the temperature and thickness of the dry-out zone do not vary with time. The temperature is assumed to be above boiling until all the water in its pores leaves the dry-out zone; however, at long time periods (>1,000 yrs) refluxing water will equal infiltration if reasonable values are selected for parameters of REFLUX2.

Water exiting the reflux cycle and entering the repository near-field environment is computed from:

$$R_{wp} = (R_I L_I + R_D L_D) A_{sa}$$ (3-36)

These values are then converted from a subarea basis to a WP basis by multiplying the flow out of the reflux zone by the fraction of the subarea associated with a WP, passed to EBSREL, to determine the amount of water that contacts the waste form and enters the WP.

3.2.3 Assumptions and Conservatism of the NFENV Approach

The near-field environment will affect the corrosion of the WP and the dissolution of SF and the amount of water entering the drift and WP. The NFENV module is an abstraction of the physical and chemical processes in the near field of the repository and includes the determination of (i) temperature of the WP and waste; (ii) evolution of near-field chemistry; and (iii) refluxing of condensed water and ground-water percolation.

NFENV has two methods, from which the user must select one, for determining the temperature of the WP and the waste form. One method determines the temperatures through an analytic conduction-only heat transfer model at the mountain-scale and an analytic, steady-state heat transfer model at the drift-scale that accounts for conduction, convection, and radiation in the vicinity of the WP. The analytic solutions simplify the problem by using uniform thermal properties and uniform heat loading at the repository level, and nine rectangular regions to represent the repository. The other method allows the user to supply WP and waste temperatures in tabular form (temperature predictions would presumably be based on the results of a more detailed model that would include conductive and advective heat transfer in rocks). Both of these methods provide a single temperature for a representative WP in each repository subarea. The use of a single temperature and a representative WP for each subarea should provide a conservative result because all WPs in a subarea will simultaneously corrode and have the potential for simultaneous release of radionuclides (actual release will depend on the number of packages that get wet).

NFENV has two methods, from which the user must select one, for determining the RH. One method determines the RH through an analytic expression based on vapor pressure as a function of temperature, whereas the other method allows the user to supply the RH for specific times in tabular form. Both these methods provide RH values for a representative WP (at the WP surface) in each repository subarea. This approach is considered reasonable and is conservative, based on the reasons discussed previously for temperature.

Water will be redistributed in the near field by evaporation caused by heat from the emplaced waste. Estimating the redistribution of water is highly uncertain because of the complicated two-phase flow processes, variations in the thermal gradient, and heterogeneities in the fracture and matrix hydrologic properties. Refluxing water is assumed to penetrate the boiling isotherm (Phillips, 1996) in the vicinity of a WP, allowing liquid water to contact the package even when temperatures exceed the boiling point of water. Reflux of water is represented by two different models (one specifies a reflux zone and the other specifies a dry-out zone and the fraction of infiltration that escapes the reflux cycle each year). These models have the flexibility to produce conditions varying from a long dry-out period to a short dry-out period. Detailed numerical analysis and heater experiments are continuing to determine the effect of the redistribution of water. The degree of conservatism of any particular approach is uncertain at this time.

The chemical composition of the water contacting the WP can affect corrosion of the WP and dissolution of the waste. The evolution of the chemistry of the water is the result of a variety of complicated processes associated with, but not limited to, mineral reactions, evaporation, and condensation. These types of complicated processes cannot be accommodated in a system code while maintaining reasonable execution times. The chemical composition of water in a partially saturated environment is affected by evaporative effects produced by the heat released from the waste. For use in NFENV, an average chemical composition of the repository near field is computed using the MULTIFLO numerical model to estimate the composition (but not the amount) of the fluid that could contact the WP (e.g., pH, chloride concentration, silica concentration, carbonate concentration, and oxygen fugacity). Because of the uncertainty in estimating the water chemistry, the parameter values are varied between bounding values, to obtain conservative estimates of performance. For example chloride concentration is varied by multiplying the tabular data by a scaling factor.

There are several specific assumptions related to thermohydrologic conditions:

- The mountain-scale conduction model assumes the ground surface is at a constant temperature and is not affected by climate change. This assumption may result in greater heat losses, and therefore may be nonconservative.

- The thermohydrologic conceptual model implemented in NFENV assumes there are both matrix and fracture flow continua. The hydrologic regime in the near field is assumed dominated by, and in equilibrium with, the thermal regime, during the thermal pulse.

- All infiltrating water participates in the reflux cycle.

- In REFLUX1, matrix waters are assumed to be immobile during the thermal pulse, so that refluxing only accounts for fracture waters. This assumption will tend to under-predict fluxes at the repository level at early and intermediate times, as thermally driven matrix water will contribute to fluxes in the fracture.

- REFLUX1 tends to return to ambient conditions almost immediately after the repository drops below boiling, with water in the halo above the boiling isotherm flushing through the system just before the boiling isotherm reaches the repository. As removal of mobilized matrix water would leave a large amount of empty matrix storage and refilling the matrix would tend to strongly attenuate fluxes as the boiling isotherm returns to the repository horizon, REFLUX1 would tend to overpredict fluxes at the repository level. The degree of conservatism or non-conservatism of this approach depends on when the WPs fail relative to the time of return to ambient conditions.

- REFLUX1 and REFLUX2 do not consider elevation of the boiling point caused by salt concentration and capillarity.

- In REFLUX2, all matrix water in the dry-out zone is vaporized.

- REFLUX2 water vaporized from the rock matrix above the WP participates in the reflux cycle. Vaporized rock water below the package does not participate in the reflux cycle.

- REFLUX2 always assumes that the near-field temperature is above boiling until all the water in the pores leaves the dry-out zone.

- In REFLUX2, it is assumed that no resaturation takes place when the pores are dry and temperature is below boiling.

There are also specific assumptions with regard to water chemistry:

- MULTIFLO results are assumed to adequately describe the behavior of the near-field chloride concentrations.

- Uncertainties and limitations of MULTIFLO chloride concentration results can be accounted for by use of a multiplication factor (Appendix A) representing differences between ground-water chemistry in the porous matrix (calculated by MULTIFLO) and chloride concentration on a WP, which could be higher through evaporation from the WP surface.

- For the computation of chloride concentration, the YM rock is modeled as pure quartz with 10 percent porosity. The assumption is considered reasonable for the purpose of calculations present here, which is to estimate the change in salinity and pH caused by evaporation and condensation processes.

- In REFLUX2, the dry-out zone thickness does not change as a function of time.

3.3 EBSFAIL MODULE DESCRIPTION

The EBS failure (EBSFAIL) module calculates the failure time of the EBS caused by various modes of degradation. Modes of WP degradation evaluated in EBSFAIL include dry air oxidation, humid air corrosion, uniform and localized (pitting and crevice) aqueous corrosion, and mechanical failure. Other degradation modes that may become important under certain conditions, such as stress corrosion cracking and microbially influenced corrosion, are not currently considered in EBSFAIL. Failures by direct disruption are calculated in FAULTO, SEISMO, and VOLCANO. Failures caused by seismically induced rockfalls calculated in SEISMO are currently considered as a process in the evolution of the repository during the simulation period. Failures calculated in VOLCANO (Section 3.11) include two categories: (i) failure from intrusive IA (contributes only to ground-water releases); and (ii) failure from extrusive IA (contributes only to ground surface releases). The disruptive event failures are relevant to TPA calculations only if the events occur before the time of failure calculated in EBSFAIL.

3.3.1 Information Flow Within TPA

3.3.1.1 Information Supplied to EBSFAIL

Inputs required by EBSFAIL provided by EXEC include the temperature and RH as a function of time and position in the EBS (i.e., subarea) and the chemical composition of the aqueous phase (e.g., pH, or chloride concentration) in contact with the WP (calculated by NFENV). All WP properties and corrosion input parameters (and other relevant input parameters) supplied to EBSFAIL are specified in the EBSFAIL section of the primary input file.

3.3.1.2 Information Provided by EBSFAIL

EBSFAIL passes WP failure time because of corrosion and mechanical failure for use in EBSREL for release calculation.

3.3.2 Conceptual Model

EBSFAIL executes the stand-alone program FAILT as a part of the EBS performance assessment code (EBSPAC) (Mohanty, et al., 1997). Three different types of WP failures in the EBS performance assessment code are: (i) initial failure (Type 1); (ii) disruptive scenario failure (Type 2); and (iii) corrosion and mechanical failures (Type 3). In Type 1 failure, a portion of the WPs in a subarea is specified to have failed at time t = 0 as a result of initial defects produced before repository closure. These WPs are assumed to have been defective or damaged before or during emplacement and are specified in the input data as a fraction of the total containers in a subarea. The number of WPs undergoing Type 1 failure is read from the input file. In Type 2 failure, WPs fail as a result of some disruptive event. The timing and number of WPs affected by Type 2 failure are calculated by consequence modules (e.g., SEISMO, FAULTO, VOLCANO). Type 2 failure is a distribution of scenario failures over time, unlike Type 1 and Type 3. To simplify radionuclide release calculations, however, the Type 2 failures are coalesced to the earliest failure time for any one of the disruptive events. The number of failed WPs attributed to Type 2 failure is the number of WPs that failed because of disruptive events before the corrosion failure of WPs. In other words, all WPs in a subarea that have not undergone Type 1 and Type 2 failures are potentially subject to Type 3 failure. This assumption implies that a Type 3 failure equally affects and fails simultaneously all WPs, in a subarea, that have not already failed under Types 1 and 2 modes.

3.3.2.1 Corrosion and Mechanical Failures

For simplicity, failure of the WP is defined as the through-wall penetration of the outer and inner overpacks by a single pit, by uniform corrosion, or by brittle fracture caused by mechanically dominated processes in the presence of residual stresses. Depending on the thermohydrological conditions and the evolution of the near-field environment, three different WP degradation processes are assumed to occur as determined by the interaction of the changing environment with the container materials. The first process is oxidation of the outer carbon steel container by interaction with gaseous oxygen in dry air at relatively elevated temperatures [100–250 °C (212–482 °F)]. The second process is humid-air corrosion of the outer container as a result of the air containing water vapor at intermediate RH values, and the third process is aqueous corrosion that occurs at even higher RHs. An additional degradation process is mechanical failure caused by brittle fracture that may occur as a result of the concurrent effect of thermal embrittlement of the steel and residual stresses produced by welding during fabrication or WP closure. No allowance for extending WP lifetimes is given to the protection ability of the inner canisters (e.g., pour canister for the vitrified defense reprocessed waste) or the fuel cladding against corrosion or mechanical failure. After the outer and inner overpacks are penetrated by corrosion or failed by fracture, the SF is considered to be either partially or fully exposed to the near-field environment.

At each time step, a calculation is performed to determine if the RH has reached a critical value. If the RH of the environment surrounding the WP is below a lower critical value, the corrosion is treated as dry air corrosion (and the penetration of the dry air oxidation front is calculated). In the same time interval, a mechanical failure test is conducted for the new thickness resulting from metal oxidation to evaluate if failure from mechanical fracture occurs. If the WP does not fail by fracture, then the time is advanced, and the RH critical value test is repeated. If RH reaches the lower critical value at any time step, then the calculation of dry-air oxidation is interrupted and the calculation of metal penetration by corrosion in humid air begins. If the RH reaches a higher threshold, the calculation of metal penetration by aqueous corrosion is initiated. The mechanical failure test is performed at all time intervals until failure occurs, regardless of the type of corrosion the WP is undergoing (e.g., dry air, humid air, or aqueous corrosion). The initial time of the time step at which full penetration occurs is computed as the failure time.

3.3.2.2 Dry-Air Oxidation

Oxidation of steel in dry air occurs at low RHs and temperatures ranging from ambient to 250 °C (482 °F). The thin oxide layers formed at such temperatures are assumed to protect the container against further oxidation. Oxide growth, however, may become localized and lead to deeper penetration of the oxidation front, adversely affecting the long-term container integrity. Localized dry-air oxidation includes internal oxidation and intergranular oxidation. In the case of internal oxidation, the oxide forms as islands in the metal underneath the uniform oxide layer. In intergranular oxidation, the oxide forms preferentially along metal grain boundaries. Localized dry-air oxidation takes place by mass transport through short-circuit diffusion paths, such as interfaces between metal and oxide, other inclusions and precipitates, or grain boundaries.

In the EBSFAIL module, it is assumed that localized dry-air oxidation occurs intergranularly by enhanced diffusion of oxygen anions along grain boundaries. For the calculations of intergranular oxide formation, a mathematical model developed by Oishi and Ichimura (1979) is used, in which oxygen diffusion in the matrix and along the grain boundary can be calculated simultaneously in an infinite 1-D body (Ahn, 1995). The main assumptions in the calculations are: (i) the effect of the external oxide layer is negligible; (ii) oxygen diffuses into metallic phases (near the interface between grain boundary oxide and metal); and (iii) oxygen also diffuses into the metallic matrix from grain boundaries. The distance of oxygen penetration in the metal at time t is:

$$Y_p = \left[\frac{4\, D_l}{r_g\, \delta\, D_g} \sum_{n=1}^{\infty} \exp\left(-\frac{D_l\, n^2\, \pi^2\, t}{r_g^2} \right) \right]^{-1/2} \tag{3-37}$$

where:

Y_p =   penetration distance by intergranular oxidation [cm]

D_l =   matrix diffusivity [cm^2/s]

D_g =   grain boundary diffusivity [cm^2/s]

δ =   the thickness of grain boundary [= 0.7×10^{-7} cm, based on Lobnig, et al., 1992]

r_g =   the grain radius [$\approx 1 \times 10^{-3}$ cm for cast steel, based on Ahn and Soo, 1983; 1984]

t =   time [s]

Using diffusivities defined by the user in the primary input file, Eq. (3-37) yields the penetration distance of oxygen along grain boundaries and is used as a surrogate for oxide formation.

3.3.2.3 Humid-Air Corrosion

Corrosion of the steel container in air at moderately high RHs, as expected under certain repository conditions after the initial dry period, bears certain similarities to atmospheric corrosion. As discussed elsewhere (Cragnolino, et al., 1998), however, significant differences exist. Atmospheric corrosion occurs when a metal surface is covered by a water film of sufficient thickness to sustain electrochemical reactions. Water can be physically adsorbed to the metal surface in molecular form or it can be chemically bonded in a dissociated form that results in the formation of metal-hydroxyl bonds (Leygraf, 1995). The thickness of the water film increases with increasing RH. The critical RH above which atmospheric corrosion of most metals occurs corresponds to the formation of a water film of multiple monolayers that displays properties similar to bulk water. Under these conditions, corrosion is governed by the same electrochemical laws applicable to corrosion of metals immersed in an aqueous electrolyte.

Atmospheric corrosion studies reveal that iron and steel exhibit a primary critical RH of around 60 percent, similar to most metals (Fyfe, 1994). Above 60 percent RH, corrosion proceeds at a slow rate, but at 75–80 percent RH, the corrosion rate sharply increases. This secondary critical RH is attributed to capillary condensation of water in the pores of the solid corrosion products. It is assumed in the EBSFAIL module that humid-air corrosion occurs above a primary critical value of RH, whereas aqueous corrosion takes place above a secondary critical value. If the RH is higher than the primary critical value, humid-air corrosion is evaluated using a constant rate set by the user in the primary input file, assumed to be independent of temperature and time.

3.3.2.4 Aqueous Corrosion

Aqueous corrosion of the outer carbon steel container could be uniform or localized, depending on the composition of the near-field environment. If the aqueous medium contacting the WP is alkaline (pH > 9), carbon steel behaves as a passive metal and the presence of Cl$^-$ in the environment may promote localized corrosion. If the corrosion potential is higher than the critical potential for the initiation of localized corrosion, the calculation of metal penetration in the form of pit growth begins immediately without an initiation or induction time. When the depth of the pit is greater than the initial thickness of the WP outer overpack, the potential of the galvanic couple formed by the outer and inner containers is calculated. If the corrosion or galvanic potential of the inner container is lower than the critical potential for localized corrosion, penetration of the inner container is computed as uniform corrosion under passive dissolution conditions. Otherwise, pit growth of the inner overpack begins and continues until the depth of the pit becomes equal to the inner overpack wall thickness.

If the RH is higher than the secondary critical value, aqueous corrosion of the steel overpack is evaluated. It is assumed that a water layer defined by an arbitrary specified thickness on the order of a few millimeters is formed and its thickness is considered to be the same regardless of the presence or absence of backfill around the WP. Water films that form on the metal surface usually contain a variety of contaminants. Soluble species such as CO_2 increase the electrical conductivity and decrease the pH of the film, leading to increased iron or steel dissolution (Leygraf, 1995). Anionic species such as Cl^- may escalate the rate of dissolution or promote a more localized form of corrosion.

In the presence of an aqueous phase, corrosion of steel is an electrochemically controlled process. This process could be in the form of uniform active dissolution at pH close to or lower than neutral, or in the form of localized passive dissolution under the alkaline conditions induced by anions present in ground water such as bicarbonate ion (HCO_3^-). Localized corrosion of the carbon steel may be promoted by the simultaneous presence of Cl^- anions at concentrations higher than a minimum critical value in the range of HCO_3^- concentrations and pH within which passivation occurs. At any given time the mode of aqueous corrosion depends on the corrosion potential and the appropriate critical potential required to initiate a particular localized (pitting or crevice) corrosion process. The corrosion potential is the mixed potential established at the metal/solution interface when a metal is immersed in a given environment. Corrosion potentials are calculated on the basis of a kinetic expression for the cathodic reduction of oxygen and water and the passive current density for the anodic oxidation of the metals.

Empirically derived equations are used in EBSFAIL for the dependence of critical potentials on environmental parameters. Pit initiation and repassivation potentials are assumed to depend only on the chloride concentration and temperature. The dependence of the critical potential on chloride concentration and temperature (computed as described in sections 3.2.2.1 and 3.2.2.2) is given by:

$$E_{crit} = E_{crit}^{\circ}(T) + B(T) \log \left[Cl^- \right] \qquad (3\text{-}38)$$

where $E_{crit}^{\circ}(T)$ and $B(T)$, which depend on the material, are linear functions of temperature. The equation is valid above a minimum Cl^- concentration required to promote localized corrosion that also depends on the material. It should be noted that $E_{crit}^{\circ}(T)$ is the value of $E_{crit}(T)$ for a Cl^- concentration equal to 1 M. Both $E_{crit}^{\circ}(T)$ and $B(T)$ were evaluated for A516 steel and Alloy 825 from initiation and repassivation potentials for both pitting and crevice corrosion (Sridhar, et al., 1993; Cragnolino, et al., 1998). Data for Alloy 625 are available only at temperatures close to the boiling point of water {95 °C (203 °F)}, because, at temperatures lower than 60 °C (140 °F) Alloy 625 is resistant to localized corrosion over a wide range of Cl^- concentrations (Gruss, et al., 1998). Because of the insufficiency of the current database, both $E_{crit}^{\circ}(T)$ and B(T) are conservatively considered to be independent of temperature until more data are available. For Alloy C-22, which exhibits even higher resistance to localized corrosion (Gruss, et al., 1998), $E_{crit}(T)$ in Eq. (3-35) does not depend on Cl^- concentration and becomes reduced to $E_{crit}^{\circ}(T)$, which represents, in this case, the potential for oxygen evolution rather than a critical potential for localized corrosion. Additional data (and eventually the application of a critical temperature criterion) will be used for a more rigorous assessment of localized corrosion of Alloy C-22.

In the EBSFAIL module, the repassivation potential, E_{rp}, is conservatively adopted as the critical potential for the initiation of localized corrosion. The same approach is applied to the outer and inner containers assuming that steel behaves as a corrosion-resistant alloy in alkaline, passivating environments. If the corrosion potential is higher than the repassivation potential, pits are assumed to grow without an initiation time. If the corrosion potential falls below the repassivation potential, previously growing pits are presumed to cease growing and the material passivates, corroding uniformly at a low rate through a passive film. The metal penetration or remaining thickness is calculated at each time step, using rates of uniform and localized corrosion, as appropriate.

The propagation of pits is considered in a simplified manner by introducing an empirical equation developed by Marsh and Taylor (1988), for carbon steel in chloride containing bicarbonate/carbonate solutions, using an extreme values statistics approach. In this equation, pit penetration is time-dependent and given by:

$$P = At^n \tag{3-39}$$

where:

P =   pit penetration [m]

t =   time [yr]

A =   experimentally determined constant specified in the primary input file (CoefForLocCorrOfOuterOverpack)

The time exponent, n, is specified in the primary input file (ExponetForLocCorrOfOuterOverpack) and is approximately equal to 0.5, consistent with a pit growth process controlled by diffusional mass transport and a rate that decreases with time. For the inner container, a constant rate of penetration ($n = 1$) was considered for pit growth and specified in the primary input file (LocalizedCorrRateOfInnerOverpack[m/yr]). Recent experimental results (Cragnolino, et al., 1998) have shown this is quite conservative since a parabolic rate law ($n = \frac{1}{2}$) was found to apply.

If the corrosion potential is lower than E_{rp}, uniform penetration caused by corrosion under passive conditions is calculated for the carbon steel container and for any of the alternative inner container materials (Alloys 825, 625, and C-22) by using experimentally available values specified in the primary input file as materials AA_1_1[C/m2/yr] and AA_2_1[C/m2/yr], respectively.

After penetration of the outer container, electrical contact of the inner and outer container through the presence of an electrolyte path (such as that provided by modified ground water) promotes galvanic coupling, assuming that metallic contact always exists between both containers. The galvanic coupling model evaluates whether penetration of the inner container by localized corrosion is possible. If not, uniform corrosion or mechanical fracture becomes the predominant failure mechanism because the inner container emerges as protected against localized corrosion.

The effect of galvanic coupling between the inner and outer overpacks on the failure time of the WP is evaluated by a simplified approach (Mohanty, et al., 1997). The corrosion potential of the inner container in the galvanic couple, E_{corr}^{wp}, formed when the wall of the outer container is penetrated by a pit, is estimated using experimentally measured values of the potential bimetallic couple, E_{couple}, for a well-defined area ratio between both components assuming perfect electrical contact (Dunn and Cragnolino, 1997; 1998). If E_{corr}^{wp} is greater than the repassivation potential of the inner overpack material, localized corrosion occurs. Otherwise, uniform corrosion takes place. The E_{corr}^{wp} is determined through a linear combination of corrosion potential, E_{corr}, of the inner overpack, in the absence of galvanic coupling at the time of the through-wall penetration of the outer overpack, and E_{couple}, according to the empirical expression:

$$E_{corr}^{wp} = (1 - \eta) \, E_{corr} + \eta \, E_{couple} \tag{3-40}$$

where $0 \leq \eta \leq 1$ is the efficiency of the galvanic coupling. The values adopted for E_{couple} for the different electrochemical parameters needed to calculate E_{corr} and for those establishing the dependence of the critical potentials with chloride concentration and temperature for the various materials considered are included in Appendix A.

3.3.2.5 Mechanical Failure

Mechanical failure of WPs in the EBSFAIL module is considered to be the result of fracture of the outer steel overpack. As a first approximation, other mechanical failure processes such as buckling or yielding are not regarded plausible for the current design of the WP because of the relatively large thickness of the container wall. Active uniform corrosion of the carbon steel overpack is not expected under the passivating conditions prevailing in the near-field environment and, therefore, failure modes such as buckling or yielding that would require significant generalized thinning of the container wall in the presence of external loads are not included.

The possibility of mechanical failure as a result of thermal embrittlement of the steel promoted by long-term exposure to temperatures above 150 °C (302 °F) is evaluated at each time step. Fracture as a result of thermal embrittlement of the steel overpack is an important failure mode to be considered for any WP design, particularly for high thermal loadings. Thermal embrittlement of low-alloy steels occurs as a consequence of prolonged exposure at elevated temperatures and results in a substantial degradation of specific mechanical properties.

One of the important mechanical properties required for a WP material is toughness, which is the ability to absorb energy in the form of plastic deformation without fracture. Toughness however, is significantly affected by thermal embrittlement, a phenomenon closely related to temper embrittlement. Thermal embrittlement is characterized by an upward shift in the ductile-brittle transition temperature, measured by the variation of the impact fracture energy for notch specimens as a function of test temperature (Vander Voort, 1990). Segregation of impurities (e.g., antimony, phosphorus, tin, and arsenic), along prior austenite grain boundaries, is the main cause of temper embrittlement (Briant and Banerji, 1983). The segregation of phosphorus, which is the dominant impurity of commercial steels, promotes fracture of notched specimens on impact and leads to a change in the low-temperature fracture mode from transgranular cleavage to intergranular fracture (Cragnolino, et al., 1996).

A simple fracture model developed on the basis of linear-elastic fracture mechanics is used in EBSFAIL. This model is based on a generalized expression for the stress intensity factor (K_I). For the case of a cylinder with a surface flaw located on its outer surface, the following equation is applicable:

$$K_I = Y\sigma(\pi a)^{0.5} \tag{3-41}$$

where:

K_I	=	stress intensity factor for the crack opening mode (I) [MPa m$^{0.5}$]
Y	=	geometry factor to account for the shape of the crack and the load configuration [unitless]
σ	=	applied stress [MPa]
a	=	depth of crack [m]

It is assumed that applied stresses are caused by only residual stresses associated with the circumferential weld used for overpack closure. It is also presumed that the maximum value attainable by residual stresses produced by welding is the yield strength of the material. The depth of the crack increases with time and is conservatively inferred to be equal to that of a pit, as a result of localized corrosion. Values of Y are calculated as $Y = M_K Q^{-0.5}$, as presented by Rolfe and Barsom (1977). The geometry factor Y corresponds to a part-through thickness thumbnail crack with a length $2c$, equal to 2 times its depth, a, for a hollow cylinder of wall thickness, t, in which the crack shape parameter, Q, is a function of $a/2c$. The magnification factor, M_K, varying from 1.0 to 1.6, is introduced for deep cracks with depths ranging from $t/2$ to t. For simplicity, the WP is considered composed of a single shell with the combined thickness of both the outer and the inner overpacks, but with the mechanical properties of the outer overpack.

In addition, a safety factor of 1.4 was applied to calculate the value of K_I by assuming that the yield strength of the material in the vicinity of the welds is higher than the base material. This value is compared at each time step with the critical stress intensity or fracture toughness of the material, K_{Ic}, to determine if failure by fracture takes place. By definition, fracture occurs instantaneously if K_I is greater than K_{Ic}. Because of the lack of data, no decrease in the value of K_{Ic}, with time, is assumed in the present analysis. If thermal embrittlement of the steel occurs through prolonged exposure (thousands of years) to temperatures above 250 °C (482 °F) however, a substantial decrease in the value of K_{Ic} may be expected (Cragnolino, et al., 1998).

3.3.3 Assumptions and Conservatism of the EBSFAIL Approach

The EBSFAIL model is an abstraction of WP failure that considers: (i) dry-air oxidation; (ii) humid-air corrosion; (iii) aqueous corrosion; and (iv) mechanical failure. Other processes and events that cause WP failure include: (i) disruptive scenario failure; and (vi) juvenile failure. In addition to the number of WPs failed, the time of failure is also estimated.

Oxidation of steel can occur in dry air at low RH and temperatures ranging from ambient to 250 °C (482 °F). It is assumed that localized dry-air oxidation occurs intergranularly by enhanced diffusion of oxygen anions along grain boundaries. Because this effect is expected only for a short time at the beginning of the post-closure period, when RH may be low during the period of high temperatures, it is not anticipated to significantly affect the WP lifetime. The inclusion of this process is considered a reasonable approach.

Corrosion of steel in air at moderately high RH, expected after the initial dry-air period, bears certain similarities to atmospheric corrosion. Humid-air corrosion is assumed to occur when the RH is above a primary RH value and below a secondary critical RH, above which aqueous corrosion takes place. Atmospheric corrosion studies for iron and steel indicate a slow rate of corrosion occurring between a primary RH of 60 percent and a secondary RH of 75–80 percent. Humid-air corrosion is not considered to have a significant effect on WP lifetime. The inclusion of this process is considered a reasonable approach.

Among dry-air oxidation, humid-air corrosion, and aqueous corrosion, degradation of the WP will be most affected by the aqueous corrosion process. Aqueous corrosion will occur when the secondary RH is exceeded, after which determination of whether localized or uniform corrosion occurs is of most importance in determining the corrosion rate (localized or pitting corrosion will breach the WP in a much shorter time than uniform corrosion). The corrosion potential is a key parameter that represents the driving force required to promote the occurrence of various corrosion processes under natural environmental conditions. Once the corrosion potential exceeds the repassivation potential, localized corrosion begins and pit growth rates are calculated by using experimentally determined expressions and parameters. Failure of the WP is defined conservatively as penetration of both overpacks by a single pit or by general dissolution. Because the repassivation potential depends on the temperature and chloride concentration, through an empirical expression based on mechanistic models, the degree of conservatism of this approach will depend on the approach used to estimate temperature and chloride concentration.

Mechanical failure of the WP is assumed to occur as a result of fracture of the outer steel overpack. Fracture of the overpack, as a result of thermal embrittlement, is an important consideration because the fracture toughness of the steel can be affected by the repository thermal environment [i.e., long-term exposure to temperatures above 150 °C (302 °F)]. Residual stress is compared with the fracture toughness of the WP to determine if failure by fracture takes place. This approach is nonconservative because the applied stress considered excludes other stresses (e.g., seismicity and faulting); however, the overall approach can be conservative depending on the value for fracture toughness specified in the input file. Fracture toughness is affected by the prolonged exposure to high temperatures and thus depends on repository design (e.g., backfill versus no backfill and thermal load).

Disruptive scenario failure of the WP is the failure of the WP from disruption caused by seismicity, faulting, or igneous activity (further details on the disruptive scenario failures are provided in their respective sections). Seismic activity can cause rockfall from the roof of the emplacement drifts, which will cause deformation and in some cases failure of the WP. Fault displacement is assumed to fail all WPs intercepted by the fault zone when a given fault displacement is exceeded. Intrusive IA is assumed to fail all WPs in an effective area intercepted by an intrusive event calculated by an intrusion orientation, length, and width (assumption assumes contact between lava and the WP results in WP failure).

Juvenile failures of the WP may occur because of manufacturing defects and emplacement accidents. The TPA Version 3.1.4 code includes a parameter to specify the fraction of WPs assumed damaged at the beginning of the simulation. There is little relevant data to determine an appropriate value for juvenile failures.

Unique failure times are assigned to each failure type: (i) juvenile failure (conservatively assumed to occur at the beginning of the simulation); (ii) disruptive scenario failure (seismicity, faulting, and intrusive igneous activity); and (iii) failure caused by degradation (dry-air oxidation, humid-air corrosion, aqueous corrosion, and mechanical failure). Because only one failure time is used for each of the three categories, the initial failure time for a particular category is used conservatively to represent the failure time of all the WPs that fail in that category. The conservative nature of this approach depends on the distribution of failures within the simulation period.

There are several specific assumptions and conservatism of the EBSFAIL module:

- Corrosion failure of all the WPs contained in a subarea occurs simultaneously and is determined by the penetration of a single pit. These assumptions do not take into consideration the full range of spatial variations in environmental, material, and electrochemical conditions that can be expected for different WPs and on different areas of a single WP.

- There is no coupling of the EBSFAIL module for mechanical failure (i.e., growth of pits or cracks with time that reduces the stresses needed for failure) with WP failure caused by disruptive events calculated outside the EBSFAIL module. No inputs from FAULTO and SEISMO modules in the form of applied stresses are used in EBSFAIL to calculate mechanical failure. Only residual stresses arising from the circumferential weld closure are considered. This lack of coupling may result in non-conservative estimates of the number of failed WPs and the time of occurrence as computed by the SEISMO module.

- Mechanical failure is calculated assuming an intact WP (i.e., no weakening of the WP caused by uniform corrosion), which may yield nonconservatively long times for mechanical failure.

- Only mechanical failure from brittle fracture of the outer steel overpack as results of thermal embrittlement is considered in EBSFAIL. This may overestimate the WP failure time but it may be partially compensated for by the lack of consideration of the contribution to mechanical integrity of the inner container, which is a very ductile material.

- For dry air oxidation of the WP, the rate of oxidation is controlled by oxygen transport into grain boundaries and the effect of the external oxide layer is negligible.

- Aqueous corrosion is understood to occur through formation of an alkaline and saline water film on the WP surface at RH. Dripping of modified ground water on the WP is not considered. This limitation may lead to nonconservative results if there are significant spatial and temporal variations in the flow and chemistry of the water.

- Simplified consideration of pit nucleation and growth in EBSFAIL may lead to rapid penetration of the outer overpack, despite its thickness. The propagation of deep pits could be limited by ohmic drop and mass transport constraints leading to the initiation of new pits rather than continuous growth of deeper pits.

3.4   SEISMO MODULE DESCRIPTION

SEISMO determines the number of WPs ruptured by seismically induced rockfalls. The number of ruptured WPs determined from this module is made available to EBSREL for release calculation.

3.4.1   Information Flow Within TPA

3.4.1.1   Information Supplied to SEISMO

The user selects the seismic disruptive process module in the primary input file. EXEC passes to SEISMO the TPA time steps and the number, magnitude, and time of the seismic events that are sampled, using the seismic hazard curve data set in the primary input file. Rock column height associated with different categories of seismic events, and rockfall area determined by joint spacing, are used to compute rockfall volume. These data are provided in *seismo.dat*. Other SEISMO input parameters are specified in the SEISMO section of the primary input file.

3.4.1.2   Information Provided by SEISMO

SEISMO calculates the cumulative fraction of WPs disturbed by the seismic event, for all subareas. These results are passed to EXEC and used in EBSREL for release calculations.

3.4.2   Conceptual Model

The SEISMO module evaluates the potential for direct rupture of WPs from rockfall induced by seismicity. The SEISMO code uses the weight of rock that is falling from the roof of the emplacement drifts to calculate the impact load on a WP. The magnitude of the impact load is assumed to be a function of the size of the falling rock block and the distance the rock falls. The size of the falling rock block is, in turn, a function of rock conditions and the magnitude of a seismic event. The SEISMO conceptual model is divided into: (i) calculation of impact load and stress; (ii) determination of WP rupture; and (iii) determination of the number of WPs ruptured. The SEISMO module accounts for variations of rock conditions, the relationship between falling rock size and magnitude of seismicity, and the timing of the seismic events.

3.4.2.1   Impact Load and Stress Calculation

The SEISMO conceptual model assumes that for impact load and stress calculation:

- No energy dissipation takes place at the point of impact because of local inelastic deformation of the WP material.

- The deformation of WPs is directly proportional to the magnitude of the dynamically applied force.

- The rock body does not fragment on impact.

- The inertia of the WP resisting an impact may be neglected.

- The WP is treated as an equivalent spring with a spring constant k_{wp} [Figure 3-11(a)].

The dynamic or impact loads determined in the SEISMO module for rock falling on a WP are approximated based on the principle of conservation of energy, using the weights of the freely falling rocks. For a rock hitting the WP, the impact load is calculated as (Popov, 1970):

$$P_{dyn} = W \left| 1 + \sqrt{1 + \frac{2h}{\Delta}} \right|$$

(3-42)

where:

P_{dyn} =   impact load [N]
W =   weight of the rock falling [N]
h =   falling distance of rocks to WPs [m]
Δ_{st} =   spring deformation [m]

(a)   (b)

Figure 3-11. (a) Waste package equivalent beam and (b) waste package vertical supports for SEISMO.

The supports for WPs are treated as flexible in the SEISMO module. In the current repository conceptual design, a WP will be supported by four equally spaced v-shaped thin beams with one vertical cylindrical bar on either side of the v-shaped beam. In the SEISMO module, only the supports at both ends of a WP are considered [Figure 3-11(b)]. In Eq. (3-42), Δ_{st} was the static deflection of a WP impacted, assuming rigid supports. To account for the deformability of WP support, Δ_{st} is made equal to:

$$\Delta_{st} = \frac{W}{k_{wp}} + \frac{W}{2N_p k_b} \tag{3-43}$$

where:

k_{wp} $\quad = \quad$ stiffness of the WP determined as unit deflection at the center of a simply supported beam when subjected to a load at the center [N/m]

N_p $\quad = \quad$ number of the supports at each end of the WP (two, in this case)

k_b $\quad = \quad$ stiffness of the vertical WP support bars [N/m]

k_{wp} can be calculated by:

$$k_{wp} = \frac{48EI}{L_{wp}^3} \tag{3-44}$$

and k_b can be calculated by:

$$k_b = \frac{AE}{L} \tag{3-45}$$

where:

A $\quad = \quad$ cross-sectional area of the vertical support bar [m^2]

E $\quad = \quad$ Young's modulus of the vertical support bar or WP [N/m^2]

L $\quad = \quad$ height of the vertical support bar [m]

I $\quad = \quad$ $\pi R_{ave}^3 t$, where t is the thickness of WP [0.12 m (4.7 in.)] and R_{ave} is the average of the outer and inner wall radii of the WP [m^4]

L_{wp} $\quad = \quad$ length of the WP [m]

No information about the shape and dimension of the bar is currently available. In the SEISMO module, the bar is assumed to be a cylinder with a radius of 0.0508 m (2 in.) and a height of 0.3048 m (1 ft).

On obtaining the impact load, the equivalent static stress resulting from the rock impact on a WP can be calculated by adopting a simple concept of two spheres in contact and by assuming the pressure is distributed over a small circle of contact (Timoshenko and Goodier, 1987). The impact stress (p) in N/m^2 can be obtained by:

$$p = \frac{3}{2\pi} \left(\frac{16P_{dyn}}{9\pi^2} \frac{1}{(C_{wp} + C_{rock})^2 R_{wp}^2} \right)^{\frac{1}{3}} \qquad (3\text{-}46)$$

where:

C_{wp} = constant for lower sphere or WP [m^2/N]

C_{rock} = constant for upper sphere or rockfall, and [m^2/N]

R_{wp} = radius of lower sphere or WP [m]

$$C_{rock} = \frac{1 - \mu_{rock}^2}{\pi E_{rock}} \qquad (3\text{-}48)$$

and:

$$C_{wp} = \frac{1 - \mu_{wp}^2}{\pi E_{wp}} \qquad (3\text{-}47)$$

where:

μ_{wp} = Poisson's ratio of WP [unitless]

E_{wp} = modulus of elasticity of WP [N/m^2]

μ_{rock} = Poisson's ratio of rockfall [unitless]

E_{rock} = modulus of elasticity of rockfall [N/m^2]

Failure Criterion

The impact stress calculated as described earlier is compared with the ultimate tensile strength of the WP material to assess whether WP failure has occurred. If the impact stress is determined to be greater than the ultimate tensile strength of the WPs, the WPs are assumed ruptured. The ultimate tensile strength, with units of N/m^2, is specified in the primary input file.

3.4.2.2 Joint Spacing and Rock Conditions in the Topopah Spring Welded Tuff Unit

It is recognized that not all rocks falling from the roof of the emplacement drifts will rupture the WPs. The effective size of the fallen rock that may impact the WPs is envisioned to be controlled by joint spacing (JS) (width and length of the rock block) and vertical extent of the rock block falling on the WP. Given the wide range distribution of JS' in the TSw2 thermal-mechanical unit, it is convenient to assume that five distinct rock conditions exist in the TSw2 unit. These rock conditions are estimated using available JS information (Brechtel, et al., 1995) for the TSw2 unit. Since each rock condition represents a range of JS, a normal distribution covering the range of the JS is assumed for the corresponding rock condition (for lack of specific information). The size of the rocks used for each rock condition is then based on the JS distribution condition. The volume of a falling rock can be calculated by JS (width) × JS (length) × vertical dimension of rock block. At this time, the SEISMO module assumes that the width of a falling rock is equal to its length.

Fractional Coverage of Rock Conditions and Determination of Number of Waste Packages Ruptured

Based on Brechtel, et al. (1995), rock condition 4 appears to contain a large portion of the TSw2 unit. About 62.9 percent of the area can be characterized as rock condition 4, and rock condition 5 occupies roughly 35.6 percent of the area. Rock conditions 1, 2, and 3 account for only 1.5 percent of the total area. Because of the lack of specific information, the 1.5 percent is equally divided into rock conditions 1, 2, and 3.

In the SEISMO module, the percentage of the total WPs in each of the rock conditions is assumed to be equal to the fractional coverage of the rock condition. For a particular seismic event, if the induced rockfalls generate an impact stress on the WPs that exceeds the ultimate material strength of the WPs, the WPs are assumed ruptured. If a seismic event triggers rockfall for a particular rock condition, rockfall is not expected to take place in the entire area under that rock condition. In fact, only a small fraction of the rock under that rock condition will fall in responding to a seismic event, because of the inherent variation associated with the rocks. A portion of the remaining intact rock may fall at a later time when another seismic event, having the same intensity, takes place. The size of the fraction may be related to the event magnitude, joint dip angles, and incidence angle of incoming seismic waves. At this time, there is little information available to determine such a relationship. In the current SEISMO module, a log-normal probability distribution function for the rockfall fraction is implemented so that users can evaluate the potential effect of variation of the rockfall fraction for a given rock condition. All WPs under that fraction of area of rockfall will be affected. If the impact stress is sufficient, all WPs under that fractional area are computed as having failed. The time history of WP failures from seismic event is used in radionuclide releases calculations in EXEC.

Seismic Hazard Parameters

The SEISMO module requires the generation of a synthetic time history of seismicity over the time period of interest. This estimate is made by using the TPA executive SAMPLER utility module. The required input for generating event history includes ground acceleration sampling points and the corresponding recurrence times, which together form a seismic hazard curve. The horizontal acceleration hazard curve provided in DOE (1995) is used to determine the seismic hazard associated with an earthquake with a given recurrence interval. This hazard curve is specifically for surface facilities. A common understanding for YM is that the seismic acceleration at the repository horizon may be half that at the ground surface (Quittmeyer, 1994). In the SEISMO module, half the surface hazard is used as a base case. The SEISMO module can be used to assess the effects of different hazard curves; however, in changing the hazard curve, the height of rockfall and associated volumes should be modified to reflect the effect of the ground acceleration magnitude.

The seismic recurrence sampling points (accelerations) and the corresponding recurrence time are provided in the primary input file. Four discrete sampling accelerations are selected for the SEISMO module. These accelerations are 0.1, 0.2, 0.3, and 0.4 g (equivalent to 0.2, 0.4, 0.6, and 0.8 g on the seismic hazard curve) (Quittmeyer, 1994). The recurrence times associated with each acceleration are 500, 2400, 8000, and 20,000 yrs.

3.4.3 Assumptions and Conservatism of the SEISMO Approach

The SEISMO module is an abstraction of the effect of seismicity on WP performance. The SEISMO module considers: (i) the frequency and magnitude of seismic events; (ii) the quantity of rockfall for each seismic event; and (iii) the effect on the WP.

The frequency and magnitude of the seismic events are based on a seismic hazard curve provided as input to the TPA Version 3.1.4 code. The seismic hazard curve provides accelerations and recurrence frequencies for seismic events over the time period of interest. The seismic hazard curve is based on historical information for surface facilities and is considered a reasonable approach for the underground facility.

The volume of rock that falls for a given event is determined based on properties related to the TSw2 thermal-mechanical unit. Five distinct rock conditions have been identified to characterize the variability in rock properties within the TSw2 unit. The abstraction for rock volume uses JS information (as width and length) and height of yield zone (as vertical extent). The height of yield zones was estimated from dynamic numerical modeling of coupled thermal-mechanical effects. The height of the yield zone has been conservatively set as an upper bound such that the volume of rockfall represents a maximum value. A heterogeneity factor is used to account for the function of drifts that experience rockfall for each seismic event. Thus, when a seismic event causes sufficient rockfall to rupture the WP, in a particular rock type, all WPs within the fraction of that rock type, specified by the heterogeneity factor, are assumed to rupture.

If a seismic event triggers a rockfall, it is assumed to equally affect the whole area of the repository, irrespective of rock conditions. In other words, all five rock conditions experience the same acceleration over the entirety of their respective areas and the fraction of each area with rockfall is the same for each rock condition (controlled by a single heterogeneity factor). In actuality, a seismic event may trigger more rockfall for one rock condition than for another. Intuitively, for a given seismic ground acceleration, a weaker rock condition should experience a relatively large area of rockfall compared with a stronger rock condition. In the current implementation, although the fractional rockfall area is a function of seismic ground acceleration, it is not a function of rock condition. Whether this assumption is conservative cannot be determined at this time.

Rockfall may affect the WP lifetime through rupture of the WP by the impact produced by the falling rock and acceleration of corrosion at the location of the impact. Only the first aspect is considered in SEISMO. The two key components of the SEISMO module include determination of the impact load and failure. In computing the impact load, the WP is treated as a simply supported beam. In converting impact load to impact stress, all energy generated through dynamic impact of the rockfall is transferred to the WP. A conservative assumption is that the WP will rupture if the impact stress exceeds the ultimate strength of the WP. The analysis assumes, nonconservatively, that rockfall occurs on an intact WP (i.e., corrosion and mechanical fracturing of the WP does not reduce WP strength, and cumulative damage caused by consecutive rockfalls is not considered). Overall, the approach is considered conservative over a 10,000-yr compliance period.

There are several specific assumptions and conservatism of the SEISMO module:

- A WP can be treated as a simply supported beam.

- No energy dissipation takes place at the point of impact because of local inelastic deformation of the WP material.

- Deformation of WPs is directly proportional to the magnitude of the impact load.

- Inertia of the WP resisting an impact may be neglected.

- Estimate of the impact stress assumes a contact area based on a spherical geometry. This assumption yields a conservative estimate because the contact area would be larger.

- In converting impact load to impact stress, falling rocks are assumed to have infinite strength (i.e., all energy generated through impact is transferred to the WPs). If rock is allowed to break, the effective impact stress to the WP should be smaller since some impact energy will be absorbed by the rock breaking.

3.5 EBSREL MODULE DESCRIPTION

The EBSREL module calculates the time-dependent release of radionuclides from the EBS to the UZ. Infiltrating water and refluxed water could reach the drift wall and subsequently the WP. If the container is breached, infiltrating water enriched with chloride and other minerals could enter the WP, coming in contact with the SF. If the cladding has failed, the infiltrating water could dissolve the SF and then carry the radionuclides out of the WP to the geosphere.

3.5.1 Information Flow Within TPA

3.5.1.1 Information Supplied to EBSREL

The values EBSREL obtains from EXEC for the representative WP in a subarea are: (i) WP temperature, by reading in a file created in EBSFAIL; (ii) volumetric flow rate of water in the near field (calculated in NFENV); and (iii) the cumulative fraction of failed WPs for all failure modes (corrosion failure, disruptive event failure, and initial defective failure). Radionuclide decay chains, half-lives, initial inventories, and molecular (isotopic) weights are passed to EBSREL by EXEC. Other values such as WP dimensions, nuclide-specific solubilities, retardation coefficients, fractions of SF wet, SF dissolution model parameters, and factors determining the amount of water entering the representative WP are specified in the primary input file.

3.5.1.2 Information Provided by EBSREL

EBSREL calculates the radionuclide releases from the EBS for all subareas and passes these results to EXEC for subsequent use in UZFT. EBSREL provides release rates as a function of time for all radionuclides specified in the primary input file for ground-water dose calculations.

3.5.2 Conceptual Model

Using WP failure information from EBSFAIL, SEISMO, FAULTO, and VOLCANO and near-field chemistry, temperature, and liquid flow rate information from NFENV, EBSREL [derived from EBSPAC (Mohanty, et al., 1996)] calculates release of radionuclides from a WP. Conceptual models used in EBSREL are described in the succeeding sections. In calculating releases, EBSREL takes into account radionuclide decay, generation of daughter products in the chains, and temporal variation in water flow. Once the container is breached, the SF may be exposed to aqueous conditions as determined by the near-field calculations. Like EBSFAIL, EBSREL looks at processes affecting an individual WP representative of all WPs in the subarea. Releases calculated by EBSREL from this breached WP are then scaled to account for the total number of WPs breached from a particular failure type (i.e., initially defective, scenario failure, and corrosion failure). Releases from all failure modes are then summed, as presented in Section 3.5.2.9.

3.5.2.1 Radionuclide Inventory and Mass Transfer

Two models for aqueous release of nuclides are available for selection by the user: the bathtub model and the flow-through model. In the bathtub model, it is assumed that at the time of failure there are at least two holes in a horizontally emplaced WP that act as inlet and outlet for water. The holes are located such that water enters through one pit and exits through the other. Another presumption is that at least one of the holes is located on the side of the WP at a level lower than that of the water entrance hole, situated at the top of the horizontally emplaced WP. After the water level in the WP rises to the specified outflow position (a sampled parameter), water begins to flow from the WP, along with the dissolved radionuclides. The flow-through model is a variant of the bathtub model, for which the flow out is assumed to be immediately equal to the flow in and the fraction of fuel wetted is not a function of water level inside the WP.

Figure 3-12 presents a schematic representation of a horizontally emplaced WP with holes representing the inlet and outlet. In this schematic, the conduit for liquid entry is shown on the upper half of the WP and the hole for the liquid exit is shown on the right side. Liquid water will accumulate in the WP until its level rises to the level of the exit hole, h. The height of the exit hole is specified in the primary input file via the SFWettedFraction parameter; 0 implies an outlet at the bottom and 1 implies an outlet at the top of the WP. In the bathtub model, all SF above this level is assumed to remain dry and does not contribute to radionuclide transfer out of the WP.

When liquid water enters the WP, after its failure, the overall mass balance model for the radionuclide inventory in liquid water contacting SF in a failed WP is:

$$\frac{dm_i}{dt} = w_{li}(t) - w_{ci}(t) - w_{di}(t) - m_i \lambda_i + m_{i-1} \lambda_{i-1} \qquad (3\text{-}49)$$

where:

m_i = amount of radionuclide i in the WP water at time t [mol]

w_{li} = rate of transfer from the SF into the resident water in the WP because of or through leaching of the SF [mol/yr]

w_{ci} = rate of advective transfer out of the WP [mol/yr]

w_{di} = rate of diffusive transfer out of the WP [mol/yr]

λ_i = decay constant of radionuclide i [1/yr]

m_{i-1} = amount of the parent at time t [mol]

λ_{i-1} = decay constant of the parent [1/yr]

The product, $m_i\lambda_i$, is the amount lost because of decay, and $m_{i-1}\lambda_{i-1}$ represents the amount generated by the decay of the parent radionuclide.

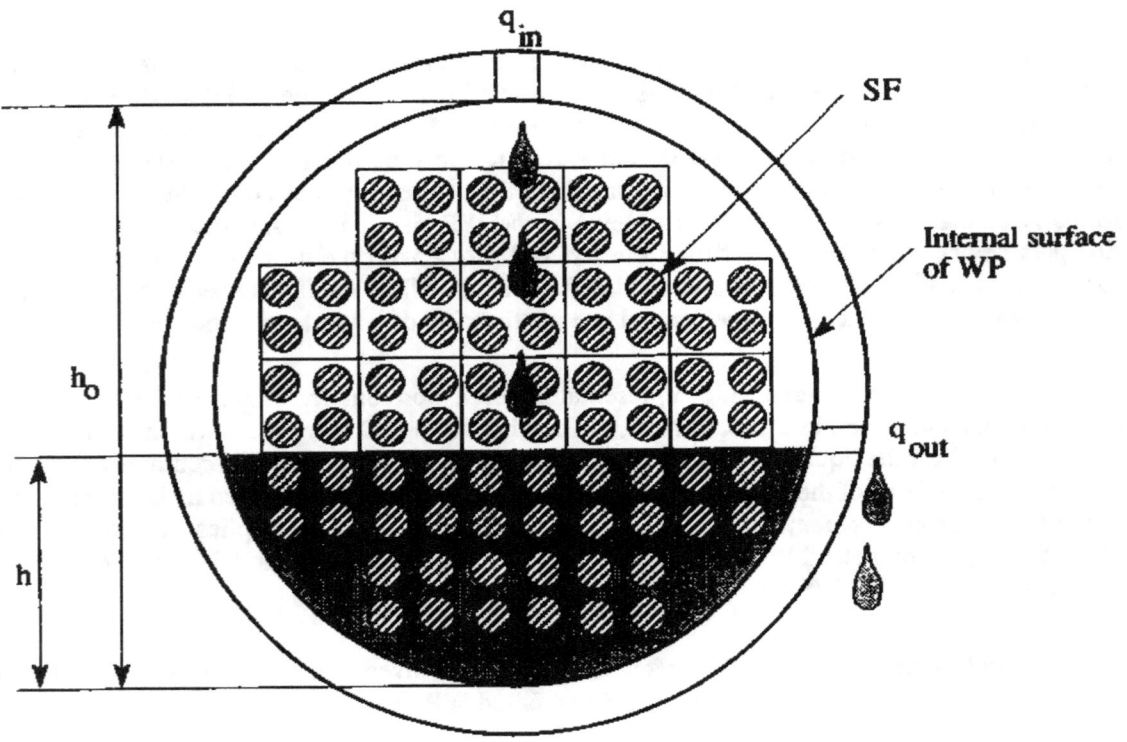

Figure 3-12. Schematic of bathtub model with incoming and outgoing water conduits.

3.5.2.2 Advective Mass Transfer

The advective mass transfer out of the WP, w_{ci}, can be represented by (Wescott, et al. 1995):

$$w_{ci}(t) = C_i(t) \, q_{out}(t) \tag{3-50}$$

where:

C_i = concentration of radionuclide i in the WP water $[mol/m^3]$

q_{out} = water leaving the WP at time t $[m^3/yr]$

The concentration C_i is determined by dividing the mass of element i, m_i, by the volume (V) of water in the WP. For the bathtub model, q_{out} is zero if the volume of water in the WP has not exceeded the WP volume below the exit hole, V_{max}, or equal to the input flow rate, q_{in}, if the volume of water exceeds V_{max}. The volume of water, V, in the WP is determined by integrating the flow rate, q_{in}, with respect to time, until it reaches V_{max}, where:

$$V = \int_0^t q_{in}(\tau)d\tau \tag{3-51}$$

For the flow-through model, the flow out of a WP is equal to the flow in for all times. Calculation of the input flow rate, q_{in}, is described later in this section.

For stability and efficiency in the numerical integration algorithm, V may not be smaller than V_{min}, which is defined as:

$$V_{min} = a \, q_{max} \tag{3-52}$$

where :

q_{max} = maximum flow rate into the WP $[m^3/yr]$

a = constant currently set at 5 yr

In the bathtub model, the fraction of fuel wetted is proportional to the water level. There is no release from the WP, however, until the actual water level exceeds the maximum water level. The flow-through model is implemented by specifying the fraction of the SF wetted and selecting the flow-through option in the primary input file. Because water does not have to fill to a certain level inside the WP, release of radionuclides occurs as soon as water enters the WP. The exit hole height parameter (SFWettedFraction) of the bathtub model is used indirectly to specify the fraction of SF wetted in the flow-through model.

Advective mass transfer is considered to occur from the inside of the WP to the EBS host rock interface instantaneously, through the nonmechanistic connection shown in Figure 3-13.

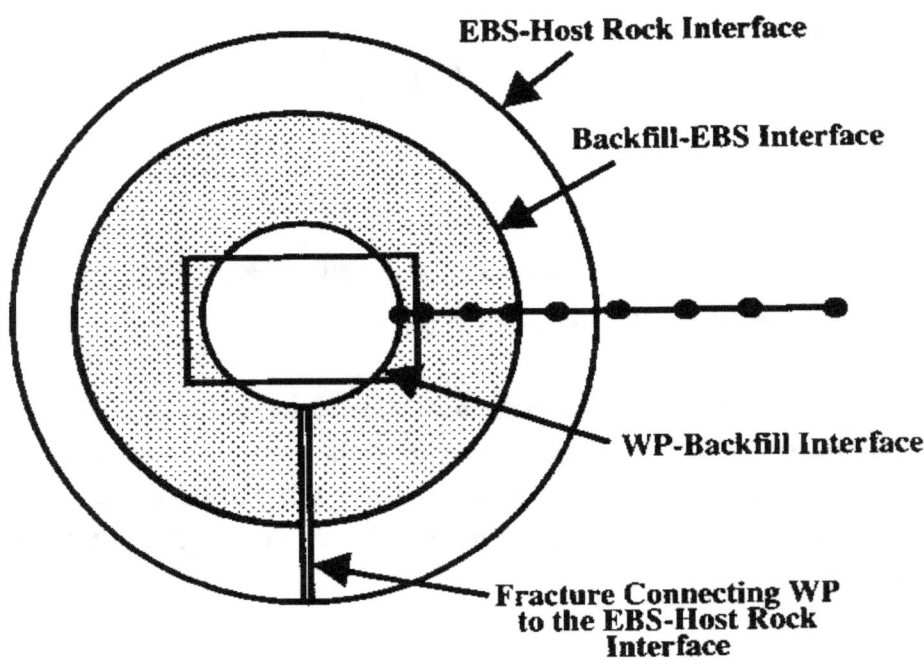

Figure 3-13. Schematic drawing for advective and diffusive mass transfer from the waste package to the host rock

3.5.2.3 Diffusive Mass Transfer

Diffusive mass transfer takes place through the medium (crushed tuff backfill properties are assumed)[1] surrounding the WP, as shown schematically in Figure 3-13. After the water leaves the WP, it is understood to envelope the whole outer surface of the WP. As a result, the radionuclides are present in uniform concentration. Then, assuming the WP can be represented by a sphere of radius r_0, the mass loss through molecular diffusion of a radionuclide into the porous medium can be represented by:

$$w_{di}(t) = 4\pi \; r^2 \; \varphi \; D \; \left. \frac{\partial C_i}{\partial r} \right|_{r = r_0} \tag{3-53}$$

---

[1] The presence of backfill, assumed only for diffusive mass transfer calculations, provides continuity in the transport pathway. Nominally, the backfill is absent.

where:

φ =   porosity of the medium surrounding the WP [unitless]

D =   diffusion coefficient of the radionuclide in the medium surrounding the WP [m²/yr]

r_0 =   radius of the equivalent WP [m]

The concentration gradient at the outer surface of the WP, $\left.\dfrac{\partial C_i}{\partial r}\right|_{r=r_0}$, can be obtained by solving the following boundary and initial value problem:

$$C_i\big|_{t=t_0} = C_{i_0} \tag{3-54}$$

$$\frac{\partial C_i}{\partial t} = \frac{D}{R_i} \frac{1}{r^2} \frac{\partial}{\partial r}\left(r^2 \frac{\partial C_i}{\partial r}\right) - \lambda C_i \tag{3-55}$$

$$C_i\big|_{r=\infty} = 0 \tag{3-56}$$

$$C_i\big|_{r=r_0} = C_i(t) \tag{3-57}$$

where:

C_i =   concentration of radionuclide i [kg m^{-3}]

t =   time [yr]

R_i =   retardation factor of radionuclide i [unitless]

λ_i =   decay constant of radionuclide i [yr^{-1}]

The retardation factor R_i is defined as:

$$R_i = 1 + \frac{1-\varphi}{\varphi}\rho_s K_{di} \tag{3-58}$$

where:

φ =   porosity of the medium [unitless]

ρ_s =   grain density [kg m^{-3}]

K_{di} =   distribution coefficient of radionuclide i [m^3 kg^{-1}]

This formulation accounts for molecular diffusion, retardation of radionuclides, and radioactive decay, but not the generation of radioactive progeny. For simplicity, it is assumed that D is independent of the species and includes the effect of tortuosity. The diffusion model is presented for a spherical geometry. Thus, the inner-boundary condition is maintained at a distance equivalent to the radius of a sphere with the same outer surface area as a single WP (Appendix A provides WP dimensions). Although the outer boundary is defined at infinity, the numerical solution of these sets of equations requires that a finite distance be assigned. A distance beyond the drift wall is considered sufficient. The spherical configuration was chosen for simplicity and lends itself to simple 1-D analytical or numerical solutions.

3.5.2.4 Waste Package Inventory

The inventory in the failed WP is calculated by a material balance, Eq. (3-49), accounting for depletion because of decay, generation of radioactive progeny, and mass depletion from diffusive and advective releases. For the case in which no mass leaves the WP, Eq. (3-49) reduces to:

$$\frac{dm_i}{dt} = -m_i \, \lambda_i + m_{i-1} \, \lambda_{i-1} \tag{3-59}$$

$$m_i\big|_{t=0} = m_{i0} \tag{3-60}$$

These differential equations are solved to determine the remaining solid mass of the radionuclides in a chain at a given time t. The analytical solution (Bateman, 1910) to this initial value problem is given by:

$$N_{ij} = N_{i0} \prod_{k=1}^{j-1} \lambda_k \sum_{L=1}^{j} \frac{e^{-\lambda_L t}}{\prod_{\substack{m=i \\ m \neq L}}^{j} (\lambda_m - \lambda_L)} \tag{3-61}$$

where:

N_{ij} =   contribution from the ith chain member to the jth chain member [mol]

N_{i0} =   initial mass of ith member of chain [mol]

λ_i =   decay coefficient of jth member of the chain [1/yr]

The total amount of the jth chain member at any time is:

$$N_j = \sum_{i=1}^{j} N_{ij} \qquad (3\text{-}62)$$

In the previous equations, $m_i = N_i$. The notation N_i is used to preserve the conventional notation for representing the solution to the Bateman equation.

3.5.2.5 Spent-Fuel Dissolution Rate

The rate of radionuclide transfer from the SF into water in the WP [w_{li} in Eq. (3-49)] is a function of the flow rate of the water, the composition of the water in which the SF is in contact, and the element solubility in that water. Fission and activation product radionuclides normally have high solubility limits in water that do not limit the release of these radionuclides from a WP. These radionuclides include technetium-99 (Tc-99), iodine-129 (I-129), cesium-135 (Cs-135), carbon-14 (C-14), and chlorine-36 (Cl-36). Neptunium-237 (Np-237) may belong to this category also, depending on the chemistry and the flow rate of water in contact with the SF. These radionuclides are understood to be released congruently with the dissolving SF matrix for immersed fuel. Gray and Wilson (1995) determined the intrinsic dissolution rate of the SF matrix from flow-through tests. The flow-through tests use artificially high flow rates to eliminate the precipitation of secondary phases on the SF surface. The secondary phases may modify the intrinsic dissolution rate of the SF matrix by altering the area of the reactive surface.

The dissolution rate in solutions containing carbonate anions, representative of altered ground water present in the near field, is expressed by Gray and Wilson (1995) as:

$$\log r = 9.310 + 0.142 \log\left[CO_3^{2-}\right] - 16.7 \log\left(p_{O_2}\right) +$$
$$0.140 \log\left[H^+\right] - \frac{2130}{T} + 6.81 \log(T) \log\left(p_{O_2}\right) \qquad (3\text{-}63)$$

where:

r	=	dissolution rate [mg m^{-2} d^{-1}]
$[CO_3^{2-}]$	=	total carbonate concentration [mol/L]
p_{O_2}	=	oxygen partial pressure [atm]
$[H^+]$	=	concentration of hydrogen ions [mol/L]
T	=	temperature [K]

Normally, the ground water at the YM repository site contains silica (Si) and calcium (Ca) ions. In the presence of these species, the dissolution rate decreases about 100 times (Gray and Wilson, 1995). The dissolution rate in the batch tests (under immersion) is about 10 times lower than that in flow-through tests (Gray, 1992). The dissolution rate in J-13 well water containing these ions can be described by:

$$r = r_0 \exp\left[-\frac{34.3}{RT}\right] \qquad (3\text{-}64)$$

where R is given in kJ/mol K and $1.4 \times 10^4 \le r_o \le 5.5 \times 10^4$. This equation was obtained from data at 25 °C (77 °F) (Gray, 1992; Gray and Wilson, 1995) and at 85 °C (185 °F) (Gray, et al., 1992). The rate magnifies by a factor of 10, as temperature increases from 25 to 85 °C (77–185 °F).

EBSREL allows the user to choose between Eqs. (3-63) and (3-64), or to assign a constant rate of dissolution via the primary input file. The $w_{li}(t)$ term in Eq. (3-49) is then calculated by multiplying the SF dissolution rates obtained from Eq. (3-63) or Eq. (3-64) (or the constant rate option) by the SF surface area.

3.5.2.6 Spent-Fuel Surface Area

Two models for determining the SF surface area are available in the primary input file. The first model determines surface area using fragmented pellets (i.e., particles) and the second model uses SF grain size criterion. The particle model assumes that SF in a WP is fragmented into small particles [~1 mm (0.04 in.) in diameter] and the intergranular porosity does not contribute to the surface area. The total surface area for the particle model is then computed as:

$$A_p = \left(\frac{M_{SF}}{M_p} \times 4\pi r_{po}^2\right) f_{wet} \qquad (3\text{-}65)$$

where:

A_p = particle surface area [m^2]

M_{SF} = SF inventory/WP [kg]

M_p = particle mass equal to $\frac{4}{3}\pi r_{po}^3 \rho_{go}$ [kg]

r_{po} = particle radius [m]

ρ_{go} = density of oxidized SF [kg m^{-3}]

f_{wet} = volume fraction of SF immersed in the WP water [unitless]

If subgranular fragmentation of the SF takes place through fuel conversion from UO_2 to $UO_{2.4}$ and U_3O_8, a smaller particle size (i.e., equivalent) can be considered in the SF particle model to represent additional exposed surface area. In the case of the second model, in which SF grains are exposed, the following expression is used to compute the total surface area:

$$A_g = 4 \times 10^{-12} \pi \left(\frac{M_{SF}}{M_p} \right) \left[(r_g - w)^2 + \frac{3}{r_g}(3r_g^2 w - 3r_g w^2 + w^3) \right] f_{wet} \quad (3\text{-}66)$$

where:

A_g = grain surface area [m^2]

r_g = SF grain radius [μm]

w = width of the oxidized zone [μm]

The surface area available for leaching is conservatively held constant throughout the leaching period, even though the radius of the unoxidized fuel grains or particles would diminish with time. In addition, preferential attacks of grain boundaries in the fuel particles were not considered.

In deriving these equations, the effects of ionizing radiation have not been considered explicitly. These effects are complex, potentially leading to higher release rates. Factors such as the age of the waste, thickness of water films, cladding protection, and protectiveness of secondary mineral layers on the fuel would have to be considered, to take ionizing radiation into account.

3.5.2.7 Cladding

SF cladding may protect the bare SF matrix from exposure to water in the WP and reduce release rates significantly. The failure mechanisms of cladding include: (i) mechanical failure by external forces such as rockfalls; (ii) localized corrosion; (iii) creep; (iv) hydrogen-induced failure; (v) splitting by matrix volume expansion; and (vi) stress corrosion cracking.

It is assumed that cladding reduces the fraction of the total SF surface exposed to water entering the WP. In the TPA Version 3.1.4 code, the fraction of SF surface area affected by cladding protection is controlled by a factor in the primary input file, which specifies the fraction of fuel that is unprotected by cladding and thus does not inhibit release (CladdingCorrectionFactor = 1 indicates no cladding protection, and 0 indicates complete cladding protection and no release). Currently, cladding protection is assumed to remain constant and unaffected by additional failure mechanisms or disruptive events.

3.5.2.8 Water Dripping Abstraction

The dripping abstraction determines the quantity of liquid water, q_{in}, eventually entering the WP. It is based on the assumption that there is a net downward percolation of meteoric water at the site after thermal reflux and water will flow in fractures within the emplacement unit.

The dripping abstraction is represented by sampled distributions rather than a model embedded in the code. This approach is computationally efficient and is more easily and transparently factored into sensitivity analysis than an embedded model. Furthermore, this approach allows incorporation of alternative conceptual flow models such as structural control of wetting along faults or fracture zones, without requiring code changes.

The WP infiltration flow rate is:

$$q_{in} = qF_{ow} \ F_{mult} \qquad (3\text{-}67)$$

where:

q = ground-water flow rate at the repository horizon after thermal reflux of water from deep percolation and reflux [m^3/yr]

F_{ow} = factor to account for the flow potentially reaching a wetted WP

F_{mult} = factor to account for near- and in-drift flow diversion

When the factor F_{ow} is multiplied by the ground-water flux q and the plan area of the WP, the resulting product defines the flow rate that potentially reaches a wetted WP. F_{ow} represents focusing of flow when it is assigned a value greater than 1 and diverging of flow when less than 1. F_{ow} is correlated with another factor, F_{wet}, which is the fraction of WP receiving a dripping flux greater than zero. No mechanistic treatment of F_{wet} has been included in the TPA Version 3.1.4 code. It is treated as a uniform random variable between 0 and 1.0. This approach entails the assumption that the number of WPs wetted is invariant in a given run and does not take into consideration that dripping locations may change with time.

The flow diversion factor, F_{mult}, which ranges in the model from 0.01 to 0.2 and is lognormally distributed, is defined as the fraction of potentially dripping water that will enter the WP and contribute to the release and transport of radionuclides. Only a fraction of the water intercepting the drifts is expected to come into direct contact with the WPs. Only a portion of the water dripping onto the WPs is expected to get inside where it can interact with the waste form. Elements considered in F_{mult} include:

- The reduction in flow to a WP because water is diverted around a drift is based on the presence of a capillary barrier. Water flow into the drift will face a capillary barrier if the fractures are small enough. Water cannot easily move across the capillary barrier and can be diverted in the fracture around the drift opening.

- Water flow crossing the capillary barrier into the drift can drip from the ceiling or from protuberances along the drift. Much of the water, however, is likely to be diverted as sheet flow along the drift walls rather than drip from the ceiling onto the containers. The closer to the crest of the tunnel, the larger would be the propensity to drip. Away from the crest, the slope of the tunnel walls would divert water to sheet flow along the walls.

- Water dripping from the ceiling would have to fall onto the WP in such a way that it could enter an open, through-going hole. If a drop of water is directly in the path of an open hole, such as when the hole results from corrosion by dripping water, then this condition would be fulfilled.

• A few drops could enter a corrosion hole, but may be unable to enter the canister because of the presence of corrosion products from the steel overpack and inner barrier. These corrosion products may be flaky, porous, or gel-like, and their density will be considerably less than the steel itself. Without a high rate of water flux, there will be no mechanism to remove them from the location where they form If the corrosion products remain in place, then water dripping into the corrosion hole would have difficulty entering the canister and would simply flow off. Holes at the crest will have a higher probability that water would enter because there would be a smaller propensity for water to flow off. Holes on the side will have higher probability that the water would flow off rather than enter.

A more detailed explanation of the derivation of initial F_{mult} (and appropriate values for it) is contained in Appendix B. Both F_{ow} and F_{mult} are specified in the primary input file as sampled parameters. F_{ow} is correlated with areal average MAI at the initial (current) climatic condition, matrix permeability (K_{sat}) in the near field, and F_{wet}. Table 3-2 presents the correlation among the dripping parameters used in the TPA Version 3.1.4 code. It is assumed that these two parameters are independent of location, temperature, and water chemistry at the repository level.

Table 3-2. Dripping Parameter Correlation Matrix

Parameter	$<MAI>$	$<K_{sat}>$	F_{wet}	F_{ow}
$<MAI>$		0	0.631	−0.224
$<K_{sat}>$	—	1	−0.623	0.13
F_{wet}	—	—	1	−0.356
F_{ow}	—	—	—	1

3.5.2.9 Computational Approach

The first step in the calculation is to determine when a liquid release starts. The release calculation includes the computation of the radionuclide inventory in the solid mass, radionuclide releases from the solid mass into the liquid surrounding the waste form, the generation of the new radionuclide inventory in the liquid because of radioactive ingrowth, advective release of mass from water leaving the WP, and diffusive losses into the medium surrounding the WP.

At each time step, the inventory of the elements is computed as the sum of the mass of all the isotopes of that element. At any given time, the concentration of an element in the WP water is calculated by dividing the element inventory in the water by the volume of water in the WP. If the concentration of that element in the WP water exceeds its solubility limit, then the calculated concentration value is discarded and the solubility limit is assigned to the concentration of the element. When release is solubility controlled, release of an individual nuclide occurs at a rate proportional to the mass fraction of all the isotopes of the element.

The number of WPs affected by each failure mode (initially defective, scenario failure, or corrosion failure) is first determined by EBSFAIL. The fraction of WPs undergoing initially defective failure is specified in the primary input file. The number of WPs actually being considered for scenario failure is determined based on the corrosion failure time and the onset of wet conditions on the WPs. If the corrosion failure occurs before the scenario failure, then the number of WPs undergoing scenario failure is set to zero (i.e., WPs can only fail once). If the corrosion failure occurs after the scenario failure and wet conditions exist prior to the time of corrosion failure, then calculations of release from scenario failure are performed using the specified number of failed WPs.

Liquid release of contaminants arises only from the wetted fraction of the SF in a WP. Radionuclides are released from the waste form at a rate proportional to the dissolution rate of the fuel (congruent release), into the volume of water present in the WP. Release of radionuclides from the WP by advection and diffusion may be limited subsequently by the elemental solubility of the radionuclides in the WP water.

At the end of this calculation, the cumulative release is recorded for each nuclide. After advancing the time, the calculation is repeated for the next time step. The calculation continues until all radionuclides are depleted from the wetted fraction of SF and the water in the WP or the end of the simulation is reached, whichever occurs first. Since release calculations focus on release from a single WP for each WP failure category, to obtain the total release in the final calculations, the release from one WP is multiplied by the total number of wetted WPs for each failure category in a subarea.

3.5.3 Assumptions and Conservatism of the EBSREL Approach

EBSREL is an abstraction of the processes that will take place in a failed waste container. The main processes that control releases of radionuclides from the SF to the boundary with the geosphere are: (i) protection of the SF by cladding; (ii) degradation of the SF by air and water vapor; (iii) contact of the SF by liquid water; (iv) mobilization of radionuclides from the SF to the liquid water; and (v) transport of dissolved or otherwise mobilized (colloids) radionuclides in the water to the outside of the WP.

There is no mechanistic model for cladding protection. Instead, cladding protection is specified as a factor between 0 and 1, representing the fraction of the surface area of the fuel exposed to dissolution. In most cases, it is conservatively assumed that no cladding protection exists. Some runs have considered substantial protection by cladding with only 0.5 percent of the available surface area of the fuel exposed. These values were derived empirically from laboratory leaching studies with deliberately failed fuel-rod samples (Wilson, 1990).

EBSREL does not explicitly consider degradation of fuel by air and water vapor. There are no mechanistic models of degradation by air and water vapor in the module. The SF is assumed in a degraded state, on contact with liquid water. It is assumed that there can be no RT from the SF until the time liquid water contacts the fuel. This is a conservative yet reasonable approach, because it is expected that most exposed fuel will degrade in the period between exposure to air and liquid water contact. The model conservatively ignores the possibly favorable effects of fuel degradation in terms of how radionuclides released from the uranium oxide waste form might be bound into low-solubility secondary phases such as uranyl silicates and hydroxides. There is a body of evidence that suggests this mechanism will significantly reduce the availability of several important radionuclides such as neptunium. Furthermore, secondary minerals may form protective layers, around the fuel, that could lead to reduction in radionuclide releases. These effects have been conservatively ignored.

The contact of exposed fuel by liquid water is presumed to be controlled by either the bathtub or flow-through models. In the bathtub model, the fraction of fuel exposed is presumed to be controlled by the water height in the container. In the flow-through model, contact is independent of the water level. It seems clear that if water were to pool in the container, the fuel that was submerged would be in contact with liquid water. Aside from submersion, contact of fuel with liquid water would be controlled by complex and ill-defined processes occurring inside the container, such as dripping, water films on the fuel controlled by surface tension, evaporation, and thermal reflux. Because the actual processes are not well-understood, the choice of coefficients defining the fraction of fuel contacted by liquid water is selected to effect conservative results.

The fraction of fuel contacted by water may or may not be important, depending on whether the radionuclides contributing most to the dose are limited by solubility. If solubility controls the release, the fraction of fuel contacted is unimportant and the flow rate of water in contact with the fuel is important. Conversely, if solubility is not the limiting factor, the fraction of fuel exposed to liquid water is important.

Released radionuclides are assumed to be transported to the geosphere by water leaving the WP and by diffusion through a layer of porous material. There is no accounting for the possibility that radionuclides in the water might be captured along the liquid-water pathway by precipitation or sorption on material inside the WP. If backfill is considered part of the EBSREL model, sorption of radionuclides from the flowing water might offer a significant reduction in releases. Molecular diffusion through backfill or rock is considered a minor factor and may be eliminated in future versions.

There are several specific assumptions related to the EBSREL module:

*   Liquid release calculations focus on release from a representative WP in each subarea. It is assumed that the release environment within the subarea is not affected by the location or the geometric configuration of other WPs. Therefore, to obtain the total release in the final calculations, the release from a representative WP is multiplied by the number of WPs for each applicable WP failure category in the subarea.

*   EBSREL considers only radionuclide releases from SF; no consideration is given to radionuclide release from glass waste form or the DOE SF.

*   Although the thermal model provides the specifics of backfill emplacement, it is conservatively assumed that the backfill is always present in the diffusive release calculations. Diffusive releases only occur if backfill is present; they are identically zero if backfill is absent. Diffusive releases are generally much smaller than advective releases.

- Colloidal transport is not explicitly considered, which could be nonconservative.

- The surface area of SF exposed to water during the leaching process is constant even if shrinking of particle size takes place with time. Preferential attacks of grain boundaries have not been considered.

- For the bathtub model, all fuel below the water level is assumed to come into contact with water uniformly.

- No new pits are formed, with time, that would alter the volume of the bathtub. Once the water reaches the level of the exit hole, it can leave the WP unimpeded.

- Advective mass transport is considered from the inside to the outside of the WP and into the UZ but not into the material surrounding it, as schematically shown in Figure 3-13 for diffusive mass transfer. Ignoring transport time through the backfill and invert is conservative.

- There is no radionuclide release from SF above the water level, which could be nonconservative.

- Formation of a subsequent release of radionuclides from secondary phases is not considered.

- The change in water chemistry in the WP caused by corrosion of container material is not explicitly considered, but can be taken into account by adjusting the parameters of the leaching rate equation.

- Gaseous release occurs only from dry SF, but is not used in the dose calculations because gaseous contribution to dose is considered negligible.

- SF leaching is based on a congruent release mode.

- For diffusive mass transport, ingrowth of radioactive progeny in the rock is ignored.

- The assumption of spherical geometry for radionuclide diffusion into a porous medium may be conservative (Wescott, et al., 1995) because:

  — The boundary condition $C|_{r = r_0}$ implies that the radionuclide concentration at the outside wall of the WP is the same as that inside the container.

  — A steeper concentration gradient results from the absence of the generation of radioactive progeny in the chain, contributing to faster diffusive release.

— The boundary condition $C|_{r=r_w} = C_0$ results in a steeper concentration gradient than expected because the contribution of other WPs is conservatively ignored.

3.6 UZFT MODULE DESCRIPTION

The UZFT module determines the release rate of radionuclides into the SZ below the repository footprint by simulating the transport of radionuclides in the UZ between the repository and the water table, taking into account fracture versus matrix flow and retardation, because of adsorption and ion exchange.

3.6.1 Information Flow Within TPA

3.6.1.1 Information Supplied to UZFT

EXEC passes to UZFT the time-varying release rates computed in EBSREL of all ground-water pathway nuclides and the time history of the volumetric flow rate from UZFLOW into the subarea. Other inputs to UZFT, such as hydrologic properties of stratigraphic units and retardation coefficients, are specified in the UZFT section of the primary input file.

3.6.1.2 Information Provided by UZFT

UZFT outputs the ground-water release rates of all ground-water pathway nuclides, as a function of time from the UZ, along with the ground-water travel time (GWTT) for the UZ. Although GWTT values are used only for purposes of reporting intermediate calculational results, the release rate histories are provided as input to SZFT.

3.6.2 Conceptual Model

The UZFT module simulates the temporal and spatial variation of deep percolation and radionuclide transport from the repository horizon to the water table. UZFT uses the stand-alone code NEFTRAN II (Olague, et al., 1991) to track contaminant transport through the UZ below the repository.

UZ flow and RT are complicated by geologic heterogeneities at a variety of spatial scales, in the hydrologic and geochemical properties of the fractures and matrix, and in temporal variation in deep percolation. Although these complications affect accurate predictions at the small scale (on the order of meters), the emphasis is on estimating performance of the overall behavior of the repository. Therefore, an approach that considers flow and transport at a larger scale (on the order of tens to hundreds of meters), which neglects many of the small-scale spatial and temporal heterogeneities in favor of average properties at a larger scale, is assumed to be appropriate. The UZFT module captures a sufficient level of detail in the properties of the UZ such that performance calculations will provide insight on how variations in hydrologic and geochemical properties of the UZ affect overall performance.

3.6.2.1 Unsaturated Zone Flow Model

UZ flow is the primary mechanism for transporting dissolved radionuclides from the repository to the water table below the repository. Fracture versus matrix flow, ground-water velocity, and moisture content are the characteristics of the UZ flow used in UZFT. Spatial and temporal variabilities of these characteristics are considered, using a simplified approach: (i) spatial variability is accounted for by using seven repository subareas where each subarea has a distinct stratigraphy and deep percolation, and each stratigraphic unit has its own hydrologic and geochemical properties (Table 3-3 provides stratigraphic unit thicknesses for each of the seven repository subareas); and (ii) temporal variation is accounted for by using the time-varying deep percolation determined in UZFLOW. This simplified approach assumes there is no lateral diversion between the repository and the water table (i.e., flow is 1-D and vertical); the UZ flow field is in equilibrium (i.e., time variations in the flow field occur rapidly in the UZ); and thermal perturbations on UZ flow do not affect the deep percolation below the repository (the waste containers are considered to be mostly intact at early times, when this assumption is most in question).

Table 3-3. Stratigraphic Thicknesses (m) for Each of the Seven Repository Subareas

Stratigraphic Unit	Subarea 1	Subarea 2	Subarea 3	Subarea 4	Subarea 5	Subarea 6	Subarea 7
Topopah Spring–welded	33	116	20	110	20	53	121
Calico Hills–vitric	0	0	0	0	113	125	0
Calico Hills–zeolitic	163	154	122	132	0	0	114
Prow Pass–welded	34	39	40	34	38	26	43
Upper Crater Flat	67	20	158	57	158	136	63
Bullfrog–welded	0	0	0	0	32	0	0
Total Thickness	297	329	340	333	361	340	341

Determination of fracture versus matrix flow is considered the most important aspect of the UZ flow model because: (i) water velocity in fractures tends to be large relative to velocity in the matrix, because of differences in porosity (generally one to two orders of magnitude); and (ii) retardation of radionuclides within fractures is generally considered to be much smaller than retardation in the matrix because of the much larger surface area available for sorption in the matrix compared with the fracture. For each stratigraphic unit in the UZ, a determination is made in the UZFT module whether fracture or matrix flow properties will be used, by comparing the deep percolation with the saturated conductivity of the matrix. Fracture flow properties are used for transporting radionuclides when the deep percolation exceeds the saturated conductivity of the matrix. This conceptualization assumes that the flow conditions (fracture versus matrix flow) remain the same within a particular stratigraphic unit and any transition between matrix and fracture flow occurs only at the interface between stratigraphic units. Thus each stratigraphic sequence has its own associated velocity and saturation characteristic of either the fracture or the matrix continuum.

Calculation of the water velocity and moisture content uses the van Genuchten (1980) characteristic curve to represent the relationship between saturation and conductivity:

$$K_r = \sqrt{S} \left[1 - (1 - S^{1/\lambda})^\lambda \right]^2 \qquad (3\text{-}68)$$

where:

K_r =   relative hydraulic conductivity [unitless]

S =   saturation [unitless]

λ =   fitting parameter [unitless]

The velocity and saturation calculation assumes that deep percolation is occurring in either the matrix or the fracture, but not both. This approach represents a simplification of the methodology used in the NRC IPA Phase 2 effort (Wescott, et al., 1995), which accounted for the partitioning of fracture and matrix flow. The simplification was done to improve efficiency in the UZ flow and transport module to accommodate other improvements considered to be more important (e.g., time-varying deep percolation and simulation periods beyond 10,000 yrs). Additionally, the partitioning of fracture and matrix flow is not considered as important to performance, given the higher estimates for deep percolation (stratigraphic units will be either primarily fracture flow or matrix flow).

As mentioned previously, time-varying deep percolation is determined in UZFLOW. A time-varying velocity is calculated for each stratigraphic unit from the time-varying deep percolation, using the formula:

$$v(t) = \frac{Q(t)}{A n S} \qquad (3\text{-}69)$$

where:

$v(t)$ =   time-varying velocity for the stratigraphic unit [m/yr]

$Q(t)$ =   time-varying deep percolation rate [m^3/yr]

n   =   porosity of the stratigraphic unit corresponding to either the matrix porosity or the fracture porosity [unitless]

S =   saturation of the stratigraphic unit corresponding to either the matrix saturation or the fracture saturation [unitless]

A =   cross-sectional area of subarea [m^2]

To aid in the computational efficiency of the NEFTRAN II code, the velocities are artificially limited to an upper bound corresponding to a 5-yr residence time for a given stratigraphic unit. This restriction should not have a significant effect on the dose calculation.

From the time-varying velocities, the time-varying residence times are determined for each stratigraphic unit, using the formula:

$$t_I(t) = \frac{x}{v(t)} \tag{3-70}$$

where:

$t_I(t)$ = time-varying residence time [yr]

x = thickness of the stratigraphic unit [m]

$v(t)$ = time-varying velocity for the stratigraphic unit [m/yr]

The mean residence time T_T for each stratigraphic unit is determined as a weighted average of the time-varying residence times over the time period of interest. A simple trapezoidal integration of the residence time is used for the calculation. In addition, a sum of the T_T from each stratigraphic unit is calculated and represents the total residence time T_{RT} for the entire UZ. The mean residence time of a stratigraphic unit is used to select which stratigraphic units satisfy a set of criteria that maintains the computational efficiency of the NEFTRAN II code. That is, stratigraphic units where T_T is less than 10 yrs or that have a T_T less than $T_{RT}/10$ are removed from consideration for the RT. These criteria should not have a significant effect on the dose calculation.

The time-varying velocities for each contributing stratigraphic unit are output to file *nefii.vel* for input to NEFTRAN II. The UZ residence time T_T for each subarea is written to the file *gwttuzsz.res*, as an intermediate result.

3.6.2.2 Unsaturated Transport

Transport of radionuclides in geologic media is often retarded relative to the water velocity because of reactions (e.g., adsorption and ion exchange) between the dissolved radionuclides and the geologic materials. Retardation factors (ratio of the velocity of a dissolved radionuclide to the water velocity) are determined for the matrix, using the following distribution coefficient, or K_d approach:

$$R_f = 1.0 + \frac{\rho(1-n)}{\theta} K_d \tag{3-71}$$

where:

R_f = retardation factor [unitless]

ρ	=	grain density of porous matrix [kg/m^3]
n	=	porosity of matrix [unitless]
θ	=	moisture content of matrix [unitless]
K_d	=	distribution coefficient (radionuclide, and stratigraphic-unit-specific) [m^3/kg]

Retardation within fractures in the UZ is considered constrained by the limited surface area available for ion exchange and adsorption. Therefore, the above K_d approach was not considered appropriate for fracture flow, and a retardation factor is used to characterize retardation in fractures in the primary input file. The use of a retardation factor for fracture transport was included in UZFT for completeness in the event that sufficient evidence becomes available that supports retardation of radionuclides within fractures. It is expected however, the retardation factor would typically be set for no retardation (value of 1.0).

NEFTRAN II

NEFTRAN II (Olague, et al., 1991) uses the following three files: (i) the standard NEFTRAN II input file that specifies the radionuclides, transport path lengths, and transport path properties (*nefii.inp*); (ii) the time history of the transport velocity developed in UZFT, using hydrologic parameters specified in the primary input file and time history of deep percolation determined in UZFLOW (*nefii.vel*); and (iii) the time history of radionuclide releases from the repository provided by EBSREL (*sotnef.dat*). NEFTRAN II uses a series of 1-D transport paths or legs (each representing a particular stratigraphic unit) that are defined by their length, velocity, saturation, porosity, dispersion length, and retardation. (Note: matrix diffusion is a process contained in the NEFTRAN II program; however, this process was not included in the UZ transport because of the uncertainty in assigning parameter values that determine the magnitude of its effect.) As described in the previous sections, UZFT calculates velocity, saturation, and retardation factors. NEFTRAN II uses this information as well as input parameters (porosity and dispersion length) to simulate advection and hydrodynamic dispersion of dissolved radionuclides using a distributed velocity method (DVM). DVM is similar to a particular tracking approach where discrete particles are used to simulate contaminant movement; however, DVM calculates a distribution of velocities for each radionuclide and each leg, based on the mean velocity and the dispersion length, and uses the velocity distribution to transport groups or packets of particles. The formulation used in calculating the velocity distribution is

$$v_j = \varepsilon_j \sqrt{\frac{2dv_m}{t\, R_f} + \frac{v_m}{R_f}} \qquad (3\text{-}72)$$

where:

v_j	=	j^{th} interval of the velocity distribution (seven intervals are used to represent the distribution) [m/yr]
ε_j	=	j^{th} abscissa from a standard normal distribution [unitless]
d	=	dispersion length [m]
v_m	=	mean velocity (velocity calculated in UZFT and used as input to NEFTRAN II is assumed to be the mean velocity) [m/yr]

R_f =   retardation factor [unitless]

t =   time to exit the leg[2] (length of leg / v_m) [yr]

Figure 3-14 illustrates the implementation of the 1-D legs or transport paths in NEFTRAN II to represent the UZs and SZs at YM. The UZFT module provides the radionuclide releases to the water table at discrete times. This information is then used by the SZFT module for the SZ transport calculation.

3.6.2.3 Efficiency of Simulating Flow and Transport

Efficiency in simulating flow and transport in the UZ was considered in development of the UZFT module, because of potentially long run times associated with long-term performance periods (much greater than 10,000 yrs) together with additional processes such as matrix diffusion. Long simulation times can occur when the transport velocity is such that the use of a small time step over the simulation period is required. A resolution to this problem was to bypass consideration of stratigraphic units that had short residence times (less than 10 yrs or less than 10 percent of the residence time of all legs). When the velocity in a particular leg results in a short residence time, the leg is not included in the NEFTRAN II simulation. If all the legs for a particular subarea are bypassed, the entire UZ transport calculation is bypassed and releases from the repository are provided directly to the SZ transport module, SZFT.

Additionally, the time intervals over which the velocity varies, which are provided by UZFLOW, can affect simulation times (e.g., short intervals resulting in long run times). Therefore, the UZFT module varies the velocity over a time interval dictated by a minimum change in the velocity field (an input parameter specifies the UZ minimum velocity change factor), where a small fractional change will result in shorter time intervals and associated longer simulation time.

3.6.3 Assumptions and Conservatism of the UZFT Approach

The UZFT module is an abstraction of the flow of ground water and the transport of radionuclides from the repository to the SZ. The main attributes of the UZ that control RT are: (i) the transport velocity of radionuclides in ground water; (ii) radionuclide sorption; (iii) matrix diffusion; and (iv) the hydrologic stratigraphy.

The transport velocity of radionuclides in ground water is influenced by the presence or absence of fracture flow. Because of differences in porosity between the fracture and matrix, transport velocities can be much larger (up to 2 to 3 orders of magnitude) in the fractures than in the matrix, for the same flux. The transport velocity within a specific hydrostratigraphic unit (e.g., Calico Hills zeolitic) is determined by assuming vertical flow below the repository and comparing the vertical flow with the saturated conductivity of the matrix (if the vertical flow exceeds the saturated conductivity any time during the simulation, fracture transport velocities are used). This approach is considered a reasonable representation for the bulk behavior

---

[2]The use of the exit time is a departure from the original NEFTRAN II formula that used the value for the time step. When used in conjunction with time-varying velocity field, the time step approach led to an unrealistic velocity distribution—the mean velocity was small, the time step was small, and a velocity distribution was dependent on the time step of the simulation rather than the physical properties of the leg. The revised formula, using the time to exit the leg, corrected these deficiencies and is considered a more appropriate approach for determining the velocity distribution.

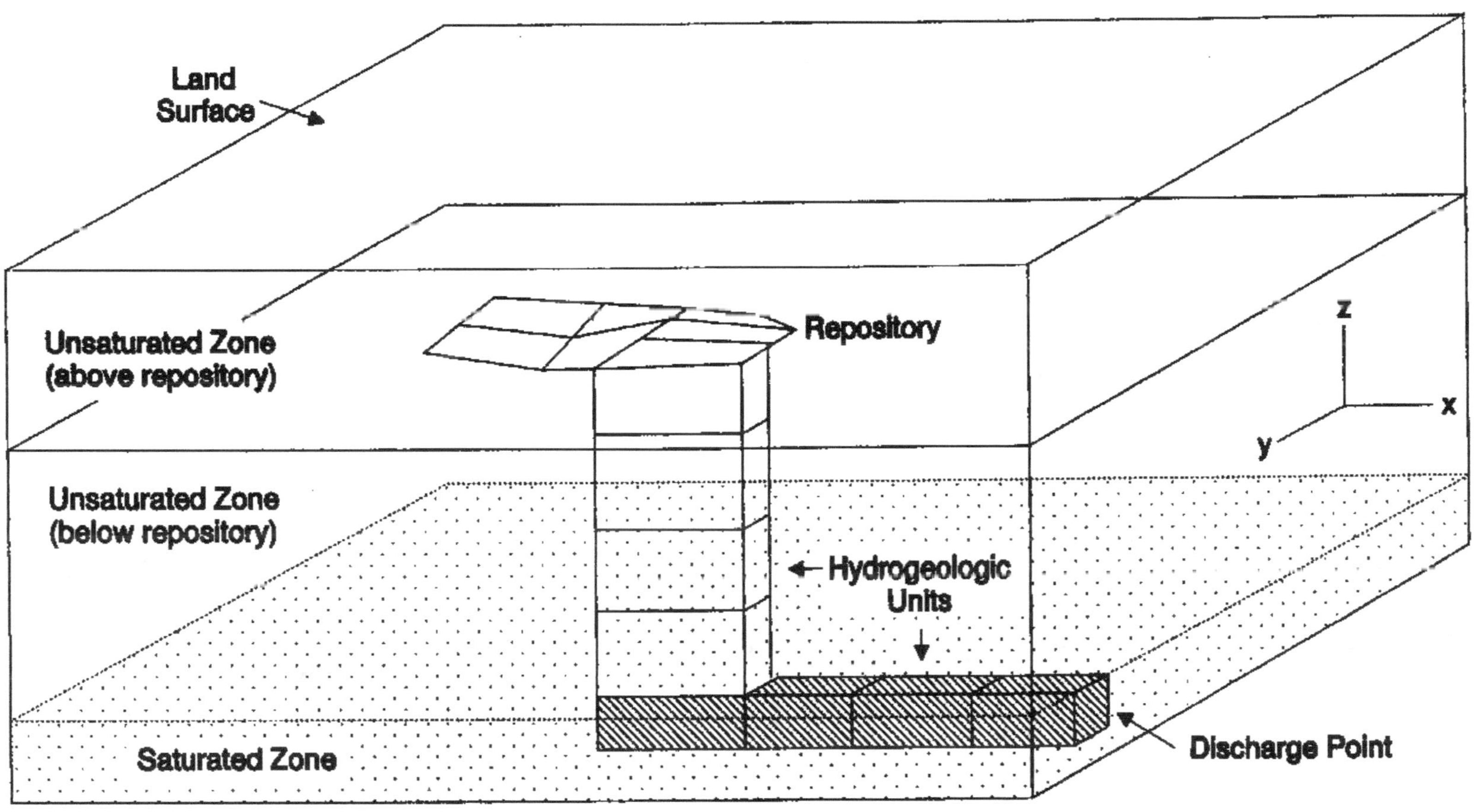

Land Surface

Unsaturated Zone (above repository)

Repository

Unsaturated Zone (below repository)

Hydrogeologic Units

Saturated Zone

Discharge Point

x
z
y

Figure 3-14. UZFT contaminant transport legs.

of a number of the specific hydrologic units at YM because saturated conductivity of the matrix is either significantly above or below the net flux (e.g., Calico Hills zeolitic, Calico Hills vitric, and Topopah Springs welded). Although this approach does not account for spatial variability of flow caused by heterogeneities in the hydrologic properties of the fractures and matrix, or the episodic nature of the infiltration, the approach generally yields short travel times to the SZ, using the current hydraulic properties and infiltration estimates. Short travel times are considered consistent with the presence of chlorine-36 at multiple locations in the Exploratory Studies Facility at YM. The degree of conservatism in these travel times cannot be determined until site investigations, including the chlorine-36 data, are complete.

Transport of radionuclides can be significantly slowed by the sorption of radionuclides on mineral and rock surfaces. Retardation of radionuclides is considered limited within fractures relative to the matrix, because retardation of radionuclides is assumed largely affected by the surface area available for sorption, whereas flow through the matrix encounters significantly more surface area than fracture flow. Although the TPA Version 3.1.4 code has the capability to represent retardation of radionuclides within fractures, fracture retardation factors have been set to 1.0 (i.e., no retardation) in the data set. This is a conservative approach considered to have a limited effect on the results given the assumption that sorption of radionuclides within fractures will be limited is true.

Transport of radionuclides within fractures can be delayed because of diffusion of radionuclides within the fractures into the matrix. The rate of diffusion of radionuclides is affected by variations in fracture and matrix flow, the concentration gradient between the fracture and matrix water, and the nature of the fracture surface (e.g., permeability of fracture coatings).Within the TPA Version 3.1.4 code, it is conservatively assumed that matrix diffusion does not occur in the UZ. Overall, this assumption in UZFT is consistent with little, if any, delay of radionuclides moving large distances in the unsaturated zone over short time scales, as evidenced by the presence of chlorine-36 at multiple locations in the Exploratory Studies Facility at YM. Moreover, observed differences in water chemistry between matrix and fracture water would not be present if matrix diffusion were significant. The degree of conservatism in this assumption depends on the ongoing site investigations.

It is obvious from the prior discussions that transport in the UZ will be sensitive to the occurrence of fracture flow. Fracture flow will dominate in a unit with low matrix conductivity (e.g., Calico Hills zeolitic) whereas matrix flow will dominate in a unit with high matrix conductivity (e.g., Calico Hills vitric). Thus, the presence or absence of particular units below the repository and the assumption that the flow is primarily vertical have a strong influence in determining the occurrence of fracture flow. The abstraction for stratigraphy uses seven subareas to account for the spatial variability in the thickness of the hydrostratigraphic units (primarily based on the spatial variability of the Calico Hills vitric and zeolitic units). This approach was used to provide a general representation of the UZ and does not represent thin stratigraphic units [i.e., on the order of a few meters (10 ft)]. Although the inclusion of some of these thin units could increase travel times, through lateral diversion or matrix-only flow, they are not included in the current approach.

There are several specific assumptions related to the UZ flow below the repository:

- Flow is downward toward the water table with no lateral diversion.

- The UZ flow field is time-varying, but in equilibrium (temporal variation in the near surface infiltration, because of climate variation, equilibrates rapidly in the UZ).

- Deep percolation between the repository and the water table is not affected by thermal reflux. During the highest thermal period (e.g., first 1,000 yr), when this assumption is most uncertain, it is assumed that few containers are failed.

- Fracture flow occurs only when the percolation exceeds the saturated conductivity of the matrix; when this occurs, it is assumed the interconnected fractures are capable of conducting the remaining flow.

- Transition between matrix and fracture flow occurs only at the interface between stratigraphic units (the amount of fracture and matrix flow is uniform within a stratigraphic unit).

There are also specific assumptions related to the UZ transport:

- RT is assumed to occur in either the fractures or matrix and is not partitioned when flow occurs in both fractures and matrix (fracture flow is used when the deep percolation exceeds the saturated matrix conductivity of the matrix); not accounting for the partitioning of flow between the fracture and the matrix will be conservative when the fracture flow is a small percentage of the matrix flow.

- Retardation within fractures does not significantly delay RT in the UZ.

- Matrix diffusion does not significantly delay RT in the UZ.

3.7   SZFT MODULE DESCRIPTION

The SZFT module describes radionuclide transport in the SZ, from the location at which radionuclides enter the water table immediately below the repository, to a receptor location. The SZ transport model consists of an array of 1-D streamtubes originating at the water table below the repository and terminating at a receptor location. RT in the SZFT module is simulated using the NEFTRAN II code (Plague, et al., 1991), which calculates the radionuclide release rate (Ci/yr) at the down-gradient receptor location. In the SZFT module, there are a number of SZ streamtubes each connecting to one or more UZ streamtubes. Figure 3-15 illustrates a streamtube assuming 1-D flow in the SZ.

3.7.1 Information Flow Within TPA

3.7.1.1 Information Supplied to SZFT

SZFT receives, from EXEC, the radionuclide release rate as a function of time from the UZ, as computed by the UZFT module. Other inputs to SZFT are specified in the SZFT section of the primary input file or in the *strmtube.dat* file. Data provided by the primary input file include hydrologic properties,

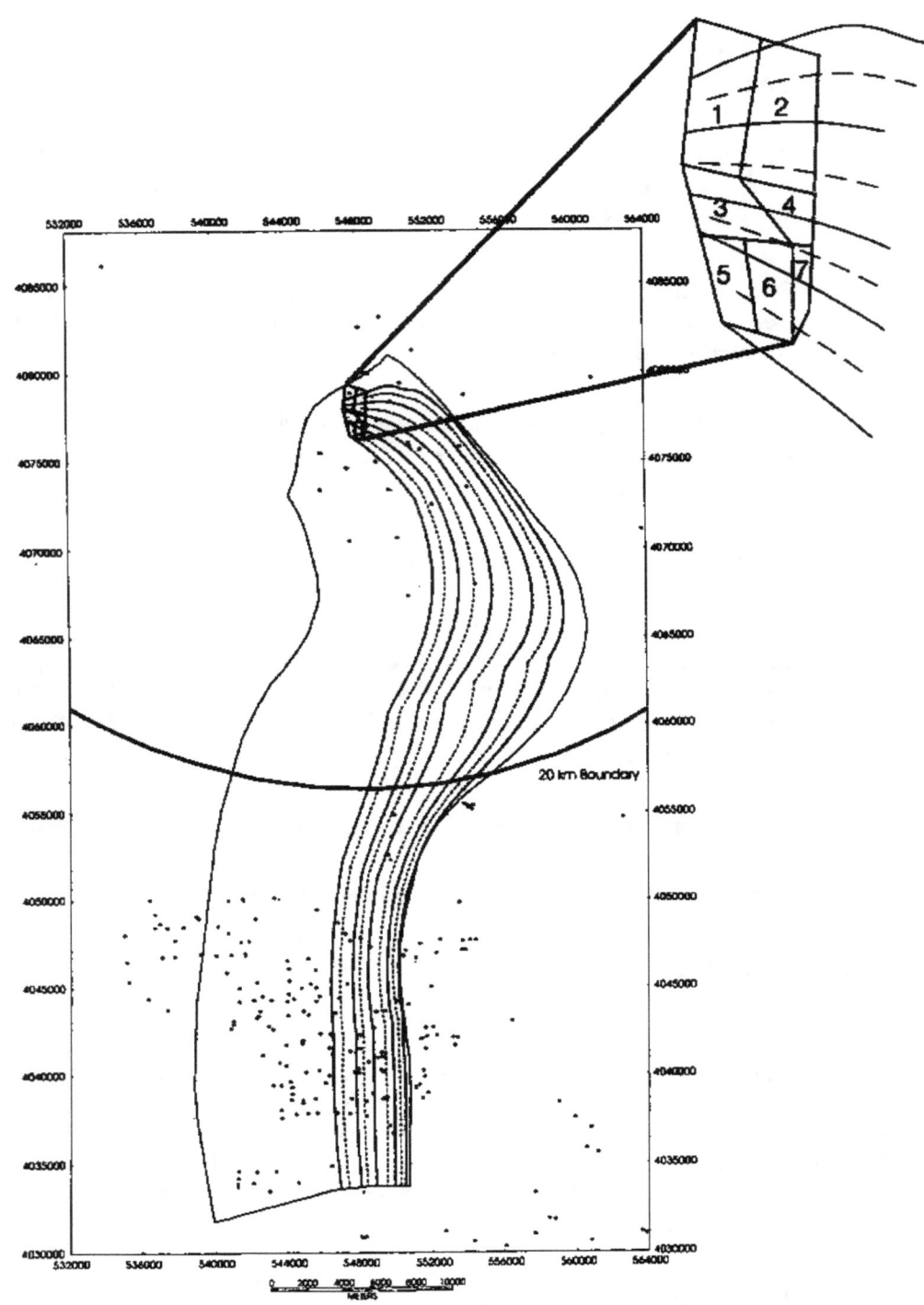

Figure 3-15. Saturated zone streamtubes model showing four streamtubes extending from just upgradient of the repository footprint southward to the major ground-water withdrawal area of Amargosa farms. The inset shows in detail the connection between the subareas and the streamtubes.

retardation factors, and matrix diffusion parameters. The distance to the receptor group determines flow unit hydrologic properties, streamtube flow rates, widths, and lengths for the tuff and alluvium specified in the *strmtube.dat* file.

3.7.1.2 Information Provided by SZFT

SZFT computes estimates of the total activity of the radionuclides transported to the receptor location as a function of time in Ci/yr. SZFT results for the time-varying release rates from the SZ are passed to EXEC. GWTT for the SZ is reported as an intermediate result for subsequent use in dose calculations.

3.7.2 Conceptual Model

The SZFT module describes the temporal and spatial variation of radionuclide mass transport from the water table at a location immediately below the repository to the receptor location. SZFT uses the stand-alone code NEFTRAN II (Olague, et al., 1991) to track radionuclide mass transport.

Transport of radionuclides in the SZ is complicated by: (i) spatial variability in the geochemical properties of the fracture surfaces and rock matrix; (ii) heterogeneity of pore-scale to formation-scale transport pathways; (iii) temporal variations in the flow field caused by climatic change and pumping for water use; and (iv) variability in the rate at which radionuclides transiting the UZ reach the water table. The extent to which these features and processes are incorporated into predictive models affect the accuracy of point estimates of radionuclide concentrations in the aquifer. The approach adopted in DCAGW (Section 3.8) for determining the radionuclide concentration of water that is consumed at the receptor location allows the SZFT model to neglect many of the high-resolution spatial and temporal variations in transport processes.

Radionuclide mass (activity) release rates at the receptor location are converted into radionuclide concentrations by multiplying the mass release rate by the fraction of mass captured by a well or well field and then dividing this product by the volume of water that is pumped. In DCAGW, simplified well-capture zone relationships are used to determine the fraction of mass captured by low-discharge wells or well fields. At high-discharge well fields used to supply irrigation water, it is assumed that all radionuclides are captured. Therefore, SZFT does not need to accurately describe the 3-D distribution of radionuclide concentrations at the receptor location. The SZFT module needs only to predict the mean trajectory and travel times of radionuclides in the SZ and take into account the variation in geochemical properties along the transport path. Although temporal changes in the flow field caused by climatic change and variations in pumping rates may affect transport, in SZFT it is assumed that a steady-state representation of the flow field is sufficient.

3.7.2.1 Saturated Zone Flow Model

The steady-state velocity field used in the SZFT module for the SZ region from below the repository to the receptor location is abstracted from a detailed 2-D model (Baca, et al., 1996). The abstraction is made by constructing four streamtubes that emanate from the vertical projection of the repository lateral boundary onto the water table and terminate at the appropriate receptor location. Differences in the mean travel time along each of the four streamtubes arise from differences in streamtube length and the sampled values of effective porosity for the tuff and alluvium units.

The boundaries of the detailed 2-D model from which the four streamtubes were abstracted, shown in Figure 3-15, are based in part on information from the 2-D subregional flow model developed by Czarnecki and Waddell (1984). The eastern and western boundaries of the flow domain shown in Figure 3-15 are coincident with a pair of streamlines from this subregional model that extend from just upgradient of the proposed YM repository southward to the major ground water withdrawal area of Amargosa Farms [Czarnecki and Waddell (1984), Plate 1]. The northwestern and southern boundaries, which were selected to be roughly coincident with head contours of 800 and 675 m (2600 and 2200 ft), were treated as Dirichlet boundaries in the detailed model. Because the western and eastern edges of the flow domain are coincident with streamlines, they were treated as no-flow boundaries in the detailed model.

To match measured head data, the domain of the detailed flow model was divided into seven zones, in which the hydraulic conductivity was assumed to be uniform. The boundaries of these seven zones were determined from available hydrostratigraphic cross-sections and from inspection of head contours. Initial estimates of hydraulic conductivity were obtained by manually calibrating the detailed flow model. These estimates were refined, using an automatic calibration routine, to fit heads predicted from the detailed model to measured head data (Baca, et al., 1996). Because the detailed model uses two Dirichlet and two no-flow boundary conditions, the values of hydraulic conductivity in the seven zones are not uniquely identifiable in the absence of prior estimates of either areal flux or hydraulic conductivity (Carrera and Neuman, 1986). Inasmuch as areal recharge within this region is minimal, fixing one hydraulic conductivity value is the best option. Zone 7, located at the southern end of the detailed model, was assigned a fixed hydraulic conductivity value of 1.7×10^{-5} m/s (Czarnecki, 1985).

3.7.2.2 Saturated Zone Transport Model

The four streamtubes abstracted from the 2-D detailed flow model describe the trajectories of the dissolved radionuclides. Mean Darcy velocities along the centerline trajectory of each streamtube are also extracted from the detailed flow model. As shown in Figure 3-15, the widths of the streamtubes vary from the repository to the receptor location. Because of the variation in streamtube width, mean centerline Darcy velocities change from 0.25 to 1.3 m/yr (0.8 to 4.3 ft/yr). Mean transport velocities along the centerline trajectories are determined by dividing the Darcy velocity by the sampled effective or kinematic porosities. As specified in the primary input file, the effective porosity varies from 0.001 to 0.01 for the portion of the streamtube that transits fractured tuff, and from 0.10 to 0.15 for the portion transiting the alluvium. Based on these ranges for effective porosity, linear transport velocities in the fractured tuff and alluvium range from 40 to 1300 and from 1.7 to 4 m/yr (130 to 4300 and from 5.6 to 13 ft/yr).

Other sampled transport parameters whose probability distribution or sampling range is defined in the primary input file include: (i) longitudinal dispersivity (α_L); (ii) retardation coefficients for each radionuclide; and (iii) parameters used to represent matrix diffusion in the fractured tuff unit. Inasmuch as the longitudinal macrodispersivity is generally assumed to increase with the scale of the contaminant plume until an asymptotic upper bound is attained (Gelhar, 1993), it is incorrect to assign a separate dispersivity value to each NEFTRAN II transport leg. Consequently, a single value is sampled that is dependent on the distance to the critical group location. The retardation coefficients set in the primary input file depend on the dominant mineralogy of the medium in the transport leg as well as the particular radionuclide being transported.

NEFTRAN II has the capability to account for the migration of dissolved contaminants from flowing pores and fractures (dynamic) into the more-or-less stagnant water within the rock matrix pores. The governing equation used in NEFTRAN II to account for the rate of change of concentration in the immobile phase is (Olague, et al., 1991):

$$\theta_s R_s \frac{\partial C_s(x,t)}{\partial t} = \beta \left[C_d(x,t) - C_s(x,t) \right] \qquad (3\text{-}73)$$

where:

θ_s	=	immobile porosity or stagnant volume fraction occupied [unitless]
R_s	=	retardation factor [unitless]
C_s	=	stagnant concentration [mass per unit volume]
x	=	distance in direction of flow [length]
t	=	time [yr]
β	=	mass transfer rate coefficient [1/yr]
C_d	=	dynamic concentration [mass per unit volume]

This process is commonly referred to as matrix diffusion. Parameters defined in the primary input file for matrix diffusion are immobile porosity and the mass transfer rate coefficient (l/yr).

3.7.3 Assumptions and Conservatism of the SZFT Approach

The abstracted model implemented in the SZFT module is assumed to adequately capture the range of processes and features that control and affect the transport of dissolved radionuclides from the SZ below the repository to the receptor location. Processes simulated by the SZFT module include: (i) advective transport through the tuff and alluvial aquifers; (ii) longitudinal dispersion during transport; (iii) chemical sorptive processes that retard the RT in the alluvial aquifer and in the matrix of the tuff aquifer; and (iv) diffusion of radionuclides from the fractures to the matrix in the tuff aquifer.

Because the estimated magnitude of water well pumping for irrigation at the receptor location is sufficient to capture most radionuclides emanating from the repository (Fedors and Wittmeyer, 1998), a relatively simple 1-D model can be used to convey radionuclides directly from the repository to the receptor location. The 1-D flow tube model used in SZFT is based, in part, on the 2-D ground-water flow model developed by Czarnecki and Waddell (1984) and, in part, on more recently obtained hydraulic conductivity and head measurements. By definition, the volumetric discharge per unit saturated thickness is constant along the flow tube; however, the Darcy velocity varies inversely with the width of the flow tube. Darcy velocities in the four flow tubes range from approximately 0.25 to 1.3 m/yr (0.8 to 4.3 ft/yr). Hydraulic conductivities obtained from pump tests performed in numerous wells in the tuff aquifer range from 1 to 10 m/day (3.3 to 33 ft/day). Using this range for hydraulic conductivity and a measured mean hydraulic gradient in the tuff aquifer of 0.00125, ambient Darcy velocities range from 0.46 to 4.6 m/yr (1.5 to 15 ft/yr), which agree reasonably well with those used in the SZFT abstraction. Transport velocities along the flow tubes are equal to the ratio of the Darcy velocity to the effective porosity, assumed to vary from 0.001 to 0.01 in the tuff aquifer, and from 0.10 to 0.15 in the alluvial aquifer.

The effects of lateral dispersion on vertical spreading of the plume are not explicitly accounted for in constructing the 2-D flow tubes. Since vertical dispersion would likely reduce *in situ* radionuclide concentrations or reduce the fraction of mass captured by the pumping well, neglecting its effects is conservative. The effects of longitudinal dispersion are included in the model; however, the value of longitudinal dispersivity used in the SZFT abstraction is conservatively assumed to be between 0.01 to 0.1 of the total transport distance.

Transport of radionuclides can be significantly retarded by sorption of radionuclides on mineral surfaces. Retardation of radionuclides is considered much more limited in fractures than in the matrix because sorption is assumed to be primarily controlled by the mineral surface area, which is much greater in the rock matrix. Although the TPA Version 3.1.4 code has the capability to represent retardation of radionuclides within fractures, fracture retardation factors have been set to 1.0 (i.e., no retardation) in the data set. This approach for fracture retardation, although conservative, will not have a significant impact on transport if indeed fracture retardation effects are small. Retardation caused by sorption of radionuclides on the abundant clays and iron minerals in the alluvium is assumed to be significant and is included in the SZFT model abstraction.

Transport of radionuclides in the tuff aquifer can be delayed by diffusion of radionuclides within the fractures into the matrix. The rate of diffusion of radionuclides is affected by variations in fracture and matrix flow, the concentration gradient between the fracture and matrix water, and the permeability of mineral coatings on the fracture surface. Within the TPA Version 3.1.4 code, it is conservatively assumed that matrix diffusion generally does not occur in the saturated tuff aquifer. Matrix diffusion can be accounted for in the tuff aquifer; however, it is conservatively limited to a small fraction of the total matrix porosity. The degree of conservatism in this assumption cannot be assessed until ongoing site investigations are complete.

There are several key assumptions regarding the SZ flow field:

- The ground-water model of Czarnecki and Waddell (1984) adequately describes the vertically averaged 2-D velocity field in the YM to Amargosa Farms area.

- Flow in the fractured tuff aquifer(s) can be adequately modeled using a vertically averaged 2-D flow model.

- Recharge is so small that it has little effect on the saturated flow regime; vertical velocity components are negligible.

- Water table exits the tuff aquifer and enters the alluvium approximately 15 km (9.3 mi) south of the repository.

- Climate change during the period of repository performance has no effect on the regional ground-water flow.

There are also key assumptions regarding transport in the SZ:

- The velocity field can be adequately modeled by dividing the region into four flow tubes whose lateral boundaries are defined by streamlines emanating from either edge of the repository.

- Within the tuff aquifer, advective transport takes place primarily through interconnected fractures.

- Radionuclides entering a flow tube are uniformly mixed across the width of the flow tube.

- The longitudinal dispersivity is related to the length of the transport path.

There is conservatism adopted in SZFT:

- Use of relatively small longitudinal dispersivities reduces mixing during transport.

- Assumption of steady-state flow precludes dispersion of the radionuclide plume caused by changes in the magnitude and direction of the mean velocity field through pumping and climatic change.

3.8 DCAGW MODULE DESCRIPTION—FARMING RECEPTOR GROUP

The DCAGW module calculates annual doses (TEDE) from exposure to radionuclide concentrations in ground water to an average member of a farming receptor group.

3.8.1 Information Flow Within TPA

3.8.1.1 Information Supplied to DCAGW

Time-varying release rates computed in SZFT for each radionuclide released via ground water are passed to DCAGW in the EXEC, from which radionuclide ground-water concentrations are calculated using the dilution volume. Time evolution of average annual precipitation and temperature from UZFLOW are passed by the EXEC to DCAGW, to determine the time when the current condition will switch to pluvial conditions. Ground-water pathway DCFs used in determining doses from radionuclide concentrations for a farming receptor located 20 km (12 mi) from YM, coincident with the location of the ground-water releases, are contained in two data files: *gw_cb_ad.dat* and *gw_pb_ad.dat*. Each file pertains to different climatic conditions (either current biosphere, designated by *cb*, or pluvial biosphere, designated by *pb* in file names). Parameters that specify the pumping rates of the receptor group, location of the receptor group, location of the conductive zone being pumped, and plume thickness are specified in the primary input file. Other input parameters are read from the DCAGW section of the primary input file.

3.8.1.2 Information Provided by DCAGW

The ground-water pathway dose for each radionuclide, as a function of time, is passed to EXEC.

3.8.2 Conceptual Model

The DCAGW module calculates dilution volume at the pumping well, using the water-use characteristics of the farming receptor group, a community at distances at least 20 km (12 mi) from YM and chosen based on location from the repository and assumptions made on the depth to the water table. Once the dilution volume is determined, the module converts the activity per unit time values calculated by SZFT to activity per unit volume of water, by dividing by the pumping volume. DCFs are then used to determine dose to the average member of the receptor group. The streamtube information used in SZFT is tied to the location of the receptor group. Therefore, any change to the location of the receptor group must be consistent with the streamtube data files.

For each time step, the product of each radionuclide concentration and DCF are summed within and among ground-water pathways and radionuclides, to calculate total doses. In addition to total doses, selected output is stratified by realization, time step, and radionuclide. The EXEC uses results from all realizations to identify and report peak doses.

3.8.2.1 Development of Radionuclide Concentrations in Water for a Farming Receptor Group

For the farming receptor group, the volume of water into which the released radionuclides are diluted is the greater of the flow rate of water within the uppermost producing horizon in the pumped aquifer and the volumetric flow rate of water pumped for household and agricultural needs of the receptor group. The flow rate of water within the uppermost horizon is set to the greater of the UZ flow rate and SZ flow rate. This condition sets a reasonable lower limit for the dilution volume that is consistent with the assumption that the combined well discharge rates at the farming location are large enough to capture all released radionuclides. Borehole concentrations at the farming location are computed by dividing the radionuclide release rate by the appropriate dilution flow rate.

For the farming receptor group location, there are no readily available data that can be used to assess the variation in hydraulic conductivity or borehole inflow with depth. The nearest wells completed in alluvium from which there is detailed lithologic information are located in the northwestern Amargosa Valley near the Beatty low-level waste facility. The soil texture varies from clay to gravel in this area; however, drillers' logs indicate that the predominant soil texture classes are sands and gravels. Data compiled from drillers' logs by Oatfield and Czarnecki (1991) show that the fraction of coarse-grained sediments in the Amargosa Farms area ranges from 0.50–0.70. Because sands and gravels are the predominant alluvial facies, it is reasonable to assume wells are screened over their entire depth. In fact, statistics on water well screen depths and lengths suggest that the typical well in the Amargosa Farms region is continuously screened over the lower 75 percent of the saturated section penetrated by the borehole. The mean and standard deviation of the thickness of the uppermost producing horizon, which are defined as extending from the water table to the bottom of the well screen, are 55 and 32 m (180 and 105 ft). The minimum and maximum values of the producing horizon thickness are 8.5 and 158 m (28 and 520 ft), respectively.

The distribution of the producing horizon thickness for the farming receptor group is positively skewed and may be suitably represented by a log-normal or gamma distribution; however, it is assumed to be uniform. For producing horizon thicknesses of 20, 80, and 140 m (66, 260, and 460 ft), the flow rates in the uppermost producing horizon in the alluvial aquifer are 40,940; 163,760; and 286,580 m^3/yr (1.4, 5.8, and 10 million ft^3/yr), respectively. Expected flow rates for farming are much greater than those for residential use. Assuming that the primary water use is for center-pivot irrigation of quarter-section alfalfa fields, a reasonable range of consumptive water use is 6,222,230–17,973,330 m³/yr (220–630 million ft³/yr) [13 and 27 quarter-section alfalfa fields; 126 irrigated acres (0.5 km²) per quarter-section; and 0.9–1.3 m (3.1–4.38 ft) of water applied each growing season].[3, 4] Clearly, the volumetric flow rate of water pumped exceeds the flow rate of water within the uppermost producing horizon.

3.8.2.2 Development of Dose Conversion Factors

The DCAGW module contains DCFs designed to convert ground-water concentrations to TEDEs for the average member of the receptor group. A farming receptor group is used because site-specific information indicates farms exist down-gradient of YM. Two climate regimes are assumed for the biosphere— one based on the present climatic conditions in the Amargosa Valley area and a second based on a future pluvial climate (i.e., a cooler and wetter climate). DCAGW uses DCFs for present climate conditions until the MAP and MAT (used as indicators for climate change) exceed a threshold value based on the Köppen-Geiger system of climate classification, at which point pluvial DCFs are to be used for the dose conversion.

In the Köppen-Geiger system of climate classification, the annual precipitation boundary between humid and arid/semiarid precipitation/climate regimes is determined for places where winter is the wet season (such as YM) by:

$$R_{\text{humid/subhumid boundary}} = 0.44(T-32) \qquad (3\text{-}74)$$

where R is the average annual boundary precipitation in inches and T is the MAT in °F. For YM, which has a MAT of about 16 °C or 61 °F, the boundary between a humid and subhumid climate would be 32 cm (12.8 in.).

The boundary in the Köppen-Geiger climate classification between arid and semiarid climates is one-half the precipitation value obtained for the humid/subhumid boundary:

$$R_{\text{semiarid/arid boundary}} = 0.22(T-32) \qquad (3\text{-}75)$$

Thus, the precipitation boundary for an arid climate classification at YM would be 16 cm (6.4 in.). In DCAGW, $R_{\text{semiarid/arid}}$ is calculated using MAT from UZFLOW. If MAP from UZFLOW is greater than $R_{\text{semiarid/arid}}$, pluvial DCFs are used.

Pluvial DCFs reflect estimated changes to the biosphere as described in LaPlante and Poor (1997). Estimated pluvial conditions did not change receptor groups or exposure pathways considered in DCF calculations (i.e., similar types of farming activities are expected under both estimated climate states).

---

[3] 0.9 m/yr (3.1 ft/yr) near Los Angeles, California (Table 14-2, page 377, Linsley and Franzini, 1979).

[4] 1.3 m/yr (4.38 ft/yr) in Mesa, Arizona (Table 2-50, page 99, van der Leeden, et al., 1990).

The exposure pathways considered in the farming scenario are depicted in Figure 3-16 and include: ingestion (of contaminated water, crops, and animal products); inhalation (from resuspension of contaminated soil); and direct exposure. The following assumptions apply to the farming receptor group. The farmer grows alfalfa (for beef and milk cow feed); vegetables, fruits, and grains for personal consumption; and feed for egg-laying hens. Drinking and irrigation water is pumped from a ground water well at the farm. Water is consumed at a rate of 2L/day (0.5 gal./day) pumped from a ground water well. Additional details of methods and assumptions used in the exposure scenarios for DCF calculations are provided in the succeeding sections and in LaPlante and Poor (1997).

3.8.3 Assumptions and Conservatism of DCAGW Approach

DCAGW is an abstraction of complex and uncertain processes occurring in the biosphere. The conversion of ground-water concentrations to receptor dose involves assumptions about: (i) the location of the receptor group; (ii) the lifestyle characteristics, of the receptor group, that form the basis for exposure pathways; (iii) processes that determine fate and transport of contaminants in the biosphere; (iv) calculation of human doses from factors that convert exposure to contaminated media to effective dose equivalents; and (v) well-pumping rates.

Receptor location is based on the assumption that present physical constraints (topography, depth to water table, and soil conditions) that limit present farming to the Amargosa Valley area will continue to limit farming south of YM to within approximately 20 km (12 mi) from the proposed repository site. Because the receptor location and lifestyle assumption relates to potential human behavior, it is speculative. The use of present knowledge to define the receptor group location helps to limit speculation and is consistent with recommendations of the National Academy of Sciences (1995).

The conversion of ground-water concentrations to receptor dose requires conceptualization of potential exposure pathways based on the site-specific characteristics of the release pathway, YM biosphere, and receptor group. When ground water is the source of contamination, the potential exposure pathways are assumed to be those resulting from pumping (including dilution) and agricultural use of water consistent with present farming conditions south of YM. These pathways include ingestion (of contaminated water, crops, and animal products); inhalation (from resuspension of soil contaminated by irrigation); and direct exposure to contaminated soil.

The assumed diet of locally produced food is important to estimating the dose for a farming receptor group. The farmer is assumed to grow alfalfa (for beef and milk cow feed); vegetables, fruits, and grains for personal consumption; and feed for egg-laying hens. The practices the farmer is presumed to engage in have been confirmed to exist in areas south of YM; however, it appears unlikely that all activities identified would be practiced by a single individual. Therefore, the doses calculated from such assumptions are expected to be greater than what might occur under current conditions. Because bartering is known to exist among community members (Eisenberg, 1996)[5], it is reasonable to project that (for the most highly exposed group) a significant portion of diet could be obtained from local sources. Thus, the assumption is considered conservative, but not excessively so. These assumptions may change when additional information on local consumption patterns is available.

---

[5]Eisenberg, N.A. 1996. *Staff Visit to Amargosa Valley: Trip Report*, Memorandum (May 14) to M. Federline, Nuclear Regulatory Commission, Washington DC: Nuclear Regulatory Commission.

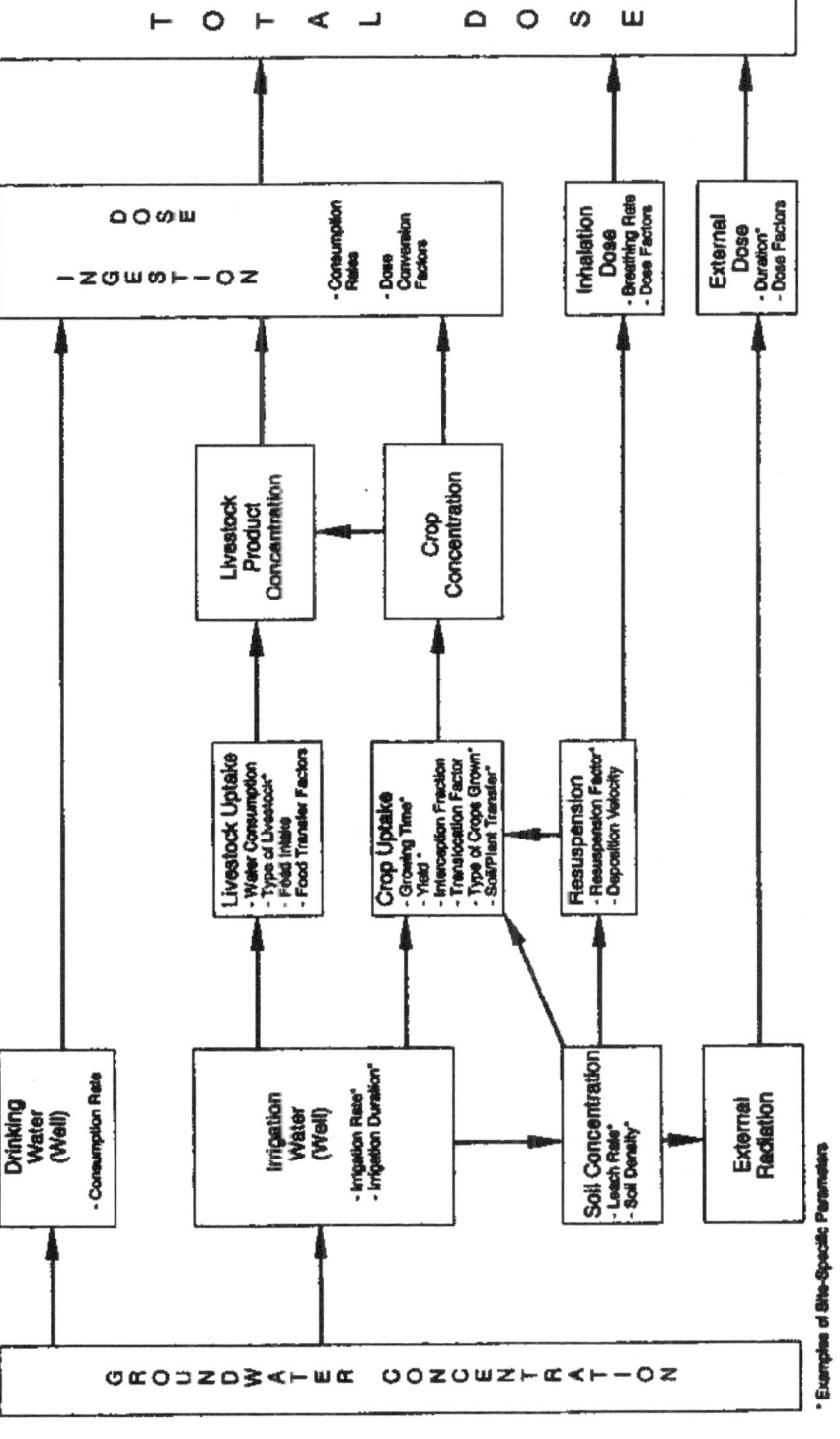

Figure 3-16. Farmer exposure pathways.

Modeling the fate and transport of radionuclides in the biosphere involves consideration of radionuclide deposition (from ground water) to soil, using local irrigation rates; determination of crop concentrations, using soil/plant uptake factors; determination of air concentrations from resuspended contamination, using a mass load model; and accounting for feed intake and transfer of contamination to livestock.

Radionuclide deposition from irrigation of crops is calculated from the irrigation rate, irrigation duration, and radionuclide concentration in ground water. The irrigation rate and duration are relevant to local conditions in the Amargosa Farms area and Nye County. Loss of contaminants to deeper soil layers from leaching is accounted for by a K_d-based model that includes factors which contribute to infiltration, such as rainfall and evapotranspiration. Rainfall and evapotranspiration rates are obtained from local or regional information sources and K_ds are relevant to the general soil texture classification reported for Amargosa farming soils. Full documentation of parameters is provided in LaPlante and Poor (1997).

Crop concentrations are based on transfer factors that represent the ratio of plant and soil concentrations for a class of crops grown in contaminated soil. Food transfer factors are generic values based primarily on published information (International Atomic Energy Agency, 1994; Baes, et al., 1982). The selected factors are expected values and have not been arbitrarily increased for conservatism.

Air concentrations from resuspended contamination are calculated by applying a mass loading model. This model uses a mass loading factor that is the ratio of airborne and surface soil radionuclide concentrations. Mass loading factors are highly uncertain and influenced by a number of processes including wind speed, surface roughness, resuspended particle size and density, and the moisture content of soil. Because the GENII-S code does not provide a mass loading model for crop deposition, a resuspension model was used with an adjusted mass loading factor.

Determination of radionuclide concentrations in livestock products is done by quantifying the intake of contaminated feed and applying a transfer coefficient. The animal transfer coefficients are based on studies that measure radionuclide concentrations in livestock, based on measured intake. The coefficients are element- and livestock-specific and based on published information (International Atomic Energy Agency, 1994; Baes, et al., 1982).

Calculation of human dose is accomplished through the use of dose factors that convert the estimated concentrations in water, food, air, and soil, together with the corresponding intakes and residence times, to TEDEs. The dosimetry models that form the basis for the internal dose factors (U.S. Environmental Protection Agency, 1988) are consistent with those described in the International Commission on Radiological Protection (ICRP) Publication 30. The dosimetry models represent state-of-the-art techniques and methodologies in existence at the time the publication was released. Refinements to dosimetry models have since been published by the ICRP, but have not yet been adopted by the U.S. Government and are therefore not used in DCAGW.

For the farming receptor group, sampled cumulative well discharge rates are based on assumed ranges of irrigated acreage (13–27 quarter-sections) and consumptive water use [0.9–1.3 m (3.1–4.38 ft) of water per year for alfalfa]. It is assumed that the radionuclide plume emanating from the repository would be completely captured by any configuration of wells whose total discharge equals the sample discharge rate.

Borehole concentrations at the farming receptor locations are calculated by dividing the radionuclide mass release rate by the cumulative well discharge rate.

More specific assumptions for DCAGW are provided in the following list and in LaPlante and Poor (1997).

- The dose calculation from consumption of crops does not account for washing fruits and vegetables before consumption and is, therefore, conservative.

- Food transfer factors are generic values based primarily on a large survey of the literature (International Atomic Energy Agency, 1994). Because of the lack of site-specific information, it is reasonable to use generic values; however, they may not be representative of arid conditions that exist in southern Nevada. The selected factors are expected values and have not been arbitrarily increased for conservatism.

- Consumption rates for food and water are based on national average values owing to lack of site-specific information. These values may not be representative of local consumption practices. Use of average values is not expected to be overly conservative. DOE has conducted a survey of local consumption practices and, when results are available, DCFs may be recalculated if the new consumption rates are found to differ significantly from the national averages used for current DCFs.

- For human dosimetry modeling, the average member of the receptor group is an adult, consistent with the EPA-proposed "Federal Radiation Protection Guidance for Exposure of the General Public" (U.S. Environmental Protection Agency, 1994). The individual is presumed to have physiology consistent with the assumptions of the reference man model (International Commission on Radiological Protection, 1975).

- Irrigation rates are assumed to be at the maximum permitted amount allowed by local water authorities in water permits (Nevada Division of Water Resources, 1995). When compared with State averages for alfalfa irrigation (U.S. Department of Commerce, 1994), the maximum permitted values are about 2 times greater. Thus, it is likely the irrigation rates used for DCF calculations are higher than amounts used locally, leading to some conservatism. These may not be unrealistically high, however, because southern Nevada is hotter and drier than the rest of the State, and water use is expected to be higher than the State-wide average.

- Inhalation and external exposure times are based on generic information provided in Kennedy and Strenge (1992). For the farming group, 55 percent of total time is assumed inside the residence, 20 percent outside of residence, 24 percent at offsite work, and 1 percent gardening and farming alfalfa, which is not a labor-intensive crop to farm. These activity times are considered to be conservative for a common residential setting (Kennedy and Strenge, 1992), but may not be for a more active farming resident.

- Resuspension of contaminated soil applies to crop deposition, inhalation, and external dose modeling. A mass loading model is applicable to a homogeneously mixed layer of contamination in the soil; therefore, a reasonable mass loading factor from the literature is used (Kennedy and Strenge, 1992). For crop deposition, the GENII-S code includes a resuspension model; therefore, the selected mass loading factor is converted to a resuspension factor, using an equation in the GENII-S user manual (Napier, et al., 1988). The mass loading and resuspension models are abstractions of complex and varying processes. Thus, considerable uncertainty exists in applying them to YM conditions.

- For pluvial climate conditions, the biosphere is assumed to be similar to that of northern Nevada and southeastern Idaho. These areas have average annual rainfall and monthly temperature profiles similar to predictions of future YM climate made from analysis of local paleoclimatic studies summarized by NRC (1997).

- Use of a concentration of radionuclides uniformly mixed in water for household and agricultural needs of the critical group is assumed appropriate for estimating doses to the average member of the critical group. Although some variation in concentrations would be expected between specific locations and well characteristics, assuming all crops within the farming community are irrigated with water containing radionuclides, a concentration uniformly mixed in the volume pumped by the farming community is considered a reasonable approach for estimating average exposure conditions.

3.9   DCAGW MODULE DESCRIPTION—RESIDENTIAL RECEPTOR GROUP

The DCAGW module calculates annual doses (TEDEs) from exposure to radionuclide concentrations in ground water to an average member of a residential receptor group. A residential receptor group located between 5 and 20 km (3.1 and 12 mi) has been included in the TPA Version 3.1.4 code, to assist timely completion of the sensitivity analyses by allowing shorter simulation periods appropriate to the shorter distances. Having the ability to evaluate a residential receptor group also provides the staff with the flexibility to address alternative assumptions for the receptor group should that be necessary.

3.9.1 Information Flow Within TPA

3.9.1.1 Information Supplied to DCAGW

Time-varying release rates computed in SZFT for each radionuclide released via ground water are passed to DCAGW in the EXEC from which radionuclide ground-water concentrations are calculated using the dilution volume and the mass fraction of the ground-water contaminant plume captured by the residential receptor group. Time evolution of average annual precipitation and temperature from UZFLOW are passed by the EXEC to DCAGW, to determine the time when the current condition will switch to pluvial conditions. Ground-water pathway DCFs used in determining doses from radionuclide concentrations for a resident receptor group, located between 5 and 20 km (3.1 and 12 mi) from YM, coincident with the location of the ground-water releases, are contained in two data files: *gw_cb_ci.dat* and *gw_pb_ci.dat*. Each file pertains to different climatic conditions (either current biosphere, designated by *cb* or pluvial biosphere, designated by *pb*, in file names). A detailed calculation is used to determine the portion of the plume that enters the receptor group water supply. Parameters that specify the pumping rates of the residential receptor group, location of the residential receptor group, location of the conductive zone being pumped, and plume thickness, are specified in the primary input file. Other input parameters are read from the DCAGW section of the primary input file.

3.9.1.2 Information Provided by DCAGW

The ground-water pathway dose for each radionuclide, as a function of time, is passed to EXEC.

3.9.2 Conceptual Model

The DCAGW module calculates dilution volume at the pumping well using the water use characteristics of the residential receptor group. The receptor group is represented by a residential community downgradient at a distance between 5 and 20 km (3.1 and 12 mi) from YM. Once the pumping volume is determined, the module converts the activity per unit time values calculated by SZFT to activity per unit volume of water, using a site-specific mass fraction of radionuclides captured by the receptor group. DCFs are then used to determine dose to the average member of the residential receptor group. The streamtube information used in SZFT is tied to the location of the receptor group. Therefore, any change to the location of the receptor group must be consistent with the streamtube data files.

For each time step, the products of each radionuclide concentration and DCF are summed within and among ground-water pathways and radionuclides, to calculate total doses. In addition to total doses, selected output is stratified by realization, time step, and radionuclide. The EXEC uses results from all realizations to identify and report peak doses.

3.9.2.1 Development of Radionuclide Concentrations in Water for a Residential Receptor Group

At the residential receptor group location [between 5 and 20 km (3.1 and 12 mi) from YM], a calculation is performed to determine the fraction of the radionuclide plume captured in the well withdrawal. Using sampled values of well discharge, aquifer thickness, and plume thickness, steady-state well hydraulics equations are solved to determine the fraction of the radionuclide plume captured. Borehole concentrations at the residential location are computed by multiplying the radionuclide release rate by the capture fraction and dividing the product by the volumetric flow rate for the well.

It is assumed the tuff aquifer from below the repository to the residential receptor location is homogenous and isotropic with a hydraulic conductivity of 1 m (3.3 ft)/day and a regional gradient of 0.00125. The radionuclide plume in the SZ below the repository is assumed to have a width equal to the width of the four streamtubes described in Section 3.7. It is also assumed that all wells have an effective radius of 0.254 m (0.8 ft).

After sampled values are obtained for the volumetric flow rate at the well and the thickness of the aquifer, an estimate of the appropriate well screen length is interpolated from a table of values determined using Thiem's equations for steady state, confined radial flow (Lohman, 1972), and Muskat's adjustment for partially penetrating wells (McWhorter and Sunada, 1977). Using the analytic element model GFLOW (Haitjema, 1995), tables of capture zone thickness and capture zone width as functions of pumping rate and aquifer thickness were constructed. After interpolated values of capture zone thickness and width are obtained from the look-up tables for the sampled pumping rate and aquifer thickness, the fraction of mass captured is determined using one of three procedures listed.

- If the screen length is greater than the sampled plume thickness or if the capture zone thickness is greater than 90 percent of the aquifer thickness, the fraction of mass captured is equal to the ratio of the capture zone width to the plume width.

- If the screen length is less than the plume thickness, but the capture zone thickness is greater than the plume thickness, the fraction of mass captured is equal to the ratio of the capture area within the plume to the cross-sectional area of the plume. The capture area within the plume is composed of two parts: (i) that portion of the plume lying above the bottom of the well screen; and (ii) that portion of the half-elliptical section of the capture zone that lies below the bottom of the well screen and above the lower vertical extent of the plume.

- If the screen length and capture zone thickness are less than the plume thickness, the fraction of the mass captured is equal to the ratio of the cross-sectional area of the capture zone to the cross-sectional area of the plume. The capture area within the plume is composed of two parts: (i) that portion of the plume lying above the bottom of the well screen; and (ii) the area of the half-elliptical section of the capture zone that lies below the bottom of the well screen.

The radionuclide concentration at the wellhead is computed by multiplying the radionuclide activity release rate by the fraction of mass captured and dividing this product by the volumetric pumping rate.

It is assumed that all water at the residential receptor group location is supplied by a single, partially penetrating well in an aquifer of specified thickness. Pumping rates are sampled from a uniform distribution with a range of 57 to 1000 m^3/day [1.5×10^4 to 2.64×10^5 gal./day (gpd)]. The minimum value for the distribution is based on a community with a population of 100 persons with a per capita daily water use of 0.6 m^3 (150 gal.), whereas the maximum value is based on a community of 880 persons with a daily per capita water use of 1.1 m^3 (300 gal.).[6,7] The range of 0.6–1.1 m^3/day (150–300 gpd) encompasses local estimates of per capita water use: (i) 0.8 m^3/day (210 gpd) for Nye County (Basse, 1990); (ii) 0.7 m^3/day (182 gpd), from 1995 metering for Beatty, Nevada (Buqo, 1996); and (iii) 1.2 m^3/day (319 gpd) by assuming 1200 m^3/yr (1 ac-ft/yr) and 2.8 people per household for Amargosa Valley, Nevada (Buqo, 1996).

Data from flowmeter surveys conducted at boreholes penetrating the saturated tuffs at YM indicate that flow is confined to a few, relatively thin, interconnected fracture zones. In conformance with the assumptions underlying the vertically averaged, 2-D flow model used to define the flow tubes in SZFT, it is assumed these highly conductive fracture zones can be represented by an equivalent aquifer. At YM, the saturated tuffs are divided into two fractured tuff aquifers separated by an aquitard composed of bedded ash-fall tuffs. The upper volcanic aquifer is composed of the Topopah Spring member of the Paintbrush

---

[6] Per capita daily water use for Tucson, Arizona, is 0.6 m^3 (150 gal.). (Table 5-16, page 319, van der Leeden, et al., 1990).

[7] Per capita daily water use for Las Vegas, Nevada, is 1.1 m^3 (300 gal.). (Table 5-16, page 319, van der Leeden, et al., 1990).

Group and is over the upper volcanic confining unit composed of the Calico Hills Formation, which is in turn over the lower volcanic aquifer composed of the Prow Pass, Bullfrog, and Tram members of the Crater Flat Group (Luckey, et al., 1996). These hydrogeologic units are separated from the underlying Paleozoic carbonate aquifer by the lower volcanic confining unit composed of the Lithic Ridge tuff and older tuffs (Luckey, et al., 1996).

As noted in Luckey, et al. (1996), the upper volcanic aquifer is present east of YM at water supply wells J-12 and J-13, where 70 percent of the aquifer is saturated; south of YM near boreholes USW WT-11 and UE-25 WT#12, where the lower 8–15 percent of the aquifer is saturated, and in Crater Flat, at VH-1, where the aquifer is fully saturated. In the region immediately below the proposed repository, only the lower volcanic aquifer is saturated. Therefore, radionuclides released from the repository will most likely be transported within the lower volcanic aquifer until they encounter major north-trending faults across which there is significant offset of hydrostratigraphic units. Although site-scale transport modeling studies (Baca, et al., 1996; Cohen, et al., 1997) indicate that faults may have a significant effect on the vertical and horizontal spreading of the radionuclide plume, these effects were not incorporated into the SZFT or DCAGW modules. It is assumed, however, that the upper and lower volcanic aquifers function as a single composite aquifer. It is also presumed that vertical mixing in this composite tuff aquifer is greater than would be expected in a relatively homogeneous sand and gravel aquifer owing to heterogeneities at a variety of scales.

Data from Luckey, et al. (1996) indicate that the composite saturated thickness of the upper and lower volcanic aquifers ranges from 337 to 719 m (1100 to 2400 ft.). It is therefore assumed that the aquifer thickness is uniformly distributed with a minimum of 300 m (980 ft) and a maximum of 700 m (2300 ft).

Compilations of dispersivity values estimated from laboratory and field data suggest that longitudinal dispersivity is approximately equal to one-tenth of the distance traveled by the contaminant plume (de Marsily, 1986). Data compiled by Gelhar, et al. (1992) suggest that horizontal transverse dispersivities are an order of magnitude smaller than longitudinal dispersivities and vertical transverse dispersivities are one to two orders of magnitude smaller than horizontal transverse dispersivities. The length of the mean particle trajectory from the repository to a 5-km (3.1-mi) boundary is approximately 7 km (4.3 mi). Using the empirical method suggested previously, the estimated vertical transverse dispersivity ranges from 7 to 70 m (23 to 230 ft). In the final report of the "Saturated Zone Flow and Transport Expert Elicitation Project" (Geomatrix Consultants, Inc., 1998), one member of the expert panel (Lynn Gelhar) estimated a mean (geometric) longitudinal dispersivity of 50 m (160 ft), but estimated that the mean (geometric) horizontal transverse dispersivity is 0.5 m (1.6 ft) and the vertical transverse dispersivity is 0.005 m (0.2 in.). Gelhar's estimates of transverse dispersivity were primarily based on data from the Cape Cod and Borden test site experiments, which were both conducted in relatively homogeneous sand and gravel aquifers (Geomatrix Consultants, Inc., 1998), and may be too small to apply to the complex fractured and faulted tuff units at YM.

Assuming a mean Darcy velocity of 0.46 m/yr (1.5 ft/yr), kinematic porosities ranging from 0.001 to 0.01, a plume trajectory 7 km (4.3 mi) in length, and vertical transverse dispersivities ranging from 7 to 70 m (23 to 230 ft), the one standard deviation vertical spreading thickness (σ_z) of a point source Gaussian plume at a 5-km (3.1-mi) receptor location varies from 104 m (340 ft) to 330 m (1100 ft).[8] Using the same calculational procedure, but substituting the vertical transverse dispersivity estimate of 0.005 m (0.2 in.)

---

[8] $3\sigma_z = (2\alpha_z v T)^{1/2}$, where α_z is the vertical dispersivity, v is the linear ground-water velocity, and T is the mean travel time.

provided by Gelhar, the one standard deviation vertical spreading thickness is 8.4 m (28 ft). Because there are no site-specific data for vertical transverse dispersivities, it is extremely difficult to select an appropriate distribution for the thickness of the plume at the receptor location. Although it may appear to be conservative to select a distribution of plume thicknesses that maximizes the likelihood that a low-discharge well captures the entire plume, even the smallest pump discharge rate considered 57 m^3 (1.5×10^4 gal.)/day creates a capture zone more than 150 m (490 ft) thick. Based partly on the bounding calculations described previously and on a need to account for variability in this important parameter, plume thickness was assumed uniformly distributed between 10 and 100 m (33 and 330 ft). Although this analysis at 5 km (3.1 mi) was used in determining the primary input file parameters, the same methodology applies to all distances of the residential group.

3.9.2.2 Development of Dose Conversion Factors

The DCAGW module contains DCFs designed to convert ground-water concentrations to TEDEs for the average member of the receptor group. Site-specific information indicates residential dwellings exist down-gradient of YM. Two climate regimes are assumed for the biosphere in the TPA Version 3.1.4 code; however, for the residential exposure group the exposure pathway (drinking water) is not expected to be affected by the predicted biosphere changes. Thus, no biosphere variations are included in the exposure pathway calculations for this receptor group. The following assumptions apply to the residential receptor group. The residential exposure pathways are limited to ground-water consumption. Water pumped from a water well is consumed at a rate of 2 L (0.5 gal.)/day. Additional details of methods and assumptions used in the exposure scenarios for DCF calculations are provided in the succeeding sections and in LaPlante and Poor (1997).

3.9.3 Assumptions and Conservatism of DCAGW Approach

The DCAGW module for the residential receptor group converts ground-water radionuclide concentrations to residential receptor dose. Therefore, the module involves assumptions about: (i) the location of the receptor group; (ii) the lifestyle characteristics of the receptor group; (iii) the conversion of intake to dose; and (iv) well pumping rates.

The residential receptor is based on an assumption that local topography and the economics of ground-water extraction allow for a potential nonfarming residential receptor group to exist closer to the potential repository site than a farming receptor group. Because this assumption relates to potential human behavior, it is speculative. The use of present knowledge to define the receptor group location helps to limit speculation and is consistent with the recommendations of the National Academy of Sciences (1995).

Dose calculation involves consideration of only the drinking water ingestion dose pathway because this pathway is the primary route of exposure for residential uses. Other household uses are not included in the dose calculation because it is assumed the ingestion dose will far exceed that caused by bathing and other potential uses.

The drinking water dose is calculated from the well head concentration, drinking water consumption rate, and the ingestion DCF. The consumption rate is assumed conservatively to be 2 L (0.5 gal.)/day, and the DCF is from Federal Guidance 11 (U.S. Environmental Protection Agency, 1988). For human dosimetry modeling, the average member of the receptor group is an adult, consistent with the EPA proposed "Federal Radiation Protection Guidance for Exposure of the General Public" (U.S. Environmental Protection Agency, 1994). The individual is presumed to have physiology consistent with the assumptions of the reference man model (International Commission on Radiological Protection, 1975).

The range of pumping rates for the residential community is based on a population assumed to vary from 100 to 880 persons having a per capita daily water use that varies from 0.6 to 1.1 m^3 (150 to 300 gal.). Corresponding daily well discharge rates for the residential receptor location vary from 57 to 1000 m^3 (15,000 to 260,000 gal.)/day. Analytical methods are used to calculate the fraction of the radionuclide plume of a given thickness that would be captured by a single well having the sampled daily discharge rate. Borehole radionuclide concentrations are calculated by dividing the product of the radionuclide mass release rate and the fraction captured by the well discharge rate. Use of a concentration of radionuclides uniformly mixed in water for household drinking water use by the critical group is assumed to be appropriate for estimating doses to the average member of the critical group. Although some variation in concentrations would be expected between specific locations and well characteristics, a concentration uniformly mixed in the volume pumped by the residential community is considered a reasonable approach for estimating average exposure conditions.

3.10 FAULTO MODULE DESCRIPTION

The FAULTO module determines the number of WP failures from direct fault disruption of the proposed repository block. The number, location, and time of WP failures are made available to EBSREL. Additional information about FAULTO and the stand-alone FAULTING code can be found in Ghosh, et al. (1997, 1998). As described in Chapter 2, a faulting event is assumed to not disturb the thermohydrological characteristics of the proposed repository.

3.10.1 Information Flow Within TPA

3.10.1.1 Information Supplied to FAULTO

The user selects the faulting disruptive event module in the primary input file. Other FAULTO input parameters, including time of the event and fault length and width, are specified in the FAULTO section of the primary input file.

3.10.1.2 Information Provided by FAULTO

FAULTO computes a time history of WP failure from faulting events for all subareas for subsequent use in radionuclide release calculations in EBSREL.

3.10.2   Conceptual Model

FAULTO evaluates the potential for direct WP rupture from fault displacement along planar decoupled faults. In the code, faults are considered as zones or bands of deformation with a finite width. WPs within these zones are assumed damaged, provided the fault slip exceeds a threshold displacement value. It is assumed that WP emplacement in the proposed repository will be appropriately set back from those faults known to present a potential hazard and that a minimum amount of fault displacement is needed to disrupt a WP. Thus, FAULTO is designed to evaluate hazards related to faults not accounted for in the proposed repository design such as: (i) new faults that may form during the period of concern; (ii) hidden faults within the proposed repository that are presently unknown and unmapped; or (iii) underestimated faults that are mapped and not considered significant during design or construction, but that pose significant risk over the lifetime of the proposed repository.

The code initiates a new fault zone within the repository based on geometric and recurrence parameters sampled from a series of probability distribution functions (PDFs). Geometric parameters include fault-zone location, orientation, strike, length, width, and displacement. Recurrence parameters include recurrence rate, time of faulting event, and cumulative displacement rate. Although in nature these geometric and recurrence properties may be related (i.e., longer faults seem to correlate with wider deformation zones or bigger faults tend to be more active), for simplicity they are specified as independent parameters in the TPA Version 3.1.4 code.

In FAULTO, parameters describing future faulting events in the proposed repository are sampled independently and the effective recurrence rate is sampled from a single exponential distribution. Calculating the effective recurrence interval within the repository boundary is not straightforward because, unlike point-source events, fault zones are tabular volumes with planar boundaries that have orientation, length, and width. Moreover, fault orientation, length, and width have their own PDFs. Recurrence rate, therefore, does not scale directly with change in simulation area. There is a critical simulation region (CSR) below which the recurrence interval does not increase in proportion to the reduction in the size of the simulation area. If the simulation region is smaller than the critical area, a fault could still initiate outside the area and have the necessary length and orientation to intersect the proposed repository. To overcome this problem, an empirical solution uses an area of interest larger than the repository footprint. This approach is described in the following paragraphs.

The CSR for faulting was developed to include the range of fault orientations and lengths (e.g., fault map in Simonds, et al., 1995). Approximately 90 percent of the faults have lengths of 30 km (19 mi) or shorter, or half lengths of 15 km (9.7 mi) or shorter. In addition, fault orientations range between N55° W (azimuth 305°) and N25° E (azimuth 25°). Given these constraints, the size of the CSR is 15.2 km × 32.8 km (9.4 × 21 mi). The ratio of the CSR area to the area used to estimate faulting recurrence from paleoseismic studies is 2.2 (498.6 divided by 225.0). Thus, the scaled recurrence of the CSR is 5909 yrs (13,000 divided by 2.2) yielding a conditional probability for faulting of 1.69×10^{-4}/yr.

Recurrence value was then scaled to only the repository footprint using the FAULTING stand-alone code (not a part of the TPA Version 3.1.4 code) to empirically estimate what percentage of faults generated in the CSR would actually intersect the proposed repository, given a 5909-yr recurrence interval for the CSR. Modeling results show that an average of 3 percent of all simulated faults intersect the repository, based on a number of different realizations between 500 and 1,000,000. Thus, the recurrence for faults within the repository itself is 197,000 yrs (5909 divided by 0.03), or an annual probability (referred to as absolute probability) of 5.0×10^{-6}.

Some parameters normally used to fully describe faulting events were not used in FAULTO, including fault-dip angle, number of slip surfaces, partition of fault slip among different slip surfaces, cumulative displacement rate, and cumulative displacement rate partitioned among slip surfaces. These parameters were excluded for the following reasons:

- Fault-dip angle is not important because the WPs are assumed to be emplaced within a single almost horizontal geological stratum. Fault dips at the surface at YM are rarely less than 45°. Therefore, the problem essentially reduces to two dimensions. Variations in fault-dip angle over the vertical thickness of the emplacement drifts will not affect the consequence.

- Number of slip surfaces of a fault does not affect performance because the area affected is constant and the entire fault has the same maximum displacement. The fault slip is either divided among the slip surfaces or allowed on one surface. Because the WPs are assumed to be uniformly distributed, the area affected (fault width times the trace length of the fault within the footprint of the proposed repository)—not the exact location of the slip surface(s)—is important.

- Cumulative displacement rates for faults at YM (see Table 2-4 of Ghosh, et al., 1997) are too small to significantly affect the performance of the proposed repository, given a 10,000-yr or even 20,000-yr period of concern.

Using the effective recurrence interval for the repository, FAULTO samples midpoint location, orientation, length, width, and displacement slip of a fault in the repository. The displacement slip is sampled from the user-supplied distribution for maximum slip and is the same for the entire fault. With the orientation and width known, FAULTO calculates the intersection area of the fault with the repository. If the sampled displacement slip exceeds the threshold displacement specified by the user, all WPs in the intersection area are reported to EBSREL as having failed. Figure 3-17 illustrates a simulated fault within the proposed repository boundary. Time for WP failure is the instant the fault is generated.

3.10.3   Assumptions and Conservatism of the FAULTO Approach

The FAULTO module is an abstracted model of the failure of WPs from instantaneous fault displacement along steeply dipping, new, or inadequately characterized faults. The FAULTO module considers: (i) distribution of WPs within the emplacement drifts; (ii) fault recurrence parameters; (iii) fault geometric parameters; and (iv) fault-slip and threshold displacements.

WPs are assumed to be uniformly distributed throughout the repository as a first-order approximation. Repository design could influence the extent to which this assumption is acceptable, but using a uniform density of emplaced WPs is acceptable to estimate the significance of fault displacement without speculating on repository design.

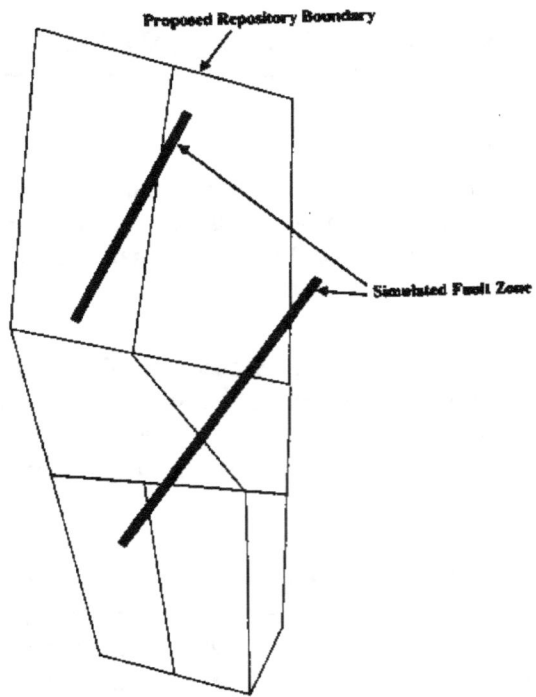

Figure 3-17. Simulated faults within the repository boundary.

The time for the next faulting event is calculated from the effective recurrence rate. The effective recurrence rate has been estimated by taking into consideration the length and orientation of the mapped faults in the YM region. The critical simulation region concept, described in Section 3.10.2, was used to estimate the annual probability of faulting in the proposed repository region at 5.0×10^{-6}. Fifty percent of faulting occurs on new faults or faults where WPs are not set back in the repository design. This assumption is conservative because most geological observations infer that large faulting events in YM will occur on known and mapped faults. For example, paleoseismic studies show that nearly all the large faults at YM produce evidence for repeated earthquake ruptures (e.g., Chapter 4, U.S. Geological Survey, 1996). Wernicke, et al. (1998) have suggested an increase in seismicity by an order of magnitude (recurrence rate $\sim 10^4$ yr). However, geologic evidence does not support this proposed increase in seismicity.[9]

The fault geometric parameters include fault midpoint location, orientation, length, and width. These parameters are generated from PDFs developed from field-measured data (Scott and Bonk, 1984; U.S. Geological Survey, 1996). Most of the faults mapped in the YMR predominantly strike in the north-northeast direction (about 95 percent) with dip angle rarely less than 45°. As a result, variations in fault-dip angle over the vertical thickness of the emplacement drifts will not significantly affect the performance. Therefore, fault-dip angle was not considered in the FAULTO module. Similarly, other parameters, such as number of slip surfaces of a fault and cumulative displacement, were ignored. It is considered a reasonable

---

[9]Connor, C.B., J.A. Stamatakos, D.A. Ferrill, and B.E. Hill. 1998. Technical Comment: Anomalous Strain Accumulation in the Yucca Mountain Area, Nevada. *Science*. In Press.

approach to drop these parameters from the module as their effects on the consequence were found insignificant. Mapped faults in the YMR tend to have widths clustered around small values, although a fault may have a large width at least on the surface. A beta distribution with small median value is considered reasonable to approximate the fault width distribution.

The fault-slip displacement is sampled from the log-normal distribution of maximum slip and is considered the same for the entire fault plane. A log-normal distribution is considered a reasonable assumption to describe the distribution of fault slips observed in the YMR. In the absence of both a model transferring forces from fault displacement to emplaced WPs and a well-developed WP design, assumptions are made about the amount of fault displacement that leads to WP failure. The threshold fault displacement directly affects the WP disruption. If the fault slip exceeds the threshold value, all WPs within the repository block along the fault zone (fault length within the repository boundary multiplied by the fault width) are considered failed. The threshold displacement is an input parameter supplied by the user that reflects lack of mechanistic understanding of the WP disruption process. Although the TPA Version 3.1.4 code does not couple the degradation of WP strength caused by corrosion and other phenomena with the threshold displacement, the approach is considered conservative because a small threshold of displacement leading to WP failure (specified in the primary input file) is believed to provide a conservative estimate of WP failure. Although WP disruption is assumed to occur only within the fault zone boundary, the assumption that all WPs within the boundary fail—if the fault displacement threshold is exceeded—is believed to provide a conservative estimate of WP failure in the absence of more sophisticated modeling.

Several specific assumptions of faulting disruption effects in the FAULTO module of the TPA Version 3.1.4 code are:

- Fifty percent of faulting is assumed to occur on new faults or faults where WPs are not set back in proposed repository design. This assumption is conservative.

- In FAULTO there is assumed to be no link between faulting, seismicity, and volcanism. In nature, volcanic eruptions are always accompanied by numerous pre- and syn-eruption earthquakes (Luhr and Simkin, 1993; Fedotov and Markhinin, 1983). Likewise, all faulting events that would affect the proposed repository would be accompanied by significant seismicity. Therefore, the current assumption of uncorrelated faulting, seismic, and volcanic phenomena is not conservative.

- The emplacement drifts are assumed to be randomly oriented. Most YM faults are oriented roughly north-south. If the actual emplacement drifts are subparallel to the fault trend, much greater number of WPs will be affected by each faulting event than currently estimated in the code. The current plan ("Civilian Radioactive Waste Management System Management and Operating Contractor," 1997) suggests that the emplacement drifts will be oriented approximately in the N 60° W direction. Therefore, the assumption of random orientation for the emplacement drifts may be nonconservative.

- The TPA Version 3.1.4 code is limited to one faulting event per realization, regardless of the selected recurrence interval. This simplification is acceptable as long as the estimate of recurrence intervals longer than or equal to 10^5 yr is correct, and the time period of interest is 10,000–20,000 yr. Otherwise, this limitation reduces the potential consequence of faulting and is, therefore, not conservative if the WP has not already failed by other mechanisms.

- There is no correlation among fault length, fault zone width, fault orientation, and probable displacement along the fault plane. In nature, longer faults have undergone larger displacements. Observations of fault zone widths in the YMR show that fault zone widths seem to be large for incipient faults and narrow as the fault zone matures. In the FAULTO module, fault zone widths are sampled from a broad range, including a few wider than 100 m (330 ft).

- Faulting could weaken WPs and make them more susceptible to corrosion over time. To compensate for not accounting for this effect, a small threshold displacement for WP failure is specified in the primary input file.

- WP failure in FAULTO is not linked to a WP mechanical failure approach, as used in EBSFAIL. This simplification is conservative because the WP is always assumed to fail when the fault displacement exceeds a threshold value.

- All WPs intersected by the fault zone are considered failed. This conservative assumption is necessary at present because the forces WPs could encounter in fault zones are poorly understood.

- The module assumes the damage zone caused by slip along a fault is entirely restricted within the fault zone boundaries (i.e., no diffuse deformation outside the fault zone). This assumption is not conservative.

- The module does not take into account the effects of slip on a new or an uncharacterized fault caused by rupture on existing faults in the repository region.

3.11   VOLCANO MODULE DESCRIPTION

The VOLCANO module provides determination of key IA processes such as: (i) timing of future igneous events; (ii) subsurface area affected by a volcanic event; (iii) type of event (i.e., intrusive only or intrusive and extrusive); and (iv) number of WPs affected by intrusions extending laterally from the volcanic conduit.

3.11.1 Information Flow Within TPA

3.11.1.1 Information Supplied to VOLCANO

The VOLCANO disruptive event module is selected by the user in the primary input file. Other VOLCANO input parameters, including time of the event, conduit diameter, and dike length, are specified in the VOLCANO section of the primary input file.

3.11.1.2 Information Provided by VOLCANO

VOLCANO computes WP failure time from a volcanic event for all subareas. VOLCANO also computes the mass of waste (in MTUs) ejected by the volcanic event for ground-surface radionuclide release calculations in ASHPLUMO, and the fraction of WPs disturbed by the volcanic dike for all subareas, for ground-water source term calculations in EBSREL..

3.11.2 Conceptual Model

Igneous activity is a disruptive scenario with a relatively low annual probability of occurrence. Event consequences can be broadly categorized as volcanic processes that disrupt WPs and directly transport HLW into the accessible environment, via the airborne pathway, and intrusive processes that enhance WP degradation, but do not directly transport HLW from the repository.

A volcanic center is allowed to form randomly within a specified area. The volcanic center is commonly restricted to the center of the proposed repository block to ensure igneous disruption is evaluated. For the currently proposed 4.8-km^2 (1.8 mi^2) occupied repository block at YM, NRC (1998) concluded that an annual probability of 10^{-7} provides a reasonably conservative value for volcanic disruption of this site. The annual probability of igneous intrusion at the proposed repository site currently is assumed to be 2 to 5 times the probability of volcanic disruption (Nuclear Regulatory Commission, 1998). Timing of the igneous event is determined by sampling a finite exponential distribution between 100-yr and 10,000-yr postclosure, using a recurrence probability of 10^{-7}.

In the TPA Version 3.1.4 code, two types of events may be modeled to occur. These are either an extrusive vent, which intersects the repository and ejects SF into the air (WP is assumed to have no effect on release) and affects other WPs through lateral intrusion, or a basaltic intrusion that disrupts WPs but does not directly exhume SF to the accessible environment. An extrusive volcanic center is always associated with a lateral intrusion, but a subsurface intrusion does not necessarily coincide with the occurrence of an extrusive event. Information required to be input into the data set by the user includes distributions for event geometries (length and width of intrusions and diameters of volcanic conduits). When an event occurs, the location for the event is identified within the repository footprint. Using parameter values sampled from the user-supplied distributions, areas for the circular conduits and linear intrusion are calculated, with the conduit area being subtracted from the total intrusion area (length times width).

Volcanic centers disrupt a finite volume of subsurface material during the emplacement and eruption of ascending basaltic magma. Work conducted at an analog volcano site in Kamchatka, Russia (Hill, 1996), demonstrated that the roughly circular conduits associated with YMR-type volcanoes may widen to a 50 ± 7 m (160 ± 20 ft) diameter during late-stage disruptive events. Several of the youngest YMR volcanoes have near-vent evidence of similar, late-stage disruptive events (Hill, 1996). Although additional work is being conducted to develop an independent technical basis for this parameter, the diameter provides a basis to calculate the number of WPs potentially disrupted by a volcanic conduit. The number of WPs disrupted, however, depends on the WP dimensions, WP spacing, and centerline spacing between emplacement drifts (Figure 3-18). Average repository waste loading values may underestimate the number of WPs disrupted by a volcanic conduit. For example, using an average waste loading of 83 MTU/acre and a 9.76-MTU WP loading, a 50 m (160 ft) diameter conduit will only disrupt 4.1 WPs. Inspection of Figure 3-18 reveals more than four WPs are disrupted by most 50 m (160 ft) diameter volcanic conduit locations.

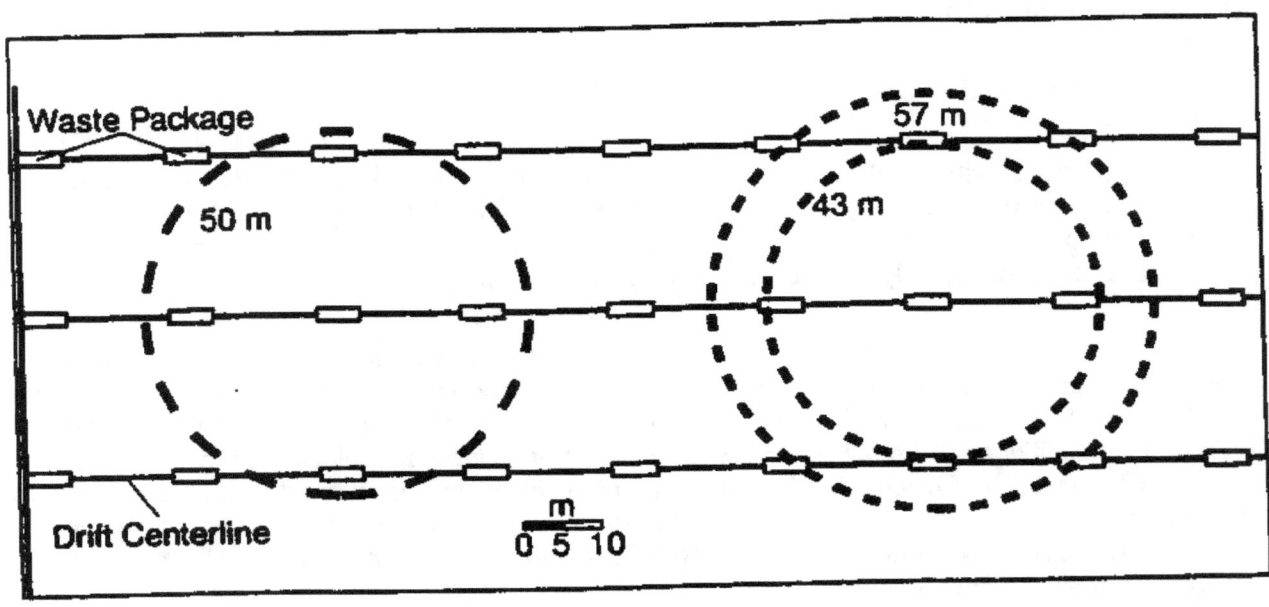

Figure 3-18. Schematic of subsurface disruption area associated with a volcanic conduit (Hill, 1996) using waste package dimension [5.68 m (18.6 ft) long, 1.8 m (5.9 ft) wide], waste package spacing [19 m (62.3 ft) center-to-center], and emplacement drift spacing [22 m (72.2 ft)] (U.S. Department of Energy, 1996).

Geometric relationships suggest that disruption of 1 to 10 WPs presents a reasonable range for volcanic conduits that may vary between 1 to 50 m (3–160 ft) in diameter.

Subsurface igneous intrusions may extend for many kilometers from the central volcanic conduit. These intrusions may penetrate farther into the repository than the area disrupted by the conduit. The effects of these intrusions are poorly constrained (e.g., Barr, et al., 1993) but likely will degrade WP performance significantly because of adverse thermal, chemical, and mechanical loading. The VOLCANO module currently calculates the number of WPs directly intersected by igneous intrusions extending from the central conduit, but does not account for additional WP failures owing to thermo-chemical effects. Area affected also does not currently account for the likelihood of lava-flow formation in open drifts. The number of WPs intersected by an intrusion is calculated by sampling uniform distributions of intrusion orientation, length, and width. This area is converted to a number of WPs using an average HLW loading for the repository.

The annual probability of volcanic disruption of the proposed repository site is bounded by 10^{-7}, which represents a reasonably conservative value based on independent analyses (Nuclear Regulatory Commission, 1998). Probabilities at least one order of magnitude higher and lower than this value can be derived from the reviewed literature (Nuclear Regulatory Commission, 1998). This version of the TPA Version 3.1.4 code assumes that the probability of igneous intrusion is equal to the probability of volcanic disruption (i.e., a single recurrence probability is specified in the primary input file). The probability of igneous intrusion, however, is likely larger by a factor of 2–5 than the probability of volcanic disruption alone (Nuclear Regulatory Commission, 1998).

3.11.3 Assumptions and Conservatism of the VOLCANO Approach

The VOLCANO module defines the source term for HLW transport by a basaltic volcanic eruption. The VOLCANO module determines: (i) timing of the volcanic event; (ii) amount of HLW available for transport by the eruption column; and (iii) area affected by igneous intrusions.

Timing of the volcanic event follows a Poisson distribution with a 10^{-7} annual probability of volcano formation for the proposed repository site. Although igneous intrusions are assumed to occur with about five times the frequency of a volcano [with a 5×10^{-7} annual probability (Nuclear Regulatory Commission, 1998)], current TPA methodology assumes it to occur with the same frequency as the volcanic event, which could be nonconservative.

The volcanic conduit ranges in size sufficient to disrupt 1 to 10 WPs, based on the shallow subsurface areas commonly disrupted by basaltic volcanic eruptions. All waste in these disrupted packages is available for transport, as staff have assumed that a WP will fail during a basaltic volcanic event.

Additional WPs are disrupted by a planar intrusion extending from the conduit. These additional WPs, however, are not available for extrusive volcanic transport, as it is unlikely that material can be transported laterally in a shallow intrusion for more than tens of meters and the SF in these WPs is available for ground-water releases.

There are several specific assumptions and conservatism in the VOLCANO module:

- Volcanic conduit is assumed to form in the center of the repository block and the stated probabilities are for volcano formation at the repository site.

- Timing of the volcanic event follows a Poisson distribution, with a 10^{-7} annual probability of volcano formation for the proposed repository site, which is a conservative value (Nuclear Regulatory Commission, 1998).

- Intrusions away from the volcanic conduit are assumed linear and have dimensions characteristic of YMR intrusions (Nuclear Regulatory Commission, 1998). The number of WPs intersected by an intrusion may not be conservative because this model does not account for magma flow into open or partially backfilled drifts. This assumption does not currently affect dose from direct releases associated with igneous events because affected

WPs do not contribute to direct release of HLW (i.e., HLW inside WPs affected by intrusion is made available for release via the ground water).

- Consequences of intrusive processes are assumed limited to the area directly occupied by an igneous intrusion. The extent of processes that will likely enhance WP degradation or affect nondisturbed repository performance, such as increasing temperatures and dry-out zones around intrusions (e.g., Connor, et al., 1997), are not considered in the current module.

- In the case of an intrusive volcanic event, the drifts are assumed to be unbackfilled. All WPs intersected by the magma dike are assumed to fail simultaneously and contribute to ground-water releases.

- The diameter of a volcanic conduit in an undisturbed geologic setting is assumed to constrain the number of WPs disrupted for direct transport of HLW via the air pathway. This assumption may not be valid because it does not evaluate the response of overpressured magma encountering the ambient pressure conditions of the 300-m (980-ft) deep repository drifts.

- The number of WPs intersected by the volcanic conduit is derived from the number of WPs intersected by a conduit ranging from 2 to 50 m (6 to 160 ft) in diameter, using WP dimensions, emplacement spacings, and interdrift spacings from the DOE advanced conceptual design. These calculations show 1 to 10 WPs can be intersected by a 2- to 50-m conduit.

- Areas affected by volcanic conduits and igneous intrusions are assumed to have simple geometries. Redistribution of rock stress because of drift emplacement will likely result in complex, irregular feature geometry. Conduits may become more elliptical and affect larger areas or may bypass drifts entirely because of drift-induced changes in local rock stress.

- Based on conservative interpretations of data and analyses presented by NRC (1998), all WPs are assumed to fail when intersected by the volcanic conduit.

3.12 ASHPLUMO MODULE DESCRIPTION

The purpose of the ASHPLUMO code module is to estimate the deposition of both tephra (i.e., ash) and incorporated SF in mass per unit area, at the point of compliance, after an extrusive volcanic event has penetrated the proposed repository. The ASHPLUMO module invokes execution of the ASHPLUME stand-alone code (Jarzemba, et al., 1997) in deterministic mode, using input from EXEC or other modules (e.g., VOLCANO) that describes the characteristics of the eruption, such as volcanic power, durations, and size distribution of the ash particle, to accomplish the aforementioned purpose.

3.12.1 Information Flow Within TPA

3.12.1.1 Information Supplied to ASHPLUMO

The mass of SF ejected in the volcanic event (i.e., ground-surface radionuclide source term), as estimated by the VOLCANO code module, is provided by EXEC. Additionally, the ASHPLUMO module uses input values from the ASHPLUMO section of the primary input file, including height of the eruption column, wind speed, and ash particulate characteristics, to determine the transport and deposition of radionuclides in volcanic ash.

3.12.1.2 Information Provided by ASHPLUMO

ASHPLUMO estimates the areal deposition of both ash and SF, at the receptor group location, that is passed to EXEC for subsequent use by the ASHRMOVO module.

3.12.2 Conceptual Model

The ASHPLUME code calculates the areal density of SF (in grams of SF per square centimeter) at points on the surface of the earth after an extrusive volcanic event penetrates the repository and exhumes SF. Using published data for wind velocity at the YM site and the estimate of pertinent volcanic parameters of events similar to those that may have occurred at YM in the past, the ASHPLUME code simulates the transport of contaminated particles (composed of SF and ash) to surface points downwind. The SF concentration from deposition is provided to the ASHRMOVO code module and subsequently to the DCAGS code module for the dose calculation.

Basaltic volcanism can encompass a variety of eruption styles, depending on the eruption energy. The energy of basaltic eruptions varies from effusive activity, where the predominant product is lava flows, to explosive activity, resulting in fragmentation of the magma into scoria fragments and transport of scoria in the atmosphere as pyroclasts. This latter style of activity generally results in the formation of cinder cones such as those found in the YMR. Explosive volcanic activity of this kind has the potential to cause dispersal of radionuclides through the biosphere. This dispersion of radionuclides resulting from volcanic activity is modeled in the ASHPLUME code, using approaches originally developed by Suzuki (1983) to model the dispersal of ash after volcanic eruptions.

Figure 3-19 shows the exposure scenario investigated with the ASHPLUME code. The exposure scenario analyzed can be divided into four subprocesses:

(i)   Magma enters the repository and becomes contaminated with SF particles.

(ii)   Tephra forms from the magma and SF is incorporated into the tephra (Jarzemba and LaPlante, 1996).

(iii)   Eruption column and contaminant plume form and produce contaminated ash fallout at various distances downwind from the volcano (Suzuki, 1983; Jarzemba, 1997).

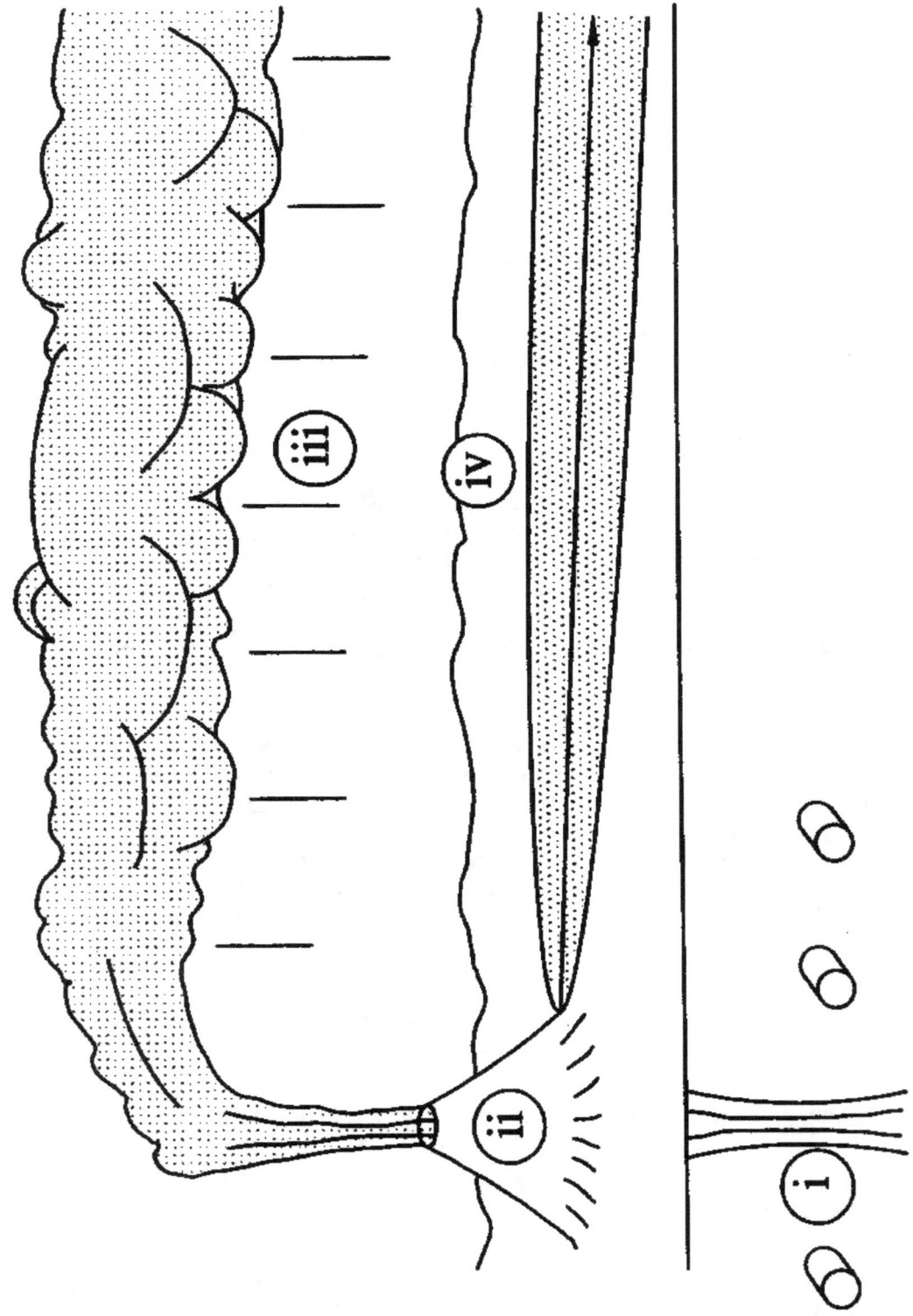

Figure 3-19. ASHPLUMO contaminant release and deposition.

(iv) Radionuclide contamination causes doses to be incurred at receptor locations (LaPlante, et al., 1995; Jarzemba and LaPlante, 1996).

Specifically, ASHPLUME addresses (ii) and (iii) from the previous list.

To assess the hypothetical radiation doses that would occur after a basaltic eruption, the distribution of SF (and hence radionuclides) in the biosphere after such an event needs to be estimated. It is assumed that the ash particles from the eruption are the carriers of the SF particles. Methods used previously to estimate radionuclide dispersal by volcanism (Wescott, et al., 1995) theorize that the ash cloud travels in a Gaussian plume, released at a stack height of one-half of the volcanic column height. Application of the Gaussian plume model presumes that a plume of contaminants travels in the same direction as the prevailing wind (x-direction), but may be somewhat depressed toward the earth surface owing to gravitational settling. Contaminants in the plume follow a Gaussian distribution in the dimensions perpendicular to the direction of travel (y- and z-directions).

The Gaussian plume model is suitable for modeling airborne and ground concentrations of contaminants for a point source release of contaminants above the surface of the earth (i.e., the stack height). A point source approximation may not be appropriate for a volcanic eruption because a volcanic eruption column is a line source of contaminants in the upward direction. Also, the Gaussian plume model does not accurately account for the effects of gravitational settling of volcanic particles with large diameters [i.e., centimeters]. This shortcoming may lead to the Gaussian plume model predicting much greater particle ranges than would be the case in reality and, hence, wider radionuclide distributions than would normally be expected after a basaltic eruption. This wider distribution of radionuclides may tend to underestimate the radiation exposure of persons in a small receptor group.

Models to predict the distribution of ash, after an eruption, have been developed with the intention of relating eruption magnitude to ash dispersion (Suzuki, 1983; Hopkins and Bridgeman, 1985; Glaze and Self, 1991). The ASHPLUME code uses the model described in Suzuki (1983) that relates eruption magnitude to ash distribution, which is modified to relate eruption magnitude to SF distribution for YM, based on a few simple assumptions. The model uses the power and duration of the eruption, along with other properties of the ash particulates, and develops an SF distribution from those sampled parameters. The SF distribution can be translated into the radionuclide distribution that can be used to model dose-to-man.

The model described in Suzuki (1983) can be summarized by the equation that describes the areal density of accumulated ash on the Earth's surface after an eruption:

$$X(x,y) = \int_{\rho=\rho_{min}}^{\rho_{max}} \int_{z=0}^{H} \frac{5QP(z) f(\rho)}{8\pi C(t + t_s)^{5/2}} \exp\left[\frac{5\{(x - ut)^2 + y^2\}}{8C(t + t_s)^{5/2}}\right] d\rho dz \qquad (3\text{-}76)$$

where:

$X(x,y)$ = mass of ash per unit area accumulated at location (x,y) [g/cm^2]

ρ_{max} = maximum value of ρ

ρ = common logarithm of particle diameter d [cm]

ρ_{min}	=	minimum value of ρ
H	=	height of the eruption column above the vent [km]
z	=	vertical distance from the ground surface [km]
Q	=	total quantity of erupted material [g]
$P(z)$	=	function for particle diffusion at height within dz about z [unitless]
$f(\rho)$	=	distribution function for particles with a log-diameter within $d\rho$ about ρ normalized per unit mass [unitless]
C	=	constant relating the eddy diffusivity and the particle fall time [$cm^2/s^{5/2}$]
t	=	particle fall time [s]
t_s	=	particle diffusion time in the eruption column [s]
x	=	x-coordinate on the surface of the earth [cm]; coordinate oriented in the same direction as the prevailing wind
u	=	wind speed [cm/s]
y	=	y-coordinate on the surface of the earth [cm]; coordinate oriented perpendicular to the direction of the prevailing wind

For calculation of dose, the necessary quantity to track is the mass of SF per unit area at the compliance point after ash released from the eruption that penetrates the proposed repository settles on the surface of the earth. To calculate this quantity, a model for SF incorporation into ash was created. This model requires the introduction of a function to determine the mass of fuel per unit mass of volcanic ash as a function of the log-diameter of the ash after the ash has been contaminated with SF [$FF(\rho^a)$]. The volcanic ash mass is assumed to be distributed log-normally:

$$f(\rho^a) = \frac{1}{\sqrt{2\pi}\,\sigma_d} \exp\left(-\frac{(\rho^a - \rho^a_{mean})^2}{2\sigma_d^2}\right) \tag{3-77}$$

where:

$f(\rho^a)$	=	normalized (per unit mass) probability distribution of ash mass as a function of ρ^a
σ_d	=	standard deviation of the log particle size
ρ^a	=	log-diameter of ash particle size [cm]
ρ^a_{mean}	=	mean of the log-diameter of ash particle size [cm]

To determine $FF(\rho^a)$, the fuel fraction (ratio of fuel mass to ash mass) as a function of ρ^a, one must consider that all fuel particles of a size smaller than ($\rho^a - \rho_c$) have the ability to simultaneously be incorporated into volcanic ash particles of size ρ^a or larger, where ρ_c is the incorporation ratio. The fuel fraction as a function of ρ^a is determined by summing all the incremental contributions of fuel mass to the volcanic ash mass from fuel sizes smaller than $(\rho^a - \rho_c)$. An expression for the fuel fraction is given as:

$$FF(\rho^a) = \frac{U}{Q} \int_{-\infty}^{\rho^a} \frac{m(\rho - \rho_c)}{1 - F(\rho)} d\rho \qquad (3\text{-}78)$$

where:

$FF(\rho^a)$	=	fuel fraction as a function of particle size [unitless]
Q	=	total mass of ash ejected in the event [g]
U	=	total mass of fuel ejected in the event [g]
$F(\rho)$	=	cumulative distribution of ash mass with ρ
$m(\rho - \rho_c)$	=	normalized probability distribution of fuel mass with $(\rho - \rho_c)$

This equation assumes the resulting contaminated particles follow the same size distribution as the original volcanic ash particles. This assumption seems reasonable since, for most events sampled in these analyses, the total mass of volcanic ash is on the order of 10^{13}—10^{15} g (2.2×10^{10}—2.2×10^{12} lb) and for most events, only several WPs are disrupted [10^7 g (2.2×10^4 lb) of fuel each]. For each simulation, Eq. (3-78) is numerically integrated to calculate the distribution of the SF and volcanic ash on the earth's surface resulting from a basaltic eruption assumed to disrupt the repository. The integrand of Eq. (3-76) is multiplied by $FF(\rho^a)$ and then evaluated to find the SF density at location (x,y).

The model developed by Suzuki (1983) is appropriate for particles of mean diameter greater than about 15–30 μm (6×10^{-4}—1.2×10^{-3} in.). This cutoff is generally accepted to be the lower limit for the importance of gravitational settling of particles (Cember, 1983; Heffter and Stunder, 1993). For particle sizes less than about 15 μm (6×10^{-4} in.), atmospheric turbulence is great enough to keep the particle aloft for a longer time than would be predicted by the model. Since the typical mean diameter of ash particles after an eruption is generally much larger than 15 μm (6×10^{-4} in.) (Suzuki, 1983), this model is useful for calculating the distribution for the vast majority of ash and, hence, radionuclides released. Jarzemba and LaPlante (1996), Jarzemba (1997), and Jarzemba, et al. (1997) contain more complete descriptions of the derivation, structure, and use of the ASHPLUME code and its algorithms.

3.12.3 Assumptions and Conservatism of the ASHPLUMO Approach

The ASHPLUMO module is an abstraction of the transport of radioactively contaminated ash (i.e., tephra) from the proposed repository to points located downwind from the event. The main processes that control the transport of SF in the ash are: (i) SF particulate incorporation in the ash; (ii) dispersion in the eruption column and plume; and (iii) transport of ash downwind from the event.

The model in ASHPLUME for SF incorporation in ash assumes that all the SF particles are eventually incorporated into larger (by a pre-set factor) ash particles. Actually, it is likely that only a percentage of the SF particles may be incorporated in this fashion, and some of the SF may be transported as bare particles. The conservatism of this approach is not yet known.

The transport model in ASHPLUME uses a dispersion-advection transfer function type solution that is empirical (e.g., eddy diffusivity constant and column shape parameter). Since the solution uses empirical constants based on observations at analogous volcanoes, it requires that future eruptions that may occur at the repository site be similar to those observed at the present time at other locations.

The transport model in ASHPLUME as implemented in the TPA Version 3.1.4 code assumes the wind speed and direction do not vary during the course of the event. This is clearly a conservative assumption, since variances in wind velocity over the event will lead to a wider distribution of SF than those predicted with a single velocity. The level of conservatism is unknown at this time, and may be averaged out when the average dose over numerous realizations is used as the performance measure.

In addition to those assumptions described before, the derivation of Eq. (4-76) assumes the following, which are reiterated here from Suzuki (1983):

- The distribution of the diameter of the released particles is well-described by a log-normal distribution.

- All particles fall at the terminal velocity and finally accumulate on the ground.

- The particles have a probability of diffusing out of the column during upward travel such that the probability density distribution, $P(z)$, has the following form.

$$P(z) = \frac{\beta W_0 Y \exp(-Y)}{V_0 H \left[1 - (1 + Y_0) \exp(-Y_0) \right]} \qquad (3\text{-}79)$$

where:

β = constant controlling the diffusion of particles out of the eruption column

W_0 = volcanic eruption velocity at the vent exit [cm/s]

Y = $\dfrac{\beta W(z)}{V_0}$

V_0 = particle terminal velocity at sea level [cm/s]

Y_0 = $\dfrac{\beta W_0}{V_0}$

$W(z)$ = particle velocity as a function of height equal to $W_0(1 - z/H)$

The terms Y and Y_0 have been modified from Suzuki (1983) where they were equal to $\beta(W(z) - V_0)/V_0$ and $\beta(W_0 - V_0)/V_0$ because with the former definition of Y, $P(z)$ has a negative value at heights approaching the top of the column. The definition of Y_0 has been altered to maintain identity as the value of $Y(z)$ at $z = 0$. The modified model has been benchmarked against actual eruptions with good agreement (Hill, et al., 1998).

- The distribution of SF particle diameter is well described by a log-triangular distribution.

- The size distribution of contaminated ash particles is unaffected by the incorporation of SF particles.

3.13   ASHRMOVO MODULE DESCRIPTION

The purpose of the ASHRMOVO module of TPA is to calculate areal radionuclide densities at the compliance point for times after a volcanic event has deposited a contaminated ash blanket. The radionuclide areal densities are greatest at the time of the event and decrease thereafter through surface erosion, leaching of radionuclides from the blanket, and radioactive decay. Ingrowth of radionuclides is also accounted for but, in general, ingrowth is more than compensated by the removal processes.

3.13.1   Information Flow Within TPA

3.13.1.1   Information Supplied to ASHRMOVO

The areal deposition of both SF and ash at the compliance point generated by ASHPLUMO are passed to ASHRMOVO from EXEC. The radionuclide inventory of the SF at times after the volcanic event is also provided by EXEC. Additionally, the ASHRMOVO code module reads input values such as ash blanket characteristics and decay parameters from the ASHRMOVO section of the primary input file.

3.13.1.2   Information Provided by ASHRMOVO

For all ground-surface radionuclides, ASHRMOVO transfers the areal radionuclide density in activity per unit area as a function of time at the compliance point back to EXEC for subsequent use in DCAGS for dose calculation.

3.13.2   Conceptual Model

The ASHRMOVO module models the time-dependent radionuclide areal densities of contaminated soil surface layers subject to removal by leaching, erosion, and radioactive decay. ASHRMOVO is used in concert with ASHPLUMO, which establishes initial radionuclide areal densities for extrusive volcanic events at the time of the event that intersects the waste repository. ASHRMOVO performs calculations to decay the inventory for succeeding times. The subsequent time history of radionuclide surficial contamination is converted to dose by the DCAGS module.

ASHRMOVO uses generalized analytical solutions to calculate the time-dependent radionuclide density in soil considering radioactive decay and ingrowth, and nonradioactive losses by leaching or erosion. For the volcanic exposure scenario previously described in ASHPLUMO, ASHRMOVO calculates the time history of radionuclide surficial contamination following the event by using analytical solutions to the following differential equations:

$$\frac{dN_i(t)}{dt} = \lambda_{i-1}^P N_{i-1}(t) - \lambda_i^P N_i(t) - \lambda_i^L N_i(t) - \lambda^B N_i(t)$$

$$= \lambda_{i-1}^P N_{i-1}(t) - \lambda_i^T N_i(t) \tag{3-80}$$

where:

$N_i(t)$	=	time-dependent areal density of radionuclide i [mol/m²]
$N_{i-1}(t)$	=	time-dependent areal radionuclide density of the parent [mol/m²]
λ_i^P	=	removal (or generation) of radionuclide i through radioactive decay [1/yr]
λ_i^L	=	relative leach rate of radionuclide i from the ash blanket [1/yr]
λ^B	=	bulk removal of the blanket through surface erosion [1/yr]
λ_i^T	=	total loss rate of radionuclide i from physical decay, leaching, and surface erosion equal to $\lambda_i^P + \lambda_i^L + \lambda^B$ [1/yr]

The initial conditions for the system of equations are the areal densities of the radionuclides at the time of the event. These densities have been estimated by ASHPLUME.

The relative leach rate of radionuclide i is based on Eq. (4.6.3) of Napier, et al. (1988). The relative leach rate has an upper limit, λ_i^{Lmax}, dependent on the solubility limit of the radionuclide and the amount of radionuclide present. When the processes are leach-rate limited, the relative leach rate of radionuclide i is given by:

$$\lambda_i^L = \lambda_i^{Lmax} \qquad\qquad N_i < \frac{S_i R_a}{\lambda_i^{Lmax}} \tag{3-81}$$

and when the leach processes are solubility-limited, it is given by:

$$\lambda_i^L = \frac{S_i R_a}{N_i} \approx \frac{S_i R_a}{\bar{N}_i} \qquad N_i > \frac{S_i R_a}{\lambda_i^{Lmax}} \tag{3-82}$$

where:

S_i = solubility limit of radionuclide i [mol/m³]

R_a = areal recharge of water [m³/(m² yr)]

$\bar{N}_i$ = average areal density over the time step of radionuclide i [mol/m²]

The quantity, λ_i^{Lmax}, is a combination of more basic parameters and is given by:

$$\lambda_i^{Lmax} = \frac{P(1 - f_e^P)f_{sat}^P + I(1 - f_e^i)f_{sat}^i}{d(1 + (\rho/\theta)K_{d_i})} \qquad (3\text{-}83)$$

where:

P = annual precipitation rate [m/yr]

f_e^P = fraction of precipitation water lost to evapotranspiration

f_{sat}^P = fraction of year blanket is saturated by precipitation

I = annual irrigation rate [m/yr]

f_e^i = fraction of irrigation water lost to evapotranspiration

f_{sat}^i = fraction of year blanket is saturated because of irrigation

d = depth of the blanket [m]

ρ = soil bulk density [g/cm³]

θ = soil volumetric water content

K_{d_i} = distribution coefficient for radionuclide i [cm³/g]

The bulk removal rate of the blanket, λ^B, is a constant value of 0.0001/yr based on preliminary estimates of ash blanket lifetimes, but may be sampled by specifying a distribution in the primary input file (Appendix A). A complete demonstration of the model in ASHRMOVO, including plots of time-dependent radionuclide areal densities for an example case, is given in Jarzemba and Manteufel (1997); however, a demonstration for the curium 245 (Cm-245) decay chain is provided here as an example. Tables 3-4 and 3-5 provide a listing of the data used in this example. Using these data and the model described in this section, the areal radionuclide densities as a function of time [$N_i(t)$] are estimated and shown graphically in Figure 3-20.

Table 3-4. A Listing of the Nonradionuclide-Specific Data Used in Estimating the Surficial Radionuclide Areal Density for Members of the Cm-245 Decay Chain

Parameter	Value
Initial SF areal density	1 kg/m^2
Annual precipitation rate	0.15 m/yr
Annual irrigation rate	1.52 m/yr
Fraction of precipitation lost by evapotranspiration	0.68
Fraction of irrigation lost by evapotranspiration	0.5
Fraction of year soil is saturated by precipitation	0.0054
Fraction of year soil is saturated by irrigation	0.2
Soil bulk density	2 g/cm^3
Soil porosity	0.4
Fractional blanket erosion rate	0.001
Depth of the rooting zone	1 m

Table 3-5. A Listing of the Radionuclide-Specific Data Used in Estimating the Surficial Radionuclide Areal Density for Members of the Cm-245 Decay Chain

Nuclide	Initial Inventory (Bq m^{-2}Kg^{-1})	Physical Decay Constant (yr^{-1})	Solubility (atoms/L)	K_d (cm^3 g^{-1})
Curium-245 (Cm-245)	4.6×10^6	8.2×10^{-5}	6.0×10^{17}	4.0×10^3
Plutonium-241 (Pu-241)	2.8×10^{12}	4.8×10^{-2}	3.0×10^{18}	5.5×10^2
Americium-241 (Am-241)	6.1×10^{10}	1.6×10^{-3}	3.0×10^{18}	1.9×10^3
Neptunium-237 (Np-237)	1.1×10^7	3.2×10^{-7}	9.5×10^{19}	5.0×10^0
Uranium-233 (U-233)	8.9×10^2	4.4×10^{-6}	2.7×10^{19}	3.5×10^1
Thorium-229 (Th-229)	5.1×10^0	9.5×10^{-5}	1.9×10^{15}	3.2×10^3

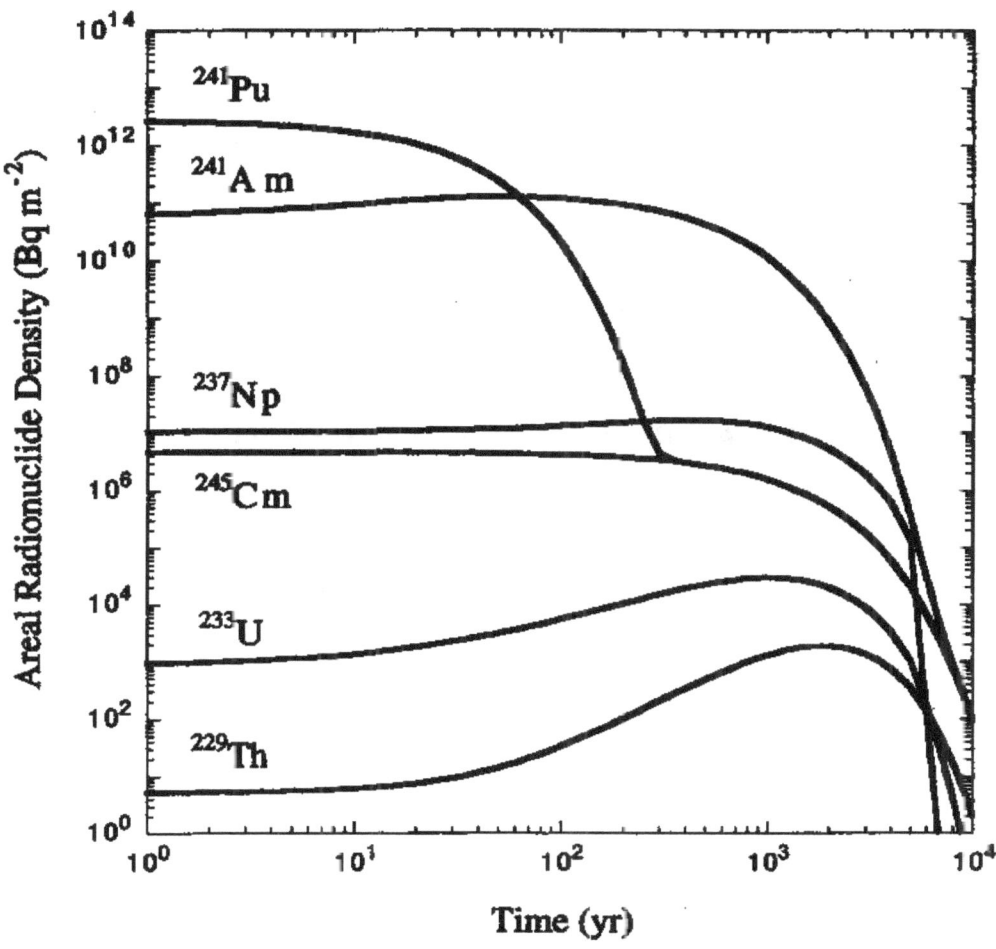

Figure 3-20. A plot of the areal radionuclide density as a function of time for radionuclides in the curium-245 (Cm-245) decay chain.

3.13.3 Assumptions and Conservatism of the ASHRMOVO Approach

The ASHRMOVO module is an abstraction of the leaching and removal of radionuclides from a contaminated soil/ash surface layer after an extrusive volcanic event has distributed radionuclides on the Earth's surface. The main processes modeled by ASHRMOVO are removal of radionuclides from the contaminated ash blanket by solubilization and leaching by infiltrating water, radioactive decay, and blanket erosion.

Infiltrating water has the ability to dissolve radionuclides in the ash blanket and transport them to deeper soil/ash layers. Since only the contamination in the top 15 cm (6 in.) of ash is considered in dose calculations, the effect of this transport is to remove leached radionuclides from the biosphere. Contamination of ground water from this source is not considered.

Radioactive decay of nuclides in the blanket is accounted for in this model. Any radioactive daughters formed from parent decay are also considered.

Blanket erosion is accounted for in the model as a first-order process. The blanket erosion term should be considered as net erosion (deposition from ash eroded from other sections of the blanket minus that eroded from the current location). Mechanical redistribution of ash is not specifically accounted for in the ASHRMOVO model.

3.14   DCAGS MODULE DESCRIPTION—FARMING RECEPTOR GROUP

The purpose of the DCAGS module is to convert the areal radionuclide activity densities (in Ci/m^2) calculated by the ASHRMOVO module into annual doses (TEDE). This section describes only the assumptions and conceptual models, beyond those already described in Section 3.8, for the DCAGW module, that are required for accomplishing this purpose.

3.14.1 Information Flow within TPA

3.14.1.1 Information Supplied to DCAGS

Time-varying areal concentrations computed in ASHRMOVO for each ground-surface radionuclide and the ash areal density calculated in ASHPLUMO are provided to DCAGS by the EXEC, from which ground-surface radionuclide dose (an annual TEDE) is obtained, using ground-surface DCFs in two data files: *gs_cb_ad.dat* and *gs_pb_ad.dat*, consistent with the lifestyle of the receptor group and the climatic conditions specified in the primary input file. The two data files are identified by the following nomenclature: *cb*—current biosphere; *pb*—pluvial biosphere; and *ad*—farming group. The time evolution of average annual precipitation and temperature from UZFLOW are passed by the EXEC to DCAGS to determine the time when the current condition will switch to pluvial conditions. Other DCAGS input parameters such as the depth to which particles can be resuspended and values for the mass load and occupancy factor are specified in the DCAGS section of the primary input file.

3.14.1.2 Information Provided by DCAGS

The ground-surface doses as a function of time for each radionuclide computed in DCAGS at every TPA time step are passed to EXEC.

3.14.2 Conceptual Model

The conceptual framework for DCAGS includes processes occurring subsequent to airborne transport and deposition of radiological contamination (released from the repository by an extrusive volcanic event) to the ground-surface where the farming receptor group is located. Processes included in the conceptual model for DCAGS at the 20-km (12-mi) receptor location include resuspension of deposited contamination, with subsequent inhalation, direct exposure, and ingestion of contaminated plant and animal products. This model is implemented in DCAGS (with the exception of the inhalation pathway, which is explained below) by using databases of DCFs that convert unit radionuclide areal densities to annual TEDEs, which is similar to the implementation in DCAGW described earlier. DCFs were calculated using a deterministic approach in the GENII-S pathway/dose assessment code (Leigh, et al., 1993; Napier, et al.,1988). These calculations were based on unit radionuclide concentrations in the soil and a combination of site-specific and generic input parameters discussed in LaPlante and Poor (1997).

For each TPA realization, DCAGS multiplies ground-surface radionuclide concentrations generated from the ASHRMOVO module by the exposure pathway/climate and radionuclide-specific DCFs. For each time step, the products of each radionuclide concentration and DCF are summed within and among exposure pathways and radionuclides to calculate total doses. The reporting of peak doses, differentiation between climates, and applicable exposure pathways are all the same as described in Section 3.8 for the DCAGW module. The only exceptions are that surface contamination, not ground water, is the source of radionuclides, and DCAGS does not include the ingestion of contaminated water.

For inhalation, a mass loading model was developed that is more applicable to expected site conditions than available models in GENII-S (Leigh, et al., 1993; Napier, et al., 1988). The mass loading model estimates the airborne concentration of radionuclides above the ash blanket after the volcanic event. The model multiplies the airborne mass load by the concentration of radionuclides in the ash being resuspended. The concentration in ash is calculated from the output of the ASHPLUMO and ASHRMOVO modules. The model also accounts for the effect of blanket thickness on the amount of material available for resuspension. The areal concentration of the ith radionuclide, $C_{a(i)}$ in Ci/m^3 is:

$$C_{a(i)} = S\eta_i f_R \qquad (3-84)$$

where:

S = airborne mass load [g/m^3]

η_i = activity of radionuclide i per mass of ash [Ci/g]

f_r = fraction of resuspended mass that emanated from the contaminated volcanic ash layer [unitless]

The activity of radionuclide i per mass of ash at the compliance point is estimated from quantities already calculated by the TPA Version 3.1.4 code modules as follows:

$$\eta_i = \frac{R_i(x,y)}{X(x,y)} \qquad (3-85)$$

where:

$R_i(x,y)$ = quantity of radioactivity per unit area accumulated at location (x,y) [Ci/m^2]

$X(x,y)$ = mass of ash per unit area accumulated at location (x,y) [g/cm^2]

The fraction of resuspended mass that emanated from the contaminated volcanic ash layer depends on the thickness of the ash blanket, T_B, and the depth to which particles can be resuspended, T_R. In other words, T_R represents the depth from the surface below which no resuspended ash can emanate over the course of 1 yr. For blankets thicker than T_R, f_r is unity. For blankets thinner than T_R, f_r is given by:

$$f_R = \frac{T_B}{T_R} \qquad (3\text{-}86)$$

Situations where f_r is equal to unity and where f_r is less than unity are shown in Figure 3-21.

The blanket thickness is a function of time and is given by:

$$T_B = T_{B\text{-}0} \exp \left[- \lambda_B (t - t_{event}) \right] \qquad (3\text{-}87)$$

where:

$T_{B\text{-}0}$ = blanket thickness at the compliance point immediately following the event (estimated from $X(x,y)$ from ASHPLUMO) [cm]

λ_B = net blanket erosion rate from ASHRMOVO [1/yr]

t_{event} = time of the volcanic event from VOLCANO [yr]

Once the airborne radionuclide concentration has been estimated, the rest of the dose conversion process is identical to what is currently implemented in DCAGW. The equation describing the rest of the dose conversion process is:

$$\dot{D}_i = BI_i C_{a(i)} f_e \qquad (3\text{-}88)$$

where:

$\dot{D}_i$ = inhalation dose rate caused by the airborne pathway for radionuclide i [rem/yr]

B = breathing rate [m³/yr]

f_e = fraction of the year the individual is exposed to contaminated air

I_i = inhalation-to-dose conversion factor in rem/Ci [(U.S. Environmental Protection Agency, 1988) choices for removal class are as listed in Pacific Northwest National Laboratory library]

The quantity BI_i is provided in the DCF tables, in the files gs_cb_ad.dat and gs_pb_ad.dat (inhalation parameter column), to the code as input data, where B is assumed to be 1.05×10^4 m³/yr (2.6 gal./min) (Bureau of Radiological Health, 1970).

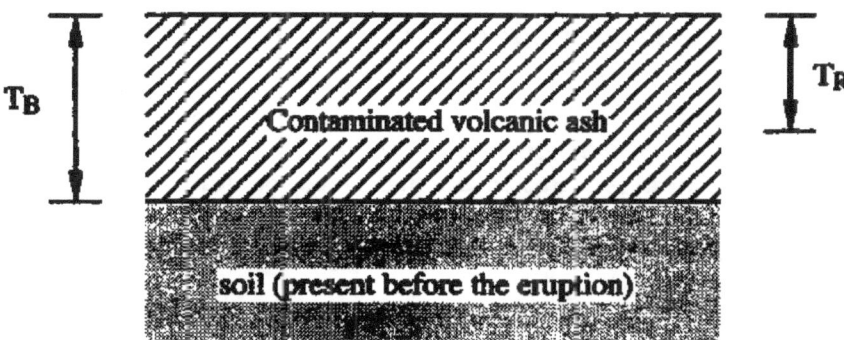

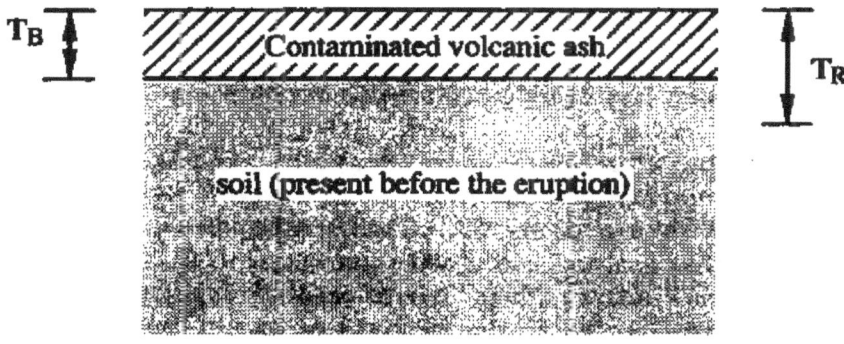

Figure 3-21. Examples where fraction of resuspended mass that emanted from the contaminated volcanic ash layer (f_R) (a) is unity and (b) is less than unity.

For each time step, the products of each radionuclide concentration and DCF are summed within and among ground-surface exposure pathways and radionuclides to calculate total doses. In addition to total doses, selected output is sorted by realization, time step, and radionuclide. The executive model uses results from all realizations to identify and report the peak dose.

3.14.3   Assumptions and Conservatism of the DCAGS Approach

DCAGS is an abstraction of complex and uncertain processes occurring in the biosphere. The conversion of soil radionuclide concentrations to receptor dose involves assumptions about: (i) the location of the receptor group; (ii) the lifestyle characteristics of the receptor group that form the basis for exposure pathways; (iii) processes that determine fate and transport of contaminants in the biosphere; and (iv) calculation of human doses from factors that convert exposure to contaminated media to effective dose equivalents.

Receptor location is based on the assumption that current physical constraints (i.e., topography, depth to water table, and soil conditions) that limit present farming to the Amargosa Valley area will continue to limit farming south of YM to within approximately 20 km (12 mi) from the proposed repository site. Because the receptor location and lifestyle assumption relate to potential human behavior, it is speculative. The use of present knowledge to define the receptor group location helps to limit speculation and is consistent with the recommendations of the National Academy of Sciences (1995).

The conversion of soil concentrations to receptor dose requires conceptualization of potential exposure pathways based on the site-specific characteristics of the release pathway, YM biosphere, and receptor group. When soil is the source of contamination, the potential exposure pathways are assumed to be those resulting from agricultural use of land, consistent with present farming conditions south of YM. These pathways include ingestion (of contaminated crops and animal products); inhalation (from resuspension of soil); and direct exposure to contaminated soil.

The assumed diet of locally produced food is important in estimating the dose for a farming receptor group. The farmer is assumed to grow alfalfa (for beef and milk cow feed); vegetables, fruits, and grains for personal consumption; and feed, for egg-laying hens. The practices the farmer is presumed to engage in have been confirmed to exist in areas south of YM; however, it appears unlikely that all activities identified would be practiced by a single individual. Therefore, the doses calculated from such assumptions are expected to be greater than what might occur under current conditions. Since bartering is known to exist among community members (Eisenberg, 1996),[10] it is reasonable to project that (for the most highly exposed group) a significant portion of diet could be obtained from local sources. Thus, the assumption is considered conservative, but not excessively so. These assumptions may change when additional information on local consumption patterns is available.

Modeling the RT in the biosphere involves determination of crop concentrations using soil/plant uptake factors, determination of air concentrations from resuspended contamination using a mass load model, and accounting for feed intake and transfer of contamination to livestock.

---

[10]Eisenberg, N.A. 1996. *Staff Visit to Amargosa Valley: Trip Report.* Memorandum (May 14) to M. Federline, Nuclear Regulatory Commission. Washington DC: Nuclear Regulatory Commission.

Crop concentrations are based on transfer factors that represent the ratio of plant and soil concentrations for a class of crops grown in contaminated soil. Food transfer factors are generic values based primarily on published information (International Atomic Energy Agency, 1994; Baes, et al., 1982). Because of the lack of site-specific information, it is reasonable to use generic values; however, they may not be representative of arid conditions that exist in Nevada. The selected factors are expected values and have not been arbitrarily increased for conservatism.

Air concentrations from resuspended contamination are calculated by applying a mass loading model. This model uses a mass loading factor that is the ratio of airborne and surface soil radionuclide concentrations. Mass loading factors are highly uncertain and influenced by a number of processes, including wind speed, surface roughness, resuspended particle size and density, and the moisture content of soil. Because the GENII-S code does not provide a mass loading model for crop deposition, a resuspension model was used with an adjusted mass loading factor. A separate mass loading model for inhalation is used in DCAGS, developed to be more applicable to site conditions than the available models in GENII-S (Leigh, et al., 1993; Napier, et al., 1988). This mass loading model estimates the airborne concentration of radionuclides above the ash blanket after the volcanic event. The model accounts for the effect of blanket thickness (as a function of time) on the amount of material available for resuspension.

Determination of radionuclide concentrations in livestock products is done by quantifying the intake of contaminated feed and applying a transfer coefficient. The animal transfer coefficients are based on studies that measure radionuclide concentrations in livestock based on measured intake. The coefficients are element- and livestock-specific and based on published information (International Atomic Energy Agency, 1994; Baes, et al., 1982).

Calculation of human dose is accomplished through the use of dose factors that convert the estimated concentrations in food, air, and soil, together with the corresponding intakes and residence times, to TEDEs. The dosimetry models that form the basis for the dose factors are consistent with those described in ICRP Publication 30 (ICRP, 1982). The dosimetry models represent state-of-the-art techniques and methodologies in existence at the time the publication was released. Refinements to dosimetry models have since been published by the ICRP, but have not yet been adopted by the U.S. Government and are therefore not used in DCAGS.

More specific assumptions for DCAGS are provided in the following list and in LaPlante and Poor (1997).

- Contamination spread on the ground surface is the sole source of contamination, as opposed to contamination dissolved in the ground water used in DCAGW. This assumption means that radionuclides leached from the ash blanket to lower soil layers and possibly the SZ are removed from the biosphere and thus do not contribute to estimated doses. Because, for this exposure pathway, the most significant contributors to dose are actinides such as plutonium; americium; and curium, and these radionuclides are not expected to be easily leached from ground surface soil/ash layers, this assumption is considered reasonable.

- The dose calculation from consumption of crops does not account for washing fruits and vegetables before consumption and is, therefore, conservative.

- Consumption rates for food are based on national average values, since there is a lack of site-specific information. These values may not be representative of local consumption practices. Use of average values is not expected to be overly conservative. DOE conducted a survey of local consumption practices and, when results are available, DCFs may be recalculated if the new consumption rates are found to differ significantly from the national averages used for current DCFs.

- For human dosimetry modeling, the average member of the receptor group is an adult consistent with the EPA proposed "Federal Radiation Protection Guidance for Exposure of the General Public" (U.S. Environmental Protection Agency, 1994). The individual is assumed to have physiology consistent with the reference man model (International Commission on Radiological Protection, 1975).

- Inhalation and external exposure times are based on generic information provided in Kennedy and Strenge (1992). For the farming group, 55 percent of total time is assumed spent inside the residence; 20 percent outside of residence; 24 percent at offsite work; and 1 percent gardening and farming alfalfa, which is not a labor-intensive crop to farm. The remainder of time is assumed to be spent away (not at work) from the residence. These activity times are considered conservative for a common residential setting (Kennedy and Strenge, 1992), but may not be for a more active farming resident.

- Resuspension of contaminated soil applies to crop deposition, inhalation, and external dose modeling. A mass loading model is applicable to a homogeneously mixed layer of contamination in the soil and, therefore, a reasonable mass loading factor from the literature is used (Kennedy and Strenge, 1992). For crop deposition, the GENII-S code includes a resuspension model; therefore, the selected mass loading factor is converted to a resuspension factor, using an equation in the GENII-S user manual (Napier, et al., 1988). A mass loading model is used to determine annual TEDEs from the inhalation pathway.

- Airborne mass loading is assumed constant over time. For times long after the time of the ash deposition, changes to soil and ash conditions (i.e., hardening of soil) will tend to reduce mass loading. However, farming activities (e.g., tilling of soil) are assumed to continually disturb the ash and soil. Provided that a reasonably conservative value for the airborne mass load is used as input, a reasonably conservative estimate of dose from the inhalation pathway for volcanism will be achieved since doses from inhalation are linearly dependent on this factor. The mass loading and resuspension models are abstractions of complex and varying processes. Thus, considerable uncertainty exists in applying them to YM conditions.

3.15   DCAGS MODULE DESCRIPTION—RESIDENTIAL RECEPTOR GROUP

The purpose of the DCAGS module is to convert the areal radionuclide activity densities (in Ci/m^2) calculated by the ASHRMOVO module into annual dose (TEDEs). This section describes only the assumptions and conceptual models beyond those already described in Section 3.9 for the DCAGW module required for accomplishing this purpose.

A residential receptor group located between 5 km (3.1 mi) and 20 km (12 mi) has been included in the TPA Version 3.1.4 code to assist timely completion of the sensitivity analyses by allowing shorter simulation periods appropriate to the shorter distances. Having the ability to evaluate a residential receptor group also provides the staff with the flexibility to address alternative assumptions for the receptor group, should that be necessary.

3.15.1   Information Flow Within TPA

3.15.1.1   Information Supplied to DCAGS

Time-varying areal ground-surface radionuclide concentrations, computed in ASHRMOVO, are provided to DCAGS by the EXEC. Ground-surface radionuclide dose (an annual TEDE) is obtained, using ground-surface DCFs in two data files: *gs_cb_ci.dat* and *gs_pb_ci.dat*, associated with the lifestyle of the receptor group and the climatic conditions specified in the primary input file. The two data files are identified by the following nomenclature: *cb*—current biosphere; *pb*—pluvial biosphere; and *ci*—residential. Other DCAGS input parameters such as the depth to which particles can be resuspended and values for the mass load and occupancy factor are specified in the DCAGS section of the primary input file. The time evolution of average annual precipitation and temperature from UZFLOW are passed by the EXEC to DCAGS to determine the time when the current condition will switch to pluvial conditions.

3.15.1.2   Information Provided by DCAGS

The ground-surface doses as a function of time for each radionuclide computed in DCAGS at every TPA time step are passed to EXEC.

3.15.2   Conceptual Model

The conceptual framework for DCAGS includes processes occurring subsequent to airborne transport and deposition of radiological contamination (released from the repository by an extrusive volcanic event) to the ground surface where the residential receptor group is located. Processes included in the conceptual model for DCAGS include resuspension of the deposited contamination at the residential receptor location and calculation of dose from inhalation of radionuclides and external exposure. This model is implemented in DCAGS (with the exception of the inhalation pathway, which is explained in Section 3.1.4.2) by using databases of DCFs that convert unit radionuclide activity areal densities to annual TEDEs. DCFs were calculated using a deterministic approach in the GENII-S pathway/dose assessment code (Leigh, et al., 1993; Napier, et al., 1988). These calculations were based on unit radionuclide concentrations in the soil and a combination of site-specific and generic input parameters discussed in LaPlante and Poor (1997).

For each TPA realization, DCAGS multiplies ground-surface radionuclide concentrations generated from the ASHRMOVO module by the exposure pathway/climate and radionuclide-specific DCFs. For each time step, the products of each radionuclide concentration and DCF are summed within and among exposure pathways and radionuclides to calculate total doses. The reporting of peak doses, the differentiation between climates, and the applicable exposure pathways are all the same as described in Section 3.9 for the DCAGS module. The only exceptions are that surface contamination, not ground water, is the source of radionuclides, and DCAGS does not include the ingestion of contaminated water.

For inhalation, a mass loading model was developed that is more applicable to expected site conditions than available models in GENII-S (Leigh, et al., 1993; Napier, et al., 1988). The mass loading model estimates the airborne concentration of radionuclides above the ash blanket after the volcanic event. The model multiplies the airborne mass load by the concentration of radionuclides in the ash being resuspended. The concentration in ash is calculated from the output from ASHPLUMO and ASHRMOVO modules. The model also accounts for the effect of blanket thickness on the amount of material available for resuspension. The areal concentration of the ith radionuclide, $C_{a(i)}$ in Ci/m^3, is computed from Eq. (3-84). The activity of radionuclide i per mass of ash at the compliance point is estimated from quantities already calculated by the TPA Version 3.1.4 code, using Eq. (3-85).

The fraction of resuspended mass emanated from the contaminated volcanic ash layer depends on the thickness of the ash blanket, T_B, and the depth to which particles can be resuspended, T_R. In other words, T_R represents the depth from the surface below which no resuspended ash can emanate over the course of 1 yr. For blankets thicker than T_R, f_r is unity. For blankets thinner than T_R, f_r is given by Eq. (3-86).

Situations where f_r is equal to or is less than unity are shown in Figure 3-21. The blanket thickness is a function of time and is given by Eq. (3-87).

Once the airborne radionuclide concentration has been estimated, the rest of the dose conversion process is identical to that currently implemented in DCAGW. Equation (3-88) describes the rest of the dose conversion process.

For each time step, the products of each radionuclide concentration and DCF are summed within and among exposure pathways and radionuclides to calculate total doses. In addition to total doses, selected output is stratified by realization, time step, and radionuclide. The EXEC module uses results from all realizations to identify and report the peak dose.

3.15.3   Assumptions and Conservatism of the DCAGS Approach

The DCAGS module for the residential receptor converts soil radionuclide concentrations to both inhalation and external doses. Therefore, the module includes assumptions about: (i) the location of the receptor group; (ii) the lifestyle characteristics of the receptor group that form the basis for exposure pathways; and (iii) the conversion of intake to dose.

The residential receptor is based on an assumption that local topography and the economics of ground-water extraction allow for a potential nonfarming residential receptor group to exist closer to the potential repository site than a farming receptor group. Although DCAGS does not include the ground-water pathways (those are addressed in DCAGW), the implicit assumption is that access to water is a precondition for the existence of residential dwellings. Because the receptor location and lifestyle assumptions relate to potential human behavior, it is speculative. The use of present knowledge to define the receptor group location helps to limit speculation and is consistent with the recommendations of the National Academy of Sciences (1995).

Because DCAGS considers the dose from airborne contaminants deposited to soil, exposure pathways are limited to inhalation and direct exposure. The inhalation dose is assumed from resuspension of contaminated soil, and external dose is primarily caused by outdoor human activities in the vicinity of contaminated soils.

Resuspension is modeled in DCAGS by use of a mass loading model developed to be more applicable to site conditions than the available models in GENII-S (Leigh, et al., 1993; Napier, et al., 1988). The mass loading model estimates the airborne concentration of radionuclides above the ash blanket after the volcanic event. The model accounts for the effect of blanket thickness (as a function of time) on the amount of material available for resuspension. The subsequent dose calculation involves the receptor breathing rate, exposure time, and inhalation-to-dose conversion factor. Exposure times for inhalation are obtained from conservative estimates in Kennedy and Strenge (1992) suggesting residents spend 4 percent of their time outdoors and 40 percent indoors (the remainder is away from residence). The inhalation-to-dose conversion factors are obtained from Federal Guidance Report No. 11 (U.S. Environmental Protection Agency, 1988) and are based on internal dosimetry models consistent with ICRP Publication 30.

Conversion of soil concentrations to external dose is a simple calculation involving the exposure time and the DCF for groundshine. The exposure times for external exposure are based on the aforementioned assumptions for inhalation exposure. The DCFs for external exposures are from Federal Guidance Report No. 12 (U.S. Environmental Protection Agency, 1993).

Following are additional details of DCAGS modeling assumptions:

- Airborne mass loading is assumed constant over time. For times long after the time of the ash deposition, changes to soil and ash conditions over time (i.e., hardening of soil) would tend to reduce mass loading of radioactive ash. Provided that a reasonably conservative value for the airborne mass load is used as input, a reasonably conservative estimate of dose from the inhalation pathway for volcanism will be achieved, since doses from inhalation are linearly dependent on this factor.

- Contamination spread on the ground-surface is the sole source of contamination, as opposed to contamination dissolved in the ground water. This assumption means that radionuclides leached from the ash blanket to lower soil layers and possibly the SZ are removed from the biosphere and thus do not contribute to estimated doses. Because, for this exposure pathway, the most significant contributors to dose are actinides such as plutonium, americium, and curium, and these radionuclides are not expected to be easily leached from ground-surface soil/ash layers, this assumption is considered reasonable. It is noted that the parameters in the base case data set that determine leaching of radionuclides out of the contaminated ash blanket are appropriate for a farming receptor group, (i.e., includes irrigation for farming) which may be nonconservative if applied to a residential group.

- For human dosimetry modeling, the average member of the receptor group is an adult, consistent with the EPA proposed "Federal Radiation Protection Guidance for Exposure of the General Public" (U.S. Environmental Protection Agency, 1994). The individual is presumed to have physiology consistent with the assumptions of the reference man model (International Commission on Radiological Protection, 1975).

4 FUTURE IMPROVEMENTS TO THE TPA CODE

Analysis of the performance of the proposed repository at YM is an important activity that assists staff in their prelicensing interactions with DOE and improves development of review plans by identifying and focusing attention on issues most important to performance. The TPA Version 3.1.4 code enhances the staff capability to review the DOE TSPA by providing staff with a code more flexible and user friendly and incorporates more detailed representations of the proposed site and engineering design relative to previously developed TPA codes. Although TPA Version 3.1.4 code represents a significant improvement over previous TPA codes, it is recognized that site characterization and engineering design are still evolving, and further refinements of this capability are anticipated. Improvements to the code are expected to incorporate new information for either the site or engineering design, improve modeling approaches (e.g., reduce conservatism, improve credibility of model), or improve the structure of the code (e.g., efficiency, flexibility, and ease of use).

Some significant improvements to TPA Version 3.1.4 have been identified based on new information. A change in the DOE material selection for the inner overpack of the WP (i.e., from Alloy 625 to Alloy C-22) occurred between the time of development and documentation of the code. Although the rate of corrosion of Alloy C-22, which results in an extended service life of Alloy C-22 compared to that for Alloy 625, has been incorporated into the current model, the current model does not account for differences in the mode of degradation (e.g., Alloy 625 degrades through pitting corrosion, whereas Alloy C-22 degrades through general passive corrosion). Modifications to the TPA Version 3.1.4 code may be needed if differences in these modes of degradation of the WP are found to have a significant effect on estimating radionuclide releases from the WP. For example, the manner in which water enters, fills, and leaves the WP through the corrosion pits of Alloy 625 could be significantly different from Alloy C-22 undergoing uniform corrosion. Therefore, the bathtub model for WP failure and the parameter that affects the amount of water entering the WP (i.e., F_{melt}) need to be evaluated considering the mode of corrosion of Alloy C-22. Additionally, the conservatism in the SEISMO module and mechanical failure of the WP, not considered significant to performance for WPs with lifetimes of 3000–5000 yrs, need to be examined for their influence on performance of a longer lived WP.

Sensitivity analyses performed at the process level (i.e., examination of the sensitivity of overall performance to variation of parameters and models within a specific process of the TPA Version 3.1.4 code, such as WP degradation and infiltration) and the system level (i.e., examination of the sensitivity to overall performance because of variation of parameters and models within the TPA Version 3.1.4 code) have been completed and are reported in Volume II of this NUREG. Initial use of the TPA Version 3.1.4 code to analyze the proposed repository has provided valuable insights assisting identification of code limitations and areas for improvement. These initial results have also aided in prioritizing code refinements based on importance to modeled repository performance. It should be noted that limitations in the analyses include identification of sources of potentially significant sensitivity that could not be evaluated in the sensitivity analysis because of their absence in the models and parameters of the code. Radionuclide transport through the emplacement drift invert is an example of a potentially significant source of sensitivity not included in the current code. In addition to the technical improvements to the modules of TPA Version 3.1.4 code, a number of additions to the input options for the code were identified to improve its flexibility and ease of use.

A number of suggestions have been made for improving the TPA Version 3.1.4 code and analyses to evaluate the need for or significance of these improvements. The suggestions for these future efforts include:

- The effect of secondary mineral formation could be added to EBSREL. The formation of secondary minerals may significantly reduce dissolution rates of the SF.

- The influence of the assumption of the presence of backfill in the diffusive release calculation could be reevaluated. Diffusive release through the backfill would not be considered if its contribution to the total release from the EBS is insignificant.

- A new module could be added to EBSREL to account for the effect of physical properties of a cementitious invert, such as porosity, permeability, and retardation (K_d) on the transport of radionuclides. The effect of this change will be radionuclide-specific, because of variation in retardation by element, and dependent on the matrix permeability of the invert material.

- Aqueous release of the inventory between the fuel pellet and cladding could be invoked in EBSREL by developing input values for the fraction of inventory in this gap (because of limitations in obtaining values for the gap fraction, aqueous release of the gap fraction is included in the current module but was not invoked by setting the gap fraction values to zero). Including the release of the gap fraction is expected to increase releases at early times.

- New models could be added to provide for more realistic treatment of the transient nature of WP and SF wetting. The current approach diverts the additional quantity of flow from increased infiltration or reflux to a fixed number of wetted WPs. In addition, mechanistic models may replace the currently used abstracted flow factors.

- New parameters account for variability in the bathtub height. A separate bathtub height could be used for each repository subarea and failure type (e.g., corrosion, initially defective, and seismic failures) as compared with the current approach, which uses a single bathtub height for subareas and failure types.

- A new relationship between acceleration and fractional area of rockfall (identified as the seismic heterogeneity factor in the TPA Version 3.1.4 code) could be developed in the SEISMO module. The number of ground acceleration magnitude classes will be increased as a refinement. Thus, the fractional area of rockfall will be a function of the magnitude of the acceleration for the seismic event, which will provide a more realistic approach.

- The height of the yield zone in SEISMO, used in determining the volume of the rockfall for impact load calculation, could be made a sampled parameter. In the TPA Version 3.1.4 code, the height of the yield zone is specified as a set of constants corresponding to acceleration values and rock type (these values are specified within an external data file for the SEISMO module). This change could provide refinement of height of the yield zone versus acceleration, allow a wider range of accelerations, and permit the user to account for uncertainty in the estimates of the height of the yield zone by specifying a probability distribution and range for the parameters for the height of the yield zone. This change is expected to reduce conservatism in the estimates of WP failures.

- Consideration of the time of WP failure from seismically induced rockfall could be improved by adding distinct time periods for grouping the failures. The TPA Version 3.1.4 conservatively uses a single time period to consolidate all the scenario failures of the WP (i.e., WP failure from seismicity, faulting, and IA are grouped together and assumed to occur at the earliest failure time). Because a number of seismic events are possible over 10,000 yr, a new approach could be developed to allow the user to specify distinct time periods for which the seismic failures are grouped within each specified time period, and failure is assumed to occur at the beginning of the respective time period. Because of their low probabilities of occurrence, faulting and igneous failures are assumed to occur once during a 10,000-yr compliance period; therefore, a single failure time for each event is sufficient. This change will reduce conservatism in the time of occurrence of the scenario failures of the WP.

- Distribution functions of particular radionuclides (e.g., uranium, neptunium, plutonium, americium, and thorium) are expected to be correlated because of similar water conditions, such as pH, that will affect sorption. Inclusion of correlation between distribution functions will improve realism of radionuclide transport. It is uncertain what effect this change will have on performance, although lower doses will tend to be lower and higher doses will tend to be higher for the radionuclides with correlated distribution coefficients.

Evaluation of the TPA Version 3.1.4 code is continuing, and additional improvements are expected as the results of the sensitivity analysis are further reviewed and additional calculations performed.

5 REFERENCES

Ahn, T.M. 1995. *Long-term Kinetic Effects and Colloid Formations in Dissolution of LWR Spent Fuels.* NUDOC Accession Number 9508030112. Washington, DC: U.S. Nuclear Regulatory Commission.

Ahn, T., and P. Soo. 1983. *Container Assessment—Corrosion Study of HLW Container Materials, Quarterly Progress Report, October–December 1983.* BNL–NUREG–34220. Upton, NY: Brookhaven National Laboratory.

Ahn, T., and P. Soo. 1984. *Container Assessment—Corrosion Study of HLW Container Materials, Quarterly Progress Report, January–March 1984, April–June 1984.* Upton, NY: Brookhaven National Laboratory.

Baca, R.G., G.W. Wittmeyer, and R.W. Rice. 1999. Analysis of contaminant dilution in groundwater. *Scoping Calculations for Revisions to the EPA Standard.* NUREG–1538. Washington, DC: U.S. Nuclear Regulatory Commission.

Baes, C.F., III, R.D. Sharp, A.L. Sjoreen, and R.W. Shor. 1982. *A Review and Analysis of Parameters for Assessing Transport of Environmentally Released Radionuclides through Agriculture.* ORNL–5876. Oak Ridge, TN: Oak Ridge National Laboratory.

Barnard, R.W., et al. 1992. *TSPA 1991: An Initial Total-System Performance Assessment for Yucca Mountain.* SAND 91-2795. Albuquerque, NM: Sandia National Laboratories.

Barr, G.E., et al. 1993. *Scenarios Constructed for Basaltic Igneous Activity at Yucca Mountain and Vicinity.* SAND 91–1653. Albuquerque, NM: Sandia National Laboratories.

Basse, B. 1990. *Water Resources in Southern Nevada.* San Antonio, TX: Center for Nuclear Waste Regulatory Analyses.

Bateman, H. 1910. The solution of a system of differential equations occurring in the theory of radioactive transformation. *Proceedings of the Cambridge Philosophical Society* 15: 473.

Bejan, A. 1988. *Advanced Engineering Thermodynamics.* New York: John Wiley and Sons.

Brechtel, C.E., M. Lin, E. Martin, and D.S. Kessel. 1995. *Geotechnical Characterization of the North Ramp of the Exploratory Studies Facility. Volume 1: Data Summary.* SAND 95–0488/1 UC–814. Albuquerque, NM: Sandia National Laboratories.

Briant, C.L., and S.K. Banerji. 1983. Intergranular fracture in ferrous alloys in nonaggressive environments. *Treatise on Materials Science and Technology. Volume 25: Embrittlement of Engineering Alloys.* C.L. Briant and S.K. Banerji, eds. New York: Academic Press.

Buqo, T.S. 1996. *Baseline Water Supply and Demand Evaluation of Southern Nye County, Nevada.* Nye County, NV: Nye County Nuclear Waste Repository Office.

Bureau of Radiological Health, ed. 1970. *Radiological Health Handbook.* Rockville, MD: U.S. Department of Health, Education, and Welfare.

Carrera, J., and S.P. Neuman. 1986. Estimation of aquifer parameters under transient and steady-state conditions, 3. Application to synthetic and field data. *Water Resources Research* 22(2): 228–242.

Carslaw, H.S., and J.C. Jaeger. 1959. *Conduction of Heat in Solids.* London: Oxford University Press.

Cember, H. 1983. *Introduction to Health Physics.* New York: Permagon Press.

Civilian Radioactive Waste Management System Management & Operating Contractor. 1997. *Repository Subsurface Layout Configuration Analyses.* BCA000000–01717–0200–00008 REV00. Las Vegas, NV: Civilian Radioactive Waste Management System Management & Operating Contractor.

Claesson, J., and T. Proberts. 1996. *Temperature Field Due to Time-Dependent Heat Sources in a Large Rectangular Grid—Derivation of Analytical Solution.* SKB 96-12. Stockholm, Sweden: Swedish Nuclear Fuel and Waste Management Co.

Codell, R.B., et al. 1992. *Initial Demonstration of the NRC's Capability to Conduct a Performance Assessment for a High-Level Waste Repository.* NUREG–1327. Washington, DC: U.S. Nuclear Regulatory Commission.

Cohen, A.J.B., et al. 1997. S^4Z: *Sub-Site Scale Saturated Zone Model for Yucca Mountain.* Berkeley, CA: Ernest Orlando Lawrence Berkeley National Laboratory. Level 4 Milestone SP25UM4.

Connor, C.B., et al. 1997. Magnetic surveys help reassess volcanic hazards at Yucca Mountain. *EOS, Transactions, American Geophysical Union* 74(7): 73–78.

Cragnolino, G.A., H.K. Manaktala, and Y-M. Pan. 1996. *Thermal Stability and Mechanical Properties of High-Level Radioactive Waste Container Materials: Assessment of Carbon and Low-Alloy Steels.* CNWRA 96-004. San Antonio, TX: Center for Nuclear Waste Regulatory Analyses.

Cragnolino, G.A., et al. 1998. Factors influencing the performance of carbon steel overpacks in the proposed high-level nuclear waste repository. *Proceedings of the CORROSION 98 Conference.* Paper No. 147. Houston, TX: NACE International.

Czarnecki, J.B., and R.K. Waddell. 1984. *Finite-Element Simulation of Groundwater Flow in the Vicinity of Yucca Mountain, Nevada–California.* WRI–84–4349. Denver, CO: U.S. Geological Survey.

Czarnecki, J.B. 1985. *Simulated Effects of Increased Recharge on the Groundwater Flow System of Yucca Mountain and Vicinity, Nevada–California*. WRI–84–4344. Reston, VA: U.S. Geological Survey.

de Marsily, G. 1986. *Quantitative Hydrogeology*. Orlando, FL: Academic Press, Inc.

Dunn, D.S., and G.A. Cragnolino. 1997. *An Analysis of Galvanic Coupling Effects on the Performance of High-Level Nuclear Waste Container Materials*. CNWRA 97-010. San Antonio, TX: Center for Nuclear Waste Regulatory Analyses.

Dunn, D.S., and G.A. Cragnolino. 1998. *Effects of Galvanic Coupling Between Overpack Materials of High-Level Nuclear Waste Containers—Experimental and Modeling Results*. San Antonio, TX: Center for Nuclear Waste Regulatory Analyses.

Fedors, R.W., and G.W. Wittmeyer. 1998. *Initial Assessment of Dilution Effects Induced by Water Well Pumping in the Amargosa Farms Area*. Letter Report. San Antonio, TX: Center for Nuclear Waste Regulatory Analyses.

Fedotov, S.A., and Ye.K. Markhinin. 1983. *The Great Tolbachik Fissure Eruption*. Cambridge. United Kingdom: Cambridge University Press.

Flint, L.E., and A.L. Flint. 1990. *Preliminary Permeability and Water-Retention Data for Nonwelded and Bedded Tuff Samples, Yucca Mountain Area, Nye County, Nevada*. Open-File Report 90-569. Denver, CO: U.S. Geological Survey.

Flint, A.L., J.A. Hevesi, and L.E. Flint. 1996. *Conceptual and Numerical Model of Infiltration for the Yucca Mountain Area, Nevada*. Milestone 3GUI623M. Las Vegas, NV: U.S. Department of Energy.

Fyfe, D. 1994. *Corrosion*. R.A. Jarman and G.T. Burstein, eds. Oxford, United Kingdom: Butterworth Heinmann.

Geldon, A.L. 1993. *Preliminary Hydrogeologic Assessment of Boreholes UE–25c#1, UE–25c#2, and UE–25c#3, Yucca Mountain, Nye County, Nevada*. WRI–92–4016. Denver, CO: U.S. Geological Survey.

Gelhar, L.W. 1993. *Stochastic Subsurface Hydrology*. Englewood Cliffs, NJ: Prentice-Hall, Inc.

Gelhar, L.W., C. Welty, and K.R. Rehfeldt. 1992. A critical review of data on field-scale dispersion in aquifers. *Water Resources Research* 28(7): 1955–1974.

Geomatrix Consultants, Inc. 1998. *Saturated Zone Flow and Transport Expert Elicitation Project*. San Francisco, CA: Geomatrix Consultants, Inc.

Ghosh, A., R.D. Manteufel, and G.L. Stirewalt. 1997. *FAULTING Version 1.0—A Code for Simulation of Direct Fault Disruption: Technical Description and User's Guide.* CNWRA 97-002. San Antonio, TX: Center for Nuclear Waste Regulatory Analyses.

Ghosh, A., et al. 1998. *Key Technical Issue Sensitivity Analysis with SEISMO and FAULTO Modules within the TPA (Version 3.1.1) Code.* Letter Report. San Antonio, TX: Center for Nuclear Waste Regulatory Analyses.

Glaze, L.S., and S. Self. 1991. Ashfall dispersal for the 16 September 1986 eruption of Lascar, Chile. Calculated by a turbulent diffusion model. *Geophysical Research Letters* 18: 1237–1240.

Gray, W.J. 1992. Dissolution testing of spent fuel. *Presentation to Nuclear Waste Technical Review Board Meeting, October 14–16, Las Vegas, Nevada.* Richland, WA: Pacific Northwest Laboratory.

Gray, W.J., H.R. Leider, and S.A. Steward. 1992. Parametric Study of LWR Spent Fuel Dissolution Kinetics. *Journal of Nuclear Material.* 190: 46–52.

Gray, W.J., and C.N. Wilson. 1995. *Spent Fuel Dissolution Studies FY 1991 to 1994.* PNNL–10540. Richland, WA: Pacific Northwest National Laboratory.

Gruss, K.A., G.A. Cragnolino, D.S. Dunn, and N. Sridhar. 1998. Repassivation potential for localized corrosion of Alloys 625 and C22 in simulated repository environments. *Proceedings of the CORROSION 98 Conference.* Paper No. 149. Houston, TX: NACE International.

Haitjema, H.M. 1995. *Analytic Element Modeling of Groundwater Flow.* New York: Academic Press.

Harrar, J.E., J.F. Carley, W.F. Isherwood, and E. Raber. 1990. *Report of the Committee to Review the Use of J-13 Well Water in Nevada Nuclear Waste Storage Investigations.* UCID–21867. Livermore, CA: Lawrence Livermore National Laboratory.

Hartman, H.L. 1991. *Mine Ventilation and Air Conditioning.* 2nd ed. J.M. Mutmansky and Y.J. Wang, co-eds. Malabar, FL: Krieger Publishing Company.

Heffter, J.L., and B.J.B. Stunder. 1993. Volcanic ash forecast transport and dispersion (VAFTAD) model. *Weather and Forecasting* 8: 533–541.

Hill, B.E. 1996. *Constraints on the Potential Subsurface Area of Disruption Associated with Yucca Mountain Region Basaltic Volcanoes.* Letter Report. San Antonio, TX: Center for Nuclear Waste Regulatory Analyses.

Hill, B.E., C.B. Connors, M.S. Jarzemba, P.C. LaFemina, M. Navarro, and W. Strauch. 1998. 1995 eruptions of Cerro Negro volcano, Nicaragua, and risk assessment for future eruptions. *Geological Society of America Bulletin* 110: 1231–1241.

Hopkins, A.T., and C.J. Bridgeman. 1985. A volcanic ash transport model and analysis of Mount St. Helens ashfall. *Journal of Geophysical Research* 90: 10,620–10,630.

Incropera, F.P., and D.P. DeWitt, 1990. *Fundamentals of Heat and Mass Transfer.* Third Edition. New York: John Wiley and Sons.

International Atomic Energy Agency. 1994. *Handbook of Parameter Values for the Prediction of Radionuclide Transfer in Temperate Environments.* Technical Report Series No. 364. Vienna, Austria: International Atomic Energy Agency.

International Commission on Radiological Protection. 1982. *Limits for Intakes of Radionuclides by Workers.* ICRP 30. 7 Volumes, including supplements. New York: Pergamon Press.

International Commission on Radiological Protection. 1975. *Reference Man: Anatomical, Physiological, and Metabolic Characteristics.* ICRP 23. Annuals of the ICRP. New York: Pergamon Press.

Jarzemba, M.S. 1997. Stochastic radionuclide distributions after a basaltic eruption for performance assessments of Yucca Mountain. *Nuclear Technology* 118: 132–141.

Jarzemba, M.S., and P.A. LaPlante. 1996. *Preliminary Calculations of Expected Dose from Extrusive Volcanic Events at Yucca Mountain.* Letter Report. San Antonio, TX: Center for Nuclear Waste Regulatory Analyses.

Jarzemba, M.S., P.A. LaPlante, and K.J. Poor. 1997. *ASHPLUME Code Version 1.0 Model Description and User's Guide.* CNWRA 97-004. San Antonio, TX: Center for Nuclear Waste Regulatory Analyses.

Jarzemba, M.S., and R.D. Manteufel. 1997. An analytically based model for the simultaneous leaching-chain decay of radionuclides from contaminated ground surface soil layers. *Health Physics* 73(6): 919–927.

Kennedy, W.E., and D.L. Strenge. 1992. *Residual Radioactive Contamination from Decommissioning: Technical Basis for Translating Contamination Levels to Annual Total Effective Dose Equivalent.* NUREG/CR–5512. Vol. 1. Washington, DC: U.S. Nuclear Regulatory Commission.

LaPlante, P.A., and K. Poor. 1997. *Information and Analyses to Support Selection of Critical Groups and Reference Biospheres for Yucca Mountain Exposure Scenarios.* CNWRA 97-009. San Antonio, TX: Center for Nuclear Waste Regulatory Analyses.

LaPlante, P.A., S.J. Maheras, and M.S. Jarzemba. 1995. *Initial Analysis of Selected Site-Specific Dose Assessment Parameters and Exposure Pathways Applicable to a Groundwater Release Scenario at Yucca Mountain.* CNWRA 95-018. San Antonio, TX: Center for Nuclear Waste Regulatory Analyses.

Leigh, C.D., et al. 1993. *User's Guide for GENII-S: A Code for Statistical and Deterministic Simulation of Radiation Doses to Humans from Radionuclides in the Environment.* SAND 91-0561. Albuquerque, NM: Sandia National Laboratories.

Leygraf, C. 1995. Atmospheric corrosion. *Corrosion Mechanisms in Theory and Practice.* P. Marcus and J. Oudar, eds. New York: Marcel Decker, Inc.: 421–455.

Lichtner, P.C., and M.S. Seth. 1996. *User's Manual for MULTIFLO: Part II—MULTIFLO 1.0 and GEM 1.0. Multicomponent-Multiphase Reactive Transport Model.* CNWRA 96-010. San Antonio, TX: Center for Nuclear Waste Regulatory Analyses.

Linsley, R.K., and J.R. Franzini. 1979. *Water-Resources Engineering.* New York: McGraw-Hill Book Company.

Lobnig, R.E., H.P. Schmidt, K.Hennesen, and H.J. Grabke. 1992. Diffusion of cations in chromia layers grown on iron-base alloys. *Oxidation of Metals* 37: 81–93.

Lohman, S.W. 1972. *Ground-Water Hydraulics.* U.S.G.S. Professional Paper 708. Washington, DC: U.S. Geological Survey.

Luckey, R.R., et al. 1996. *Status of Understanding of the Saturated-Zone Ground-Water Flow System at Yucca Mountain, Nevada, as of 1995.* WRI–96–4077. Denver, CO: U.S. Geological Survey.

Luhr, J.F., and T. Simkin. 1993. *Parícutin, The Volcano Born in a Mexican Cornfield.* Phoenix, AZ: Geoscience Press, Inc.

McWhorter, D.B., and D.K. Sunada. 1977. *Ground-Water Hydrology and Hydraulics.* Fort Collins, CO: Water Resources Publications.

Manteufel, R.D. 1997. Effects on ventilation and backfill on a mined waste disposal facility. *Nuclear Engineering and Design* 172(2): 205–220.

Manteufel, R.D., and N.E. Todreas. 1994. Effective thermal conductivity and edge conductance model for a spent fuel assembly. *Nuclear Technology* 105(3): 421–440.

Marsh, G., and K. Taylor. 1988. An assessment of carbon steel containers for radioactive waste disposal. *Corrosion Science* 28: 289–320.

Mohanty, S., et al. 1996. *Engineered Barrier System Performance Assessment Code: EBSPAC Version 1.0 BETA, Technical Description and User's Manual.* CNWRA 96–001. San Antonio, TX: Center for Nuclear Waste Regulatory Analyses.

Mohanty, S., et al. 1997. *Engineered Barrier System Performance Assessment Code: EBSPAC Version 1.1 Technical Description and User's Manual.* CNWRA 97–006. San Antonio, TX: Center for Nuclear Waste Regulatory Analyses.

Moran, J.J., and H.N. Shapiro. 1992. *Fundamentals of Engineering Thermodynamics.* New York: John Wiley and Sons.

Napier, B.A., R.A. Peloquin, D.L. Strenge, and J.V. Ramsdell. 1988. *GENII: The Hanford Environmental Radiation Dosimetry Software System; Volumes 1, 2, and 3: Conceptual Representation, User's Manual, Code Maintenance Manual.* PNL–6584. Richland, WA: Pacific Northwest Laboratory.

National Academy of Sciences. 1995. *Technical Bases for Yucca Mountain Standards.* Washington, DC: National Academy Press.

National Climatic Data Center. 1984 to 1994. *WBSAN Hourly Surface Observations.* Asheville, NC: National Oceanic and Atmospheric Administration.

Nevada Division of Water Resources. 1995. *Preliminary Special Hydrographic Abstract for Valley Basin No. 230 from the Nevada Division of Water Resources Water Rights Database.* Carson City, NV: Nevada Division of Water Resources.

Oatfield, W.J., and J.B. Czarnecki. 1991. Hydrogeologic inferences from drillers' logs and from gravity and resistivity surveys in the Amargosa Desert, southern Nevada. *Journal of Hydrology* 124(1-2): 131–158.

Oishi, Y., and H. Ichimura. 1979. Grain-boundary enhanced interdiffusion in polycrystalline CaO-stabilized Zirconia system. *Journal of Chemical Physics* 71(12): 5134–5139.

Olague, N.E., D.E. Longsine, J.E. Campbell, and C.D. Leigh. 1991. *User's Manual for the NEFTRAN II Computer Code.* NUREG/CR–5618. Washington, DC: Nuclear Regulatory Commission.

Perfect, D.L., C.C. Faunt, W.C. Steinkampf, and A.K. Turner. 1995. *Hydrochemical Data Base for the Death Valley Region, California and Nevada.* USGS Open-File Report 94–305. Denver, CO: U.S. Geological Survey.

Phillips, O.M. 1996. Infiltration of a liquid finger down a fracture into superheated rock. *Water Resources Research* 32(6): 1665–1670.

Popov, E.P. 1970. *Mechanics of Materials.* New York: Prentice Hall.

Quittmeyer, R.C. 1994. *Seismic Design Inputs for the Exploratory Studies Facility at Yucca Mountain.* BAB000000–01717–5705–00001 REV00. Las Vegas, NV: Civilian Radioactive Waste Management System Management & Operating Contractor.

Rautman, C.A., L.E. Flint, A.L. Flint, and J.D. Istok. 1995. *Physical and Hydrologic Properties of Rock Outcrop Samples from a Nonwelded to Welded Tuff Transition, Yucca Mountain, Nevada.* WRI–95–4061. Denver, CO: U.S. Geological Survey.

Rolfe, S.T., and J.M. Barsom. 1977. *Fracture and Fatigue Control in Structures.* Englewood Cliffs, NY: Prentice Hall.

Sagar, B., ed. 1997. *NRC High-Level Radioactive Waste Program Annual Progress Report: Fiscal Year 1996.* NUREG/CR–6513, No. 1. Washington, DC: U.S. Nuclear Regulatory Commission.

Schenker, A.R., et al. 1995. *Stochastic Hydrogeologic Units and Hydrogeologic Properties Development for Total System Performance Assessments.* SAND 94–0244. Albuquerque, NM: Sandia National Laboratories.

Scott, R.B., and J. Bonk. 1984. *Preliminary Geologic Map (1:12,000 scale) of Yucca Mountain, Nye County, Nevada, with Geologic Cross Sections.* U.S. Geological Survey Open-File Report 84–494. Denver, CO: U.S. Geological Survey.

Seth, M.S., and P.C. Lichtner. 1996. *User's Manual for MULTIFLO: Part 1 METRA 1.0 β Two-Phase Nonisothermal Flow Simulator.* CNWRA 96–005. San Antonio, TX: Center for Nuclear Waste Regulatory Analyses.

Simonds, W.F., et al. 1995. *Map of Fault Activity of the Yucca Mountain Area, Nye County, Nevada.* U.S. Geological Survey Miscellaneous Investigations Series Map, 1-2520. Denver, CO: U.S. Geological Survey.

Sridhar, N., G.A. Cragnolino, and D.S. Dunn. 1993. *Experimental Investigations of Localized Corrosion of High-Level Waste Container Materials.* CNWRA 93–004. San Antonio, TX: Center for Nuclear Waste Regulatory Analyses.

Stothoff, S.A., H.M. Castellaw, and A.C. Bagtzoglou. 1997. *Simulating the spatial distribution of infiltration at Yucca Mountain, Nevada.* Letter Report. San Antonio, TX: Center for Nuclear Waste Regulatory Analyses.

Suzuki, T. 1983. *A Theoretical Model for Dispersion of Tephra. Arc Volcanism: Physics and Tectonics.* Tokyo, Japan: Terra Scientific Publishing: 95–113.

Thompson, B.G.J. 1988. *A Method for Overcoming the Limitation of Conventional Scenario-Based Assessments by Using Monte Carlo Simulation of Possible Future Environmental Changes.* PAAG/DOC/88/11. Paris, France: Nuclear Energy Agency/Organization for Economic Cooperation and Development.

Thompson, B.G.J., and B. Sagar. 1993. The development and application of integrated procedures of post-closure assessment, based upon Monte Carlo simulation: The probabilistic systems assessment (PSA) approach. *Reliability Engineering and System Safety* 42: 125–160.

Timoshenko, S.P., and J.N. Goodier. 1987. *Theory of Elasticity.* New York: McGraw-Hill.

TRW Environmental Safety Systems, Inc. 1995. *Total System Performance Assessment—1995: An Evaluation of the Potential Yucca Mountain Repository.* B00000000–01717–2200–00136, Rev. 01. Las Vegas, NV: TRW Environmental Safety Systems, Inc.

TRW Environmental Safety Systems, Inc. 1996. *Total System Performance Assessment—Viability Assessment (TSPA-VA).* B00000000–01717–2200–00179. Las Vegas, NV: TRW Environmental Safety Systems, Inc.

Turner, D.R. 1998. *Radionuclide Sorption in Fractures at Yucca Mountain, Nevada: A Preliminary Demonstration of Approach for Performance Assessment.* Letter Report. San Antonio, TX: Center for Nuclear Waste Regulatory Analyses.

U.S. Department of Commerce. 1994. *Census of Agriculture: Farm and Ranch Irrigation Survey.* Washington, DC: U.S. Department of Commerce.

U.S. Department of Energy. 1995. *Seismic Design Methodology for a Geologic Repository at Yucca Mountain. Topical Report.* YMP–0030NP. Las Vegas, NV: U.S. Department of Energy.

U.S. Department of Energy. 1996. *Mined Geologic Disposal System Advance Conceptual Design Report.* B00000000–01717–5705–00027, Rev. 00. Washington, DC: U.S. Department of Energy.

U.S. Environmental Protection Agency. 1988. *Limiting Values of Radionuclides Intake and Air Concentration and Dose Conversion Factors for Inhalation, Submersion, and Ingestion.* EPA–520/1–88–020. Washington, DC: U.S. Environmental Protection Agency.

U.S. Environmental Protection Agency. 1993. *Federal Guidance Report No. 12: External Exposure to Radionuclides in Air, Water, and Soil.* EPA 402–R–93–081. Washington, DC: U.S. Environmental Protection Agency, Office of Radiation and Indoor Air.

U.S. Environmental Protection Agency. 1994. *Notice of Proposed Federal Radiation Protection Guidance for Exposure of the General Public.* Federal Register 59:31618. Washington, DC: U.S. Government Printing Office.

U.S. Geological Survey. 1996. *Seismotectonic Framework and Characterization of Faulting at Yucca Mountain, Nevada.* J.W. Whitney, coordinator. Denver, CO: U.S. Geological Survey.

U.S. Nuclear Regulatory Commission. 1997. *Issue Resolution Status Report on Methods to Evaluate Climatic Change and Associated Effects at Yucca Mountain.* Washington, DC: U.S. Nuclear Regulatory Commission.

U.S. Nuclear Regulatory Commission. 1998. *Issue Resolution Status Report Key Techncial Issue: Igneous Activity.* Revision 0. Washington, DC: U.S. Nuclear Regulatory Commission.

van der Leeden, F., F.L. Troise, and D.K. Todd. 1990. *The Water Encyclopedia.* Chelsea, MI: Lewis Publishers.

Vander Voort, G.F. 1990. Embrittlement of steels. *Metals Handbook, Vol. 1, 10th edition. Properties and Selection: Irons, Steels, and High-Performance Alloys.* Materials Park, OH: ASM International.

van Genuchten, R. 1980. A close-form equation for predicting the hydraulic conductivity of unsaturated soils. *Soil Science Society of America Journal* 44(4): 892–898.

Van Wylen, G.J., and R.E. Sonntag. 1978. *Fundamentals of Classical Thermodynamics.* New York: John Wiley and Sons.

Wernicke, B., et al. 1998. Anomalous strain accumulation in the Yucca Mountain area, Nevada. *Science* (279): 2096–2100.

Wescott, R.G., et al. 1995. *NRC Iterative Performance Assessment Phase 2: Development of Capabilities for Review of a Performance Assessment for a High-Level Waste Repository.* NUREG–1464. Washington, DC: U.S. Nuclear Regulatory Commission.

Wilson, C.N. 1990. *Results from NNWSI Series 3 Spent Fuel Dissolution Tests.* PNL–7170. Richland, WA: Pacific Northwest Laboratory.

Wilson, M.L., et al. 1994. *Total-System Performance Assessment for Yucca Mountain-SNL Second Iteration (TSPA-93).* SAND 93–2675, Vols. 1 and 2. Albuquerque, NM: Sandia National Laboratory.

Winograd, I.J., and W. Thordarson. 1975. *Hydrogeologic and Hydrochemical Framework, South-Central Great Basin, Nevada-California, with Special Reference to the Nevada Test Site.* U.S.G.S. Professional Paper 712-C. Washington, DC: U.S. Geological Survey.

Wittwer, C., et al. 1995. *Preliminary Development of the LBL/USGS Three-Dimensional Site Scale Model of Yucca Mountain, Nevada.* LBL–37356, UC–814. Berkeley, CA: Lawrence Berkeley Laboratory.

Wolery, T.W. 1992. *EQ3/6, A Software Package for Geochemical Modeling of Aqueous Systems. Package Overview and Installation Guide Version 7.0.* UCRL–MA110662. Livermore, CA: Lawrence Livermore National Laboratory.

APPENDIX A
REFERENCE DATA SET

The base case (reference) data set (i.e., primary input file) is presented in the following tables. This input data set includes information provided by various key technical issue (KTI) teams. It is expected the base case data set will evolve further as the KTI teams continue to use the code and analyze results.

**

title

Input file *tpa.inp* as supplied with TPA Version 3.1 Code.

Base case data set Rev 3.1.4 4/06/98

```
**
**   ***>>> Disruptive Scenario flags <<<***
**
```

iflag	VolcanismDisruptiveScenarioFlag(yes=1,no=0)	self-explanatory	0	—	**
iflag	FaultingDisruptiveScenarioFlag(yes=1,no=0)	self-explanatory	0	—	**
iflag	SeismicDisruptiveScenarioFlag(yes=1,no=0)	self-explanatory	0	—	**

NUREG–1668

```
**             ***>>> Problem size specification  <<<***

** Number and Location of SubAreas[m] Based on Fig 3.4-1 in TSPA95

**subarea

**]

**ZONE T="ONE RECTANGULAR ZONE Subarea", F=POINT

**      547405.7    4076362.2
**      548469.3    4076362.2
**      548469.3    4079237.8
**      547405.7    4079237.8
**      547405.7    4076362.2

subarea

7

ZONE T="Subarea 1",I=5,F=POINT
547472.0, 4079323.7
548069.2, 4079136.5
547847.3, 4077816.2
547318.4, 4077934.0
547472.0, 4079323.7

ZONE T="Subarea 2",I=5,F=POINT
548069.2, 4079136.5
548609.7, 4078968.6
548547.9, 4077654.1
547847.3, 4077816.2
548069.2, 4079136.5

ZONE T="Subarea 3",I=5,F=POINT
547318.4, 4077934.0
547847.3, 4077816.2
548322.7, 4077192.2
547474.7, 4077281.6
547318.4, 4077934.0
```

ZONE T="Subarea 4",I=5,F=POINT
547847.3, 4077816.2
548547.9, 4077654.1
548504.8, 4077170.0
548322.7, 4077192.2
547847.3, 4077816.2
ZONE T="Subarea 5",I=5,F=POINT
547474.7, 4077282.6
547887.3, 4077238.1
547995.0, 4076338.9
547670.4, 4076435.5
547474.7, 4077282.6
ZONE T="Subarea 6",I=5,F=POINT
547887.3, 4077238.1
548322.7, 4077192.2
548319.5, 4076220.2
547995.0, 4076338.9
547887.3, 4077238.1
ZONE T="Subarea 7",I=5,F=POINT
548322.7, 4077192.2
548504.8, 4077170.0
548473.1, 4076533.7
548319.5, 4076220.2
548322.7, 4077192.2
**
iconstant
StartAtSubarea
1
**
** Stop value should be zero to perform all remaining subareas.
**

NUREG-1668

iconstant

StopAtSubarea

0

**

** Number and Names of Nuclides to be Tracked for Aqueous Pathway

** NOTE: The order of the nuclide names below should not be changed.

**   This order is used in NEFTRAN for decay chain information.

aqueousnuclides

20

Cm246

U238

Cm245

Am241

Np237

Am243

Pu239

Pu240

U234

Th230

Ra226

Pb210

Cs135

I129

Tc99

Ni59

C14

Se79

Nb94

Cl36

constant	SeedForRandomNumber	Seed for random number generator	1889104452.0	---	**
iflag	LatinHypercubeSampling(yes=1,no=0)	Latin Hypercube Sampling flag	1	---	**
iconstant	NumberOfRealizations	Number of realizations	1	---	**
iconstant	StartAtRealization	Index of first realization to be executed.	1	---	**
iconstant	StopAtRealization	Index of last realization to be executed. Should be zero to perform all remaining realizations.	0	---	**
constant	MaximumTime[yr]	Simulation time period	1.0e4	---	**
iconstant	NumberOfTimeSteps	Number of time steps for simulation time period	201	---	**
constant	RatioOfLastToFirstTimeStep	Ratio of last to first time step. When the ration is set as 1, then all time intervals are equal. When the ratio is greater than 1, the time intervals geometrically progress until the maximum time.	100.0	---	**

**
** ***>>> APPEND OPTIONS <<<***
**

iconstant	OutputMode(0=None,1=All,2=UserDefined)	Flag for appending intermediate output to files (0 = no files appended, 1 = all files appended, and 2 = all files for specified realizations)	0	---	**
iconstant	UserDefinedLowerRealizationAppended	For output mode = 2, append files are written beginning with this realization.	2	---	**
iconstant	UserDefinedUpperRealizationAppended	For output mode = 2, append files are written ending with this realization.	3	---	**

uniform	ArealAverageMeanAnnualInfiltrationAtStart[mm/yr]	Areally averaged mean annual infiltration for the initial (current) climate (mm/yr)	1.0, 10.0	Current best estimate by Unsaturated and Saturated Flow under Isothermal Conditions (USFIC) team [see the Issue Resolution Status Report (IRSR) on shallow infiltration, Nuclear Regulatory Commission, 1997)].	**
uniform	MeanAveragePrecipitationMultiplierAtGlacialMaximum	Mean average precipitation (MAP) multiplier at glacial maximum	1.5, 2.5	Best estimate by USFIC team (see the IRSR on climate, Nuclear Regulatory Commission, 1997).	**
uniform	MeanAverageTemperatureIncreaseAtGlacialMaximum[deg C]	Mean average temperature (MAT) increase at glacial maximum (°C)	-10, -5	Best estimate by USFIC team (see IRSR on climate, Nuclear Regulatory Commission, 1997).	**
constant	TimeStepForClimate[yr]	Time step used in the climate model	500.0	Value selected on the basis of efficiency of the TPA Version 3.1.4 code and the rate for climate change	**
constant	StandardDeviationOfMAPAboutMeanInOneTimePeriod[mm/yr]	Standard deviation of MAP about mean in one time period (mm/yr)	0.0	Deviation about mean is not considered necessary at this point and allows efficient execution of the code.	**
constant	StandardDeviationOfMATAboutMeanInOneTimePeriod[degC]	Standard deviation of MAT about mean in one time period (°C)	0.0	Deviation about mean is not considered necessary at this point and allows efficient execution of the code.	**
constant	CorrelationBetweenMAPAndMAT	Correlation between MAP and MAT	-0.8	Best estimate by the USFIC team. Derived from climate expert elicitation (DeWispelare, et al., 1993)	**
iconstant	ClimatePerturbationSet	Options: 1 and 2; currently both sets are identical.	1	-	**

Type	Name	Description	Value	Source	
iflag	TabularTemperatureRHFlag(yes=1,no=0)	Flag to use *tefkti.inp* file containing temperature and relative humidity (RH) data computed external to the TPA Version 3.1.4 code.	0	---	**
iconstant	nsetUsedToPickTempRHDataSet	Selects the data set in *tefkti.inp* that has up to four sets of data (options 1 to 4).	1	---	**
iconstant	UseReflux2	Flag to select reflux model (reflux1 = 0 and reflux2 = 1)	1	---	**
constant	LengthOfRefluxZone[m]	Length of reflux zone (m) in reflux1 model	20	Thermal Effects on Flow (TEF) team best estimate	**
constant	MaximumFluxInRefluxZone[m/s]	Maximum flux in reflux zone (m/s) in reflux1 model	1.0e-9	Center for Nuclear Waste Regulatory Analyses (CNWRA) best estimate	**
constant	PerchedBucketVolumePerSAarea[m3/m2]	Perched bucket volume per subarea (m³/m²) in reflux1 model	0.5	TEF team best estimate	**
constant	Reflux2Thickness	Thickness of dry-out zone (m) in the reflux2 model	100.0	100 m (328 ft) consistent with MULTIFLO simulations of the dry-out zone	**
constant	Reflux2Porosity	Porosity of rock in the reflux zone in the reflux2 model	0.14	CNWRA best estimate	**
constant	Reflux2SatInit	Initial water saturation in the reflux zone in the reflux2 model	0.9	CNWRA best estimate	**
constant	Reflux2SatResid	Residual water saturation in the reflux zone in the reflux2 model	0.1	CNWRA best estimate	**
constant	Reflux2Period	Reflux cycle in the reflux2 model; when this factor has a value of 1, dry-out water recycles each year	100.0	Best estimate, upper value from MULTIFLO runs	**

Type	Variable	Description	Value	Source	
constant	Reflux2LossI	Fraction of infiltration-derived water that escapes each year	0.1	TEF team best estimate	**
constant	Reflux2LossD	Fraction of dry-out zone derived water that escapes the reflux cycle each year	0.1	TEF team best estimate	**
constant	WPLength[m]	Outer length of waste package (WP) (m)	5.682	Doering (1995)	**
constant	WPDiameter[m]	Outer diameter of WP (m)	1.802	Doering (1995)	**
constant	EmplacementDriftDiameter[m]	Emplacement drift diameter (m)	5.0	Civilian Radioactive Waste Management System (CRWMS) (1996)	**
constant	WPSpacingAlongEmplacementDrift[m]	WP spacing (m)	19.0	TRW Environmental Safety Systems, Inc. (1995)	**
constant	ArealMassLoading[MTU/acre]	Areal mass loading (AML) in metric tons of uranium (MTU)/acre	83.0	Civilian Radioactive Waste Management System Management and Operating Contractor (CRWMS M&O) (1994)	**
constant	WastePackagePayload[MTU]	WP payload (MTU/package)	9.76	U.S. Department of Energy (1987)	**
constant	AgeOfWaste[yr]	Age of waste at the time of emplacement	26.0	—	**
constant	AmbientRepositoryTemperature[C]	Ambient repository temperature (°C)	20.0	Assumed in TRW Environmental Safety Systems, Inc. (1995)	**
constant	MassDensityofYMRock[kg/m^3]	Mass density of rock (kg/m³)	2580.0	U.S. Department of Energy (1990)	**
constant	SpecificHeatofYMRock[J/(kg-K)]	Specific heat of rock (J/kg-K)	840.0	U.S. Department of Energy (1990)	**
uniform	ThermalConductivityofYMRock[W/(m-K)]	Thermal conductivity of rock (W/m-K)	1.8, 2.2	U.S. Department of Energy (1993)	**
constant	EmissivityOfDriftWall[-]	Emissivity of drift wall	0.8	Incropera & DeWitt (1995)	**
constant	EmissivityOfWastePackage[-]	Emissivity of WP	0.7	Incropera & DeWitt (1995)	**

constant	ThermalConductivityOfFloor[W/(m C)]	Thermal conductivity of floor (W/m-°C)	0.6	Incropera & DeWitt (1995)	**
constant	EffectiveThermalConductivityOfUnbackfill edDrift[W/(m C)]	Effective thermal conductivity of unbackfilled drift (W/m-°C)	0.90	Manteufel (1997)	**
constant	TimeOfBackfillEmplaced[yr]	Time of backfill emplacement (yr)	10001.0	Civilian Radioactive Waste Management System Management and Operating Contractor (1996)	**
constant	EffectiveThermalConductivityOfBackfill[W/(m C)]	Effective thermal conductivity of backfill (W/m-°C)	0.60	U.S. Department of Energy (1990)	**
constant	ThermalConductivityOfInnerStainlessSteel Wall[W/m C]	Thermal conductivity of inner stainless steel container wall (W/m-°C)	15.0	Incropera & DeWitt (1992)	**
constant	ThermalConductivityOfOuterCarbonSteel Wall[W/m C]	Thermal conductivity of outer carbon steel wall (W/m-°C)	50.0	Incropera & DeWitt (1992)	**
constant	EffectiveThermalConductivityOfBasket&S FinWP[W/(m C)]	Effective thermal conductivity of basket & spent fuel (SF) in WP (W/m-°C)	1.0	Assumed based on Manteufel and Todreas (1994)	**
constant	ElevationOfRepositoryHorizon[m]	Elevation of repository horizon above sea level (m)	1072.0	Civilian Radioactive Waste Management System (1996)	**
constant	ElevationOfGroundSurface[m]	Elevation of ground surface above sea level (m)	1400.0	Civilian Radioactive Waste Management System (1996)	**

constant	OuterWPThickness[m]	Thickness of the outer overpack (m)	0.1	Doering (1995)	**
constant	InnerWPThickness[m]	Thickness of the inner overpack (m)	0.02	Doering (1995)	**
constant	MetalGrainRadius[micrometer]	Average radius of the metal grains constituting the WP outer overpack (μm)	13.75	Container Life and Source Term (CLST) meetings 04/11/97, 04/14/97, and 04/17/97	**
constant	GrainBoundaryThickness[micrometer]	Thickness of grain boundary used in the model for calculating coupled oxygen diffusion along grain boundaries in metal (μm)	7.0e-4	Range about value assumed in Mohanty, et al. (1996); Lobnig, et al. (1992)	**
constant	DryOxidationConstant	Constant relating matrix and grain boundary oxygen diffusivities in metal	0.00001	Assumed in Mohanty, et al. (1996)	**
constant	CriticalRelativeHumidityHumidAirCorrosion	Critical relative humidity above which humid-air corrosion may initiate	0.55	CLST meeting 04/24/97	**
normal	CriticalRelativeHumidityAqueousCorrosion	Critical relative humidity above which aqueous corrosion may initiate	0.75, 0.85	CNWRA staff best estimate	**
uniform	ThicknessOfWaterFilm[m]	Thickness of water film on WP surface (m)	0.001, 0.003	Assumed in Mohanty, et al. (1996)	**
constant	BoilingPointOfWater[C]	Boiling point of water at Yucca Mountain (YM) (°C)	97.0	Manteufel (1997)	**
constant	OuterOverpackErpIntercept	Outer overpack E_{rp} intercept in mV	-620.3	Cragnolino, et al. (1998)	**
constant	TempCoefOfOuterPackErpIntercept	Temperature coefficient of outer overpack E_{rp}	0.47	Cragnolino, et al. (1998)	**
constant	OuterOverpackErpSlope	Outer overpack E_{rp} slope in mV	-95.2	Cragnolino, et al. (1998)	**
constant	TempCoefOfOuterPackErpSlope	Temperature coefficient of outer overpack E_{rp} slope	0.88	Cragnolino, et al. (1998)	**

uniform	InnerOverpackErpIntercept	Inner overpack E_{rp} intercept in mV$_{she}$	1040.0, 1240.0	Sridhar, et al. (1995). Values listed are for Alloy C-22. For Alloy 625, range is 48.5, 148.5. For Alloy 825, range is 372.8, 472.8.	**
constant	TempCoefOfInnerPackErpIntercept	Temperature coefficient of inner overpack E_{rp} intercept in mV/(°C)	0.0	Sridhar, et al. (1995). Values listed are for Alloy 625 and Alloy C-22. For Alloy 825, value is −4.1.	**
constant	InnerOverpackErpSlope	Inner overpack E_{rp} slope in mV	0.0	Sridhar, et al. (1995) Value listed is for Alloy C-22. For Alloy 625, value is -160.8. For Alloy 825, value is -64.0.	**
constant	TempCoefOfInnerPackErpSlope	Temperature coefficient of inner overpack E_{rp} slope in mV/(°C)	0.0	Sridhar, et al. (1995) Values listed are for Alloy C-22 and Alloy 625. For Alloy 825, value is -0.80	**
constant	OuterWPBetaKineticsParameterforOxygen	Transfer coefficient for oxygen reduction reaction (β_{O_2}) for the WP outer overpack	0.75	Assumed, based on Calvo and Schriffrin (1988)	**
constant	OuterWPBetaKineticsParameterforWater	Transfer coefficient for water reduction reaction (β_{H_2O}) for the WP outer overpack	0.5	Assumed, based on Bockris and Reddy (1970)	**
constant	InnerWPBetaKineticsParameterforOxygen	Transfer coefficient for oxygen reduction reaction (β_{O_2}) for the WP inner overpack	0.75	Assumed, based on Calvo and Schriffrin (1988)	**
constant	InnerWPBetaKineticsParameterforWater	Transfer coefficient for water reduction reaction (β_{H_2O}) for the WP inner overpack	0.5	Assumed, based on Bockris and Reddy (1970)	**
constant	OuterWPRateConstantforOxygenReduction[coulomb-m/mole/yr]	Rate constant for oxygen reduction for the WP outer overpack (C-m/yr/mol)	3.8e12	Assumed, based on Bockris and Reddy (1970); Calvo (1979)	**
constant	OuterWPRateConstantforWaterReduction[coulomb-m/m^2/yr]	Rate constant for water reduction for the WP outer overpack (C/m²/yr)	1.6e-1	Assumed, based on Turnbull and Gardner (1982)	**

Type	Variable	Description	Value	Source	
constant	OuterWPActivationEnergyforOxygenReduction[J/mole]	Activation energy for oxygen reduction for WP outer overpack (J/mol)	37300.0	Assumed, based on Calvo (1979)	**
constant	OuterWPActivationEnergyforWaterReduction[J/mole]	Activation energy for water reduction for WP outer overpack (J/mol)	25000.0	Assumed, based on Heusler (1976)	**
constant	InnerWPRateConstantforOxygenReduction[coulomb-m/mole/yr]	Rate constant for oxygen reduction for WP inner overpack (C-m/yr/mol)	3.0e10	Assumed, based on Bockris and Reddy (1970); Calvo (1979)	**
constant	InnerWPRateConstantforWaterReduction[coulomb-m/m^2/yr]	Rate constant for water reduction for WP inner overpack (C/yr/mol)	3.2	Assumed, based on Turnbull and Gardner (1982)	**
constant	InnerWPActivationEnergyforOxygenReduction[J/mole]	Activation energy for oxygen reduction reaction for WP inner overpack (J/mol)	40000.0	Assumed, based on Calvo (1979)	**
constant	InnerWPActivationEnergyforWaterReduction[J/mole]	Activation energy for water reduction reaction for WP inner overpack (J/mol)	25000.0	Assumed, based on Heusler (1976)	**
constant	AA_1_1[C/m2/yr]	Passive current density for WP outer overpack (C/m²/yr)	3.15e5	Assumed, based on Alvarez and Galvele (1984)	**
uniform	AA_2_1[C/m2/yr]	Passive current density for WP outer overpack (C/m²/yr)	2.0e4, 6.3e4	CNWRA experimental data. Values for Alloys C-22, 625, and 825	**
constant	MeasuredGalvanicCouplePotential	Experimentally measured galvanic couple potential carbon steel and Alloy 625	-0.46	Scully and Hack, 1984	**
uniform	CoefForLocCorrOfOuterOverpack	Coefficient for localized corrosion rate of outer overpack	8.66e-4, 8.66e-3	Marsh and Taylor (1988)	**
constant	ExponetForLocCorrOfOuterOverpack	Exponent for localized corrosion rate of outer overpack	0.45	Marsh and Taylor (1988)	**
constant	HumidAirCorrosionRate[m/yr]	Humid air corrosion rate (m/yr)	1.16e-5	Assumed	**
constant	LocalizedCorrRateOfInnerOverpack[m/yr]	Corrosion rate for localized corrosion of inner overpack	2.5e-4	Assumed in Mohanty, et al. (1997). 1997 Sensitivity Analysis indicates low sensitivity.	**
constant	FractionalCouplingStrength	Efficiency factor, varying from 0 to 1, representing galvanic coupling between the outer and inner overpack not complete sent	0.0	Switch: full credit = 1; no credit = 0. 1997 Sensitivity Analysis indicates high sensitivity. Continue evaluating.	**

constant	FactorForDefiningChoiceOfCritPotential	Factor for defining choice of critical potential (initiation or repassivation)	0.0	Mohanty, et al. (1997)	**
constant	CritChlorideConcForFirstLayer[moL/L]	Critical chloride concentration for localized corrosion of outer overpack	3.0e-4	Galvele (1978)	**
constant	CritChlorideConcForSecondLayer[moL/L]	Critical chloride concentration for localized corrosion of outer overpack	1.0	Sridhar, et al. (1995). Value listed is for Alloy C-22. Value for Alloy 625 is 3.0e-2. Value for Alloy 825 is 2.0e-3	**
uniform	ChlorideMultFactor	Factor for changing chloride concentration	1.0, 30.0	Assumed. Note solubility limit for NaCl = 6 mol/liter. 1997 Sensitivity Analysis indicates high sensitivity. Continue evaluating.	**
constant	ReferencepH	pH of water (does not change).	9.0	Assumed in Mohanty, et al. (1997); based on MULTIFLO calculations.	**
constant	WPsurfaceScaleThickness[m]	Scale deposit on WP surface	0.0	A value of 0.0 implies no scale deposit.	**
constant	TortuosityOfScaleonWP	Tortuosity of a porous layer scale deposited on the WP. Thickness does not change with time.	1.0	In the case of no deposit, the value is 1.0.	**
constant	PorosityOfScaleonWP	Porosity of the layer deposited on the WP. Porosity does not change with time.	1.0	In the case of no deposit, the value is 1.0.	**
constant	YieldStrength[MPa]	Yield strength of outer overpack	205.0	American Society for Testing Material (1995)	**
constant	SafetyFactor	Safety factor for residual stresses	1.4		**
constant	FractureToughness[MPa-m**0.5]	Fracture toughness of outer overpack	250.0	Assumed	**

NUREG-1668

Type	Name	Description	Value	Notes	
hazardcurve	SeismicHazardCurveforSEISMO	Minimum peak ground acceleration for bins, return period (yr); number of magnitudes for recurrence of seismic events, magnitude, and recurrence time	4 0.1 500.0 0.2 2400.0 0.3 8000.0 0.4 20000.0	Wong, et al. (1995), Figure 5, p. 59	**
constant	WeightPercentageOfRockFallThatHitsWPforSEISMO	Weight percentage of rock fall that hits WP	1.0	Not used	**
constant	WeightOfWPforSEISMO[N]	Weight of WP used in impact calculation during free fall (N)	1.27D05	Not used	**
constant	WPStiffnessforSEISMO[Pa*m]	WP stiffness used for rockfall impact calculation (Pa·m)	1.21D10	Calculated based on a simple supported beam with load applied at the center of the beam using the following equation: $K = 48\ E\ I/L^3$ where: K is the stiffness of the WPs; $I = \pi\ R^3t$; E is the Young's modulus of the WPs; L is the length of the WPs; R is the average of inner and outer radii of the WPs; t is the thickness of the WPs.	**
constant	WPModulusOfElasticityforSEISMO[Pa]	WP modulus of elasticity (Pa)	2.07D11	Popov (1970), Appendix Table 1, p. 423	**

normal	RockModulusOfElasticityforSEISMO[Pa]	Rock modulus of elasticity (Pa)	2.76D10, 4.14D10	Brechtel, et al. (1995), Table 2-4, p. 2-12	**
constant	WPPoissonRatioforSEISMO[]	WP Poisson ratio	0.2D0	Popov (1970), p. 36	**
normal	RockPoissonRatioforSEISMO[]	Rock Poisson ratio	0.15, 0.25	Brechtel, et al. (1995), Table 2-4, p. 2-12 supplies the mean value for the parameter. The 25% variation for the assumed normal distribution of this parameter is adopted from Table 5.1, p. 5-19 of the DOE Site Characterization Plan (U.S. Department of Energy, 1988).	**
constant	RockFallingDistanceforSEISMO[m]	Rock falling distance gap between the top of WP and the drift crown (m)	2.0D0	Estimated based on the diameter of the emplacement drift minus the diameter of the WPs minus the height of the pedestal	**
constant	WPFallingDistanceforSEISMO[m]	WP falling distance for free fall calculations (m)	0.3D0	Not currently used in this version of TPA	**
iconstant	WPNumberofSupportPairforSEISMO	Number of support pairs on the WP pedestal	2.0	The conceptual SEISMO module is based on an assumption that WPs can be represented as simply supported beams. Consequently, two support beams are assumed.	**
constant	WPSupportStiffnessforSEISMO[pa m]	Stiffness of the pedestal	5.5D09	Calculated using the following equation: $K = A\,E/L$, where: K is the support stiffness; A is the cross-sectional area of the support; E is the Young's modulus of the support; L is the height of the support.	**
constant	DistributionJointSpacing1forSEISMO	Fractional areal coverage for rock condition 1	5.0D-03	Based on Brechtel, et al. (1995), Fig. 7-3, p. 7-9, rock conditions 1, 2, and 3 take up only 1.5% of the area in total and are each equally subdivided into 0.5% of the repository.	**
constant	DistributionJointSpacing2forSEISMO	Fractional areal coverage for rock condition 2	5.0D-03	Based on Brechtel, et al. (1995), Fig. 7-3, p. 7-9, rock conditions 1, 2, and 3 take up only 1.5% of the area in total and are each equally subdivided into 0.5% of the repository.	**

constant	DistributionJointSpacing3forSEIS MO	Fractional areal coverage for rock condition 3	5.0D-03	Based on Brechtel, et al. (1995), Fig. 7-3, p. 7-9, rock conditions 1, 2, and 3 take up only 1.5% of the area in total and are each equally subdivided into 0.5% of the repository.	**
constant	DistributionJointSpacing4forSEIS MO	Fractional areal coverage for rock condition 4	0.629D0	Based on Brechtel, et al. (1995), Fig. 7-3, p. 7-9, about 62.9% of the repository can be characterized as rock condition 4.	**
constant	DistributionJointSpacing5forSEIS MO	Fractional areal coverage for rock condition 5	0.356D0	Based on Brechtel, et al. (1995), Fig. 7-3, p. 7-9, rock condition 5 occupies about 35.6% of the repository.	**
normal	SEISMOJointSpacing1[m]	Joint spacing (JS) for rock condition 1	0.466, 0.600	Not all rocks falling from the roof of the emplacement drift will impact WPs. The effective size of the rock that impacts WPs will be controlled by JS. For abstraction into the TPA Version 3.1.4 code, the TSw2 thermomechanical unit was subdivided into five distinct rock conditions. The rock conditions are estimated using available JS (Brechtel, et al., 1995) information for the TSw2 provided in Brechtel, et al. (1995), which summarizes data collected from NRG holes. In the report, the JS information is presented in terms of rating of JS (Hoek and Brown, 1982). Three JS ratings are observed in the TSw2 Unit: 5, 8, and 10. JS rating of 5 indicates a JS that is smaller than 0.06 m; JS rating of 8 is for JS between 0.06 m to 0.2 m; and JS rating of 10 represents JS of 0.2 to 0.6 m. For the rock condition scheme used in the SEISMO module, rocks related to JS rating of 10 are further subdivided into three conditions: conditions 1, 2, and 3 respectively. This subdivision is considered reasonable because: (i) the range for JS rating of 10 between 0.2 m to 0.6 m is large; and (ii) larger JS will induce greater impact failures on WPs. A normal distribution of JS is assumed in each rock condition category because no additional information about the distributions of JS' was compiled when the SEISMO module was coded. Future work will examine the validity of this assumption.	**

| normal | SEISMOJointSpacing2[m] | JS for rock condition 2 | 0.333, 0.466 | Not all rocks falling from the roof of the emplacement drift will impact WPs. The effective size of the rock that impacts WPs will be controlled by JS. For abstraction into the TPA Version 3.1.4 code, the TSw2 thermomechanical unit was subdivided into five distinct rock conditions. The rock conditions are estimated using available JS (Brechtel, et al., 1995) information for the TSw2 provided in Brechtel, et al. (1995), which summarizes data collected from NRG holes. In the report, the JS information is presented in terms of rating of JS (Hock and Brown, 1982). Three JS ratings are observed in the TSw2 Unit: 5, 8, and 10. JS rating of 5 indicates a JS that is smaller than 0.06 m; JS rating of 8 is for JS between 0.06 to 0.2 m; and JS rating of 10 represents JS of 0.2 to 0.6 m. For the rock condition scheme used in the SEISMO module, rocks related to JS rating of 10 are further subdivided into three conditions: conditions 1, 2, and 3 respectively. This subdivision is considered reasonable because: (i) the range for JS rating of 10 between 0.2 m to 0.6 m is large; and (ii) larger JS will induce greater impact failures on WPs. A normal distribution of JS is assumed in each rock condition category, because no additional information about the distributions of JS was compiled when the SEISMO module was coded. Future work will examine the validity of this assumption. | ** |

| normal | SEISMOJointSpacing3[m] | 0.20, 0.333 | Not all rocks falling from the roof of the emplacement drift will impact WPs. The effective size of the rock that impacts WPs will be controlled by JS. For abstraction into the TPA Version 3.1.4 code, the TSw2 thermomechanical unit was subdivided into five distinct rock conditions. The rock conditions are estimated using available JS information for the TSw2 provided in Brechtel, et al. (1995), which summarizes data collected from NRG holes. In the report, the JS information is presented in terms of rating of JS (Hoek and Brown, 1982). Three JS ratings are observed in the TSw2 Unit: 5, 8, and 10. JS rating of 5 indicates a JS that is smaller than 0.06 m; JS rating of 8 is for JS between 0.06 to 0.2 m; and JS rating of 10 represents JS of 0.2 to 0.6 m. For the rock condition scheme used in the SEISMO module, rocks related to JS rating of 10 are further subdivided into three conditions: conditions 1, 2, and 3, respectively. This subdivision is considered reasonable because: (i) the range for JS rating of 10 between 0.2 m to 0.6 m is large; and (ii) larger JS will induce greater impact failures on WPs. A normal distribution of JS is assumed in each rock condition category because no additional information about the distributions of JS was compiled when the SEISMO module was coded. Future work will examine the validity of this assumption. | ** |
| normal | SEISMOJointSpacing4[m] | 0.06, 0.20 | Not all rocks falling from the roof of the emplacement drift will impact WPs. The effective size of the rock that impacts WPs will be controlled by JS. For abstraction into the TPA Version 3.1.4 code, the TSw2 thermomechanical unit was subdivided into five distinct rock conditions. The rock conditions are estimated using available JS information for the TSw2 provided in Brechtel, et al. (1995), which summarizes data collected from NRG holes. In the report, the JS information is presented in terms of rating of JS (Hoek and Brown, 1982). Three JS ratings are observed in the TSw2 Unit: 5, 8, and 10. JS rating of 5 indicates a JS that is smaller than 0.06 m; JS rating of 8 is for JS between 0.06 to 0.2 m; and JS rating of 10 represents JS of 0.2 to 0.6 m. For the rock condition scheme used in the SEISMO module, condition 4 is assumed to be the rocks with JS rating of 8. A normal distribution of JS is assumed in each rock condition category, because no additional information about the distributions of JS was compiled when the SEISMO module was coded. Future work will examine the validity of this assumption. | ** |

normal	SEISMOJointSpacing5[m]	JS for rock condition 5	0.03, 0.06	Not all rocks falling from the roof of the emplacement drift will impact WPs. The effective size of the rock that impacts WPs will be controlled by JS. For abstraction into the TPA Version 3.1.4 code, the TSw2 thermomechanical unit was subdivided into five distinct rock conditions. The rock conditions are estimated using available JS information for the TSw2 provided in Brechtel, et al. (1995), which summarizes data collected from NRG holes. In the report, the JS information is presented in terms of rating of JS (Hoek and Brown, 1982). Three JS ratings are obsérvied in the TSw2 Unit: 5, 8, and 10. JS rating of 5 indicates a JS that is smaller than 0.06 m; JS rating of 8 is for JS between 0.06 to 0.2 m; and JS rating of 10 represents JS of 0.2 to 0.6 m. For the rock condition scheme used in the SEISMO module, condition 5 is assumed to be equivalent to the rock with JS rating of 5. A normal distribution of JS is assumed in each rock condition category because no additional information about the distributions of JS was compiled when the SEISMO module was coded. Future work will examine the validity of this assumption.	**
normal	WPUltimateStrength[N/m²]	Ultimate strength of WP (N/m²)	1.103448D9, 1.655172D9	Assumed based on the strength of SS carbon steel (3.5% Ni and 0.4% C). The mean value of this parameter is obtained from the Appendix, Table 1, p. 423 of Popov (1970). The 20-percent variation for the assumed normal distribution is based on the expert judgment of the authors.	**
constant	GrainDensityforTSw2SEISMO[g/cm²]	Grain density for Topopah Springs—welded (g/cm³)	2.55	Brechtel, et al. (1995) Table 2-4, p. 2-12	**
lognormal	SeismicHeterogeneityFactor	Factor that accounts for the potential effect of heterogeneity in each rock condition	1.0D-5, 0.5	CNWRA best estimate	**

					**
iflag	FlowThroughModelFlag	iflag = 0 for bathtub and iflag=1 for flow-through	0		**
lognormal	FowFactor	Flow convergence/divergence factor	0.01, 3.0	See Appendix B.	**
lognormal	FmultFactor	Flow multiplication factor	0.01, 0.2	See Appendix B.	**
uniform	SubAreaWetFraction	Subarea wet fraction	0.0, 1.0	See Appendix B.	**
uniform	DefectiveFractionOfWPs/cell	Fraction of total WPs in an SA that fails at time t = 0	1.0e-4, 1.0e-2	No available literature; based on staff discussions with Electric Power Research Institute (EPRI) and National Institute of Standards and Technology (NIST).	**
constant	WPInternalVolume[m3]	Internal WP volume where water can reside (m³)	4.83	Assumed in Mohanty, et al. (1996)	**
constant	FlowOnsetTemperature[C]	Flow onset temperature	999	Deliberately set at a large value to allow flow into the WP, starting at time 0.0 if water is available.	**
constant	SFDensity[kg/m3]	Spent fuel density (kg/m³)	10600	Wilson (1990); based on UO_2	**
iconstant	SurfaceAreaModel	Selection of model for computing surface area of SF (options 1 and 2)	1	Mohanty, et al. (1997). 1997 Sensitivity Analysis indicates high sensitivity. Continue evaluating.	**
iconstant	IModel	Selection of model for computing SF dissolution (1=absence of Ca and Si, 2=presence of Ca and Si, 3=user-specified).	2	Mohanty, et al. (1997). 1997 Sensitivity Analysis indicates high sensitivity. Continue evaluating.	**
constant	OxygenPartialPressure[atm]	Oxygen partial pressure (over pressure) (atm)	0.21	Assumed in Mohanty, et al. (1996). 1997 Sensitivity Analysis indicates high sensitivity. Continue evaluating. Based on atmospheric composition. Stockman (1997).	**

constant	NegativeLog10CarbonateConcentration[mol/L]	Negative $\log_{10}$ of carbonate concentration in surrounding water (mol/L)	6.71	Equilibrated with $CaCO_3$ and based on MULTIFLO calculations; no literature data available. 1997 Sensitivity Analysis indicates low sensitivity.	**
constant	UserLeachRate[kg/yr/m2]	User-provided leaching rate (kg/yr/m²) value 3×10^{-3} only if IModel = 3	3.0e−3	Based on a value for the oxidation rate at the Peña Blanca natural analog (Murphy, et al., 1997) scaled to the mass of uranium in the repository, basecase surface area per container, and number of containers in the repository.	**
uniform	SFWettedFraction	Water level inside the WP expressed as a fraction of the WP internal diameter	0.0, 1.0	Assumed in Mohanty, et al. (1996)	**
constant	RD_Backfill_Cm	Retardation factor of Cm in backfill surrounding the WP	10.0	Assumed in Nuclear Regulatory Commission (1995), p. A-22	**
constant	RD_Backfill_Pu	Retardation factor of Pu in backfill surrounding the WP	10.0	Assumed in Nuclear Regulatory Commission (1995), p. A-23	**
constant	RD_Backfill_U	Retardation factor of U in backfill surrounding the WP	1.0	Assumed in Nuclear Regulatory Commission (1995), p. A-22	**
constant	RD_Backfill_Am	Retardation factor of Am in backfill surrounding the WP	20.0	Assumed in Nuclear Regulatory Commission (1995), p. A-23	**
constant	RD_Backfill_Np	Retardation factor of Np in backfill surrounding the WP	1.0	Assumed in Nuclear Regulatory Commission (1995), p. A-23	**
constant	RD_Backfill_Th	Retardation factor of Th in backfill surrounding the WP	10.0	Assumed in Nuclear Regulatory Commission (1995), p. A-23	**
constant	RD_Backfill_Ra	Retardation factor of Ra in backfill surrounding the WP	30.0	Assumed in Nuclear Regulatory Commission (1995), p. A-23	**
constant	RD_Backfill_Pb	Retardation factor of Pb in backfill surrounding the WP	5.0	Assumed in Nuclear Regulatory Commission (1995), p. A-23	**
constant	RD_Backfill_Cs	Retardation factor of Cs in backfill surrounding the WP	8.0	Assumed in Nuclear Regulatory Commission (1995), p. A-23	**

constant	RD_Backfill_I	Retardation factor of I in backfill surrounding the WP	1.0	Assumed in Nuclear Regulatory Commission (1995), p. A-23	**
constant	RD_Backfill_Tc	Retardation factor of Tc in backfill surrounding the WP	1.0	Assumed in Nuclear Regulatory Commission (1995), p. A-23	**
constant	RD_Backfill_Ni	Retardation factor of Ni in backfill surrounding the WP	2.0	Assumed in Nuclear Regulatory Commission (1995), p. A-23	**
constant	RD_Backfill_Cl	Retardation factor of Cl in backfill surrounding the WP	1.0	Based on retardation factor for I	**
constant	RD_Backfill_C	Retardation factor of C in backfill surrounding the WP	1.0	Assumed in Nuclear Regulatory Commission (1995), p. A-23	**
constant	RD_Backfill_Se	Retardation factor of Se in backfill surrounding the WP	1.0	Assumed in Nuclear Regulatory Commission (1995), p. A-23	**
constant	RD_Backfill_Nb	Retardation factor of Nb in backfill surrounding the WP	1.0	Assumed in Nuclear Regulatory Commission (1995), p. A-23	**
constant	GapFractionForCM246	Cm-246/kg of SF in grain and pellet cladding gap	0.0	Not used	**
constant	GapFractionForU238	U-238/kg of SF in grain and pellet cladding gap	0.0	Not used	**
constant	GapFractionForCM245	Cm-245/kg of SF in grain and pellet cladding gap	0.0	Not used	**
constant	GapFractionForAM241	Am-241/kg of SF in grain and pellet cladding gap	0.0	Not used	**
constant	GapFractionForNP237	Np-237/kg of SF in grain and pellet cladding gap	0.0	Not used	**
constant	GapFractionForAM243	Am-243/kg of SF in grain and pellet cladding gap	0.0	Not used	**
constant	GapFractionForPU239	Pu-239/kg of SF in grain and pellet cladding gap	0.0	Not used	**
constant	GapFractionForPU240	Pu-240/kg of SF in grain and pellet cladding gap	0.0	Not used	**

constant	GapFractionForU234	U-234/kg of SF in grain and pellet cladding gap	0.0	Not used	**
constant	GapFractionForTH230	Th-230/kg of SF in grain and pellet cladding gap	0.0	Not used	**
constant	GapFractionForRA226	Ra-226/kg of SF in grain and pellet cladding gap	0.0	Not used	**
constant	GapFractionForPB210	Pb-210/kg of SF in grain and pellet cladding gap	0.0	Not used	**
constant	GapFractionForCS135	Cs-135/kg of SF in grain and pellet cladding gap	0.0	Not used	**
constant	GapFractionForI129	I-129/kg of SF in grain and pellet cladding gap	0.0	Not used	**
constant	GapFractionForTC99	Tc-99/kg of SF in grain and pellet cladding gap	0.0	Not used	**
constant	GapFractionForNI59	Ni-59/kg of SF in grain and pellet cladding gap	0.0	Not used	**
constant	GapFractionForCL36	Cl-36/kg of SF in grain and pellet cladding gap	0.0	Not used	**
constant	GapFractionForC14	C-14/kg of SF in grain and pellet cladding gap	0.0	Not used	**
constant	GapFractionForSE79	Se-79/kg of SF in grain and pellet cladding gap	0.0	Not used	**
constant	GapFractionForNB94	Nb-94/kg of SF in grain and pellet cladding gap	0.0	Not used	**
normal	InitialRadiusOfSFParticle[m]	Initial radius of UO_2 particle (m)	7.0e-4, 3.0e-3	Guenther, et. al. (1991); Belle (1961). 1997 Sensitivity Analysis indicates high sensitivity. Continue evaluating.	**
constant	RadiusOfSFGrain[m]	Radius of UO_2 grain (m)	1.25e-5	Einziger, et al. (1992); Belle (1961). 1997 Sensitivity Analysis indicates low sensitivity.	**

constant	CladdingCorrectionFactor	Cladding correction factor	1.0		**
normal	SubGrainFragmentRadiusAfterTransFrac[m]	Subgrain fragment radius of UO$_2$ particle after transgranular fracture (m); used only if fuel conversion takes place from UO$_2$, to UO$_{2.4}$ and U$_3$O$_8$; used only by the SF dissolution models that are dependent on exposed surface area.	5.0e-7, 2.0e-6	Mohanty, et al. (1997). Distribution based on distribution of grains or particles. 1997 Sensitivity Analysis indicates high sensitivity. Continue evaluating.	**
constant	ThicknessOfCladding[m]	Thickness of cladding (m)	6.1e-4	Smith & Baldwin (1989)	**
constant	SFC-14InventoryPerKgSF[ci]	(not used)	7.2e-4	Assumed in Mohanty, et al. (1996)	**
constant	CladC-14InventoryPerKgSF[ci]	(not used)	4.89e-4	Park (1992)	**
constant	ZyrOxideAndCrudC-14InvPerKgSF[ci]	(not used)	2.48e-5	Park (1992)	**
constant	GapAndGrainBoundaryInventoryPerKgSF[ci]	(not used)	6.2e-6	Park (1992)	**
uniform	SolubilityAm[kg/m3]	Solubility limit for Am	2.4e-8, 2.4e-4	TRW Environmental Safety Systems, Inc. (1995), Table 6.3-1, based on expert elicitation described in Gauthier (1993), and Nitsche, et al. (1993). 1997 Sensitivity Analysis indicates high sensitivity. Continue evaluating.	**
logtriangular	SolubilityNp[kg/m3]	Solubility limit for Np	1.2e-3, 3.4e-2, 2.4e-1	TRW Environmental Safety Systems, Inc. (1995), Table 6.3-1, based on expert elicitation described in Gauthier (1993); Nitsche, et al. (1993); and Dyer (1993).* The beta distribution recommended in TSPA-95 is approximated by a triangular distribution. 1997 Sensitivity Analysis indicates high sensitivity. Continue evaluating.	**
constant	Solubility_I[kg/m3]	Solubility limit for I	1.29e2	TRW Environmental Safety Systems, Inc. (1995), p. 6-7. Based on the assumption that there is no solubility controlling phase for iodine. Conservatively assume 1.0 mol/L.	**

constant	SolubilityTc[kg/m3]	9.93e1	Based on CNWRA assumption that there is no solubility controlling phase for technetium. Conservatively assume 1.0 mol/L.	**
constant	SolubilityCl[kg/m3]	3.6e1	TRW Environmental Safety Systems, Inc. (1995), p. 6-7. Based on assumption that there is no solubility controlling phase for iodine. Conservatively assume 1.0 mol/L.	**
constant	Solubility_C[kg/m3]	1.4e1	TRW Environmental Safety Systems, Inc. (1995), p. 6-7. Based on assumption that there is no solubility controlling phase for carbon (Gauthier, 1993). Conservatively assume 1.0 mol/L.	**
constant	Solubility_U[kg/m3]	7.6e-3	TRW Environmental Safety Systems, Inc. (1995), Table 6.3-1, based on expert elicitation described in Gauthier (1993), and Wanner and Forest (1992). 1997 Sensitivity Analysis indicates low sensitivity. The apex of the logtriangular distribution (2.4e-6 to 2.4e0) used in the sensitivity studies was chosen.	**
constant	SolubilityCm[kg/m3]	2.4e-4	TRW Environmental Safety Systems, Inc. (1995), Table 6.3-1, based on expert elicitation described in Gauthier (1993). Assumed that curium behaves similarly to Am^{3+}. (See Fuger, 1992). 1997 Sensitivity Analysis indicates low sensitivity. The maximum value of the uniform distribution (2.4e-8 to 2.4e-4) used in the sensitivity studies was chosen.	**
uniform	SolubililityPu[kg/m3]	2.4e-6, 2.4e-4	TRW Environmental Safety Systems, Inc. (1995), Table 6.3-1, based on expert elicitation described in Gauthier (1993); Niusche, et al. (1993); and Dyer (1993).[1] 1997 Sensitivity Analysis indicates high sensitivity. Continue evaluating.	**

constant	SolubilityTh[kg/m3]	2.3e-4	Based on solubility data in carbonate-containing systems (Östhols, et al. (1994), Table 3 and Fig. 6; Rai, et al. (1995), Figures 1 & 3 (and appendix). 1997 Sensitivity Analysis indicates low sensitivity. The mean value of the loguniform distribution (2.3e-7 to 2.3e-1) used in the sensitivity studies was chosen.	**
constant	SolubilityRa[kg/m3]	2.3e-5	TRW Environmental Safety Systems, Inc. (1995), Table 6.3-1, based on expert elicitation described in Gauthier (1993), and Kerrisk (1984). 1997 Sensitivity Analysis indicates low sensitivity. The apex of the distribution of the logtriangular distribution (2.3e-7 to 2.3e-3) used in the sensitivity studies was chosen.	**
constant	SolubilityPb[kg/m3]	6.6e-5	TRW Environmental Safety Systems, Inc. (1995), Table 6.3-1, based on expert elicitation described in Gauthier (1993); Andersson (1988); and Pei-Lin, et al. (1985). 1997 Sensitivity Analysis indicates low sensitivity. The apex of the logtriangular distribution (2.1e-6 to 2.1e-3) used in the sensitivity studies was chosen.	**
constant	SolubilityCs[kg/m3]	1.35e2	Based on CNWRA assumption that there is no solubility-controlling phase for cesium. Conservatively assume 1.0 mol/L.	**
constant	SolubilityNi[kg/m3]	1.1e-1	TRW Environmental Safety Systems, Inc. (1995), Table 6.3-1, based on expert elicitation described in Gauthier (1993); Andersson (1988); and Siegel, et al. (1993). 1997 Sensitivity Analysis indicates low sensitivity. The apex of the logtriangular distribution (5.9e-5 to 5.9e0) used in the sensitivity studies was chosen.	**

| constant | SolubilitySe[kg/m3] | Solubility limit for Se | 7.9e1 | Based on CNWRA assumption that release of selenium is controlled by waste form dissolution. High value of 1.0 mol/L is used. | ** |
| constant | SolubilityNb[kg/m3] | Solubility limit for Nb | 9.3e-7 | TRW Environmental Safety Systems, Inc. (1995), Table 6.3-1, based on expert elicitation described in Gauthier (1993), and Andersson (1988). 1997 Sensitivity Analysis indicates low sensitivity. The mean value of the loguniform distribution (9.3e-8 to 9.3e-6) used in the sensitivity studies was chosen. | ** |

[1] Dyer, J.R. 1993. *Radionuclide Solubility Working Group (SolWOG) Meeting Report (SCP:N/A),* Letter (August 2) to Lawrence Livermore National Laboratory and Los Alamos National Laboratory. Washington, DC: U.S. Department of Energy.

**
**   ***>>> UZFT <<<***
**

constant	UnsaturatedZoneMinimumVelocityChangeFactor[Fraction]	Same velocity is used unless the velocity changes by this fraction of the initial velocity. However, the minimum time span over which velocity does not change is 500 yr.	0.4	Same as in base case. Current best estimate by the unsaturated and saturated zone flow under isothermal condition team	**
constant	MatrixLongitudinalDispersivity[FractionOfLayer]	Maximum matrix longitudinal dispersivity specified as a fraction of layer thickness	0.1	Same as in base case. Current best estimate by the USFIC team	**
constant	FractureLongitudinalDispersivity[FractionOfLayer]	Maximum fracture longitudinal dispersivity specified as a fraction of layer thickness	0.1	Same as in base case. Current best estimate by the USFIC team	**
loguniform	MatrixKD_TSw_Am[m3/kg]	Matrix K_d of Am for Topopah Springs—welded [m³/kg] (UZ)	0.100, 2.0	Triay, et al. (1997)	**
loguniform	MatrixKD_CHnvAm[m3/kg]	Matrix K_d of Am for Calico Hills—nonwelded vitric [m³/kg] (UZ)	0.100, 1.0	Triay, et al. (1997)	**
loguniform	MatrixKD_CHnzAm[m3/kg]	Matrix K_d of Am for Calico Hills—nonwelded zeolitic [m³/kg] (UZ)	0.100, 1.0	Triay, et al. (1997)	**
loguniform	MatrixKD_PPw_Am[m3/kg]	Matrix K_d of Am for Prow Pass—welded [m³/kg] (UZ)	0.100, 2.0	Triay, et al. (1997)	**
loguniform	MatrixKD_UCF_Am[m3/kg]	Matrix K_d of Am for Upper Crater Flat [m³/kg] (UZ)	0.100, 2.0	Triay, et al. (1997)	**
loguniform	MatrixKD_BFw_Am[m3/kg]	Matrix K_d of Am for Bullfrog—welded [m³/kg] (UZ)	0.100, 2.0	Triay, et al. (1997)	**
loguniform	MatrixKD_UFZ_Am[m3/kg]	Matrix K_d of Am for Unsaturated Fracture Zone [m³/kg]	0.100, 2.0	Triay, et al. (1997)	**
loguniform	MatrixKD_TSw_Np[m3/kg]	Matrix K_d of Np Topopah Springs—welded [m³/kg] (UZ)	6.0e-8, 0.006	Triay, et al. (1997). A value 5 orders of magnitude smaller than the maximum value was used as the lower bound.	**

loguniform	MatrixKD_CHnvNp[m3/kg]	Matrix K_d of Np for Calico Hills—nonwelded vitric [m³/kg] (UZ)	1.5e-7, 0.015	Triay, et al. (1997). A value 5 orders of magnitude smaller than the maximum value was used as the lower bound.	**
loguniform	MatrixKD_CHnzNp[m3/kg]	Matrix K_d of Np for Calico Hills—nonwelded zeolitic [m³/kg] (UZ)	3.0e-8, 0.003	Triay, et al. (1997). A value 5 orders of magnitude smaller than the maximum value was used as the lower bound.	**
loguniform	MatrixKD_PPw_Np[m3/kg]	Matrix K_d of Np for Prow Pass—welded [m³/kg] (UZ)	6.0e-8, 0.006	Triay, et al. (1997). A value 5 orders of magnitude smaller than the maximum value was used as the lower bound.	**
loguniform	MatrixKD_UCF_Np[m3/kg]	Matrix K_d of Np for Upper Crater Flat [m³/kg] (UZ)	6.0e-8, 0.006	Triay, et al. (1997). A value 5 orders of magnitude smaller than the maximum value was used as the lower bound.	**
loguniform	MatrixKD_BFw_Np[m3/kg]	Matrix K_d of Np for Bullfrog—welded [m³/kg] (UZ)	6.0e-8, 0.006	Triay, et al. (1997). A value 5 orders of magnitude smaller than the maximum value was used as the lower bound.	**
loguniform	MatrixKD_UFZ_Np[m3/kg]	Matrix K_d of Np for Unsaturated Fracture Zone [m³/kg] (UZ)	6.0e-8, 0.006	Triay, et al. (1997). A value 5 orders of magnitude smaller than the maximum value was used as the lower bound.	**
constant	MatrixKD_TSw_I[m3/kg]	Matrix K_d of I for Topopah Springs—welded [m³/kg] (UZ)	0.0	Triay, et al. (1997)	**
constant	MatrixKD_CHnvI[m3/kg]	Matrix K_d of I for Calico Hills—nonwelded vitric [m³/kg] (UZ)	0.0	Triay, et al. (1997)	**
constant	MatrixKD_CHnzI[m3/kg]	Matrix K_d for I for Calico Hills—nonwelded zeolitic [m³/kg] (UZ)	0.0	Triay, et al. (1997)	**
constant	MatrixKD_PPw_I[m3/kg]	Matrix K_d of I for Prow Pass—welded [m³/kg] (UZ)	0.0	Triay, et al. (1997)	**
constant	MatrixKD_UCF_I[m3/kg]	Matrix K_d of I for Upper Crater Flat [m³/kg] (UZ)	0.0	Triay, et al. (1997)	**
constant	MatrixKD_BFw_I[m3/kg]	Matrix K_d of I for Bullfrog—welded [m³/kg] (UZ)	0.0	Triay, et al. (1997)	**

constant	MatrixKD_UFZ_I[m3/kg]	Matrix K_d of I for Unsaturated Fracture Zone [m^3/kg] (UZ)	0.0	Triay, et al. (1997)	**
constant	MatrixKD_TSw_Tc[m3/kg]	Matrix K_d of Tc for Topopah Springs—welded [m^3/kg] (UZ)	0.0	Triay, et al. (1997)	**
constant	MatrixKD_CHnvTc[m3/kg]	Matrix K_d of Tc for Calico Hills—nonwelded vitric [m^3/kg] (UZ)	0.0	Triay, et al. (1997)	**
constant	MatrixKD_CHnzTc[m3/kg]	Matrix K_d of Tc for Calico Hills—nonwelded zeolitic [m^3/kg] (UZ)	0.0	Triay, et al. (1997)	**
constant	MatrixKD_PPw_Tc[m3/kg]	Matrix K_d of Tc for Prow Pass—welded [m^3/kg] (UZ)	0.0	Triay, et al. (1997)	**
constant	MatrixKD_UCF_Tc[m3/kg]	Matrix K_d of Tc for Upper Crater Flat [m^3/kg] (UZ)	0.0	Triay, et al. (1997)	**
constant	MatrixKD_BFw_Tc[m3/kg]	Matrix K_d of Tc for Bullfrog—welded [m^3/kg] (UZ)	0.0	Triay, et al. (1997)	**
constant	MatrixKD_UFZ_Tc[m3/kg]	Matrix K_d of Tc for Unsaturated Fracture Zone [m^3/kg] (UZ)	0.0	Triay, et al. (1997)	**
constant	MatrixKD_TSw_Cl[m3/kg]	Matrix K_d of Cl for Topopah Springs—welded [m^3/kg] (UZ)	0.0	Triay, et al. (1997)	**
constant	MatrixKD_CHnvCl[m3/kg]	Matrix K_d of Cl for Calico Hills—nonwelded vitric [m^2/kg] (UZ)	0.0	Triay, et al. (1997)	**
constant	MatrixKD_CHnzCl[m3/kg]	Matrix K_d of Cl for Calico Hills—nonwelded zeolitic [m^3/kg] (UZ)	0.0	Triay, et al. (1997)	**
constant	MatrixKD_PPw_Cl[m3/kg]	Matrix K_d of Cl for Prow Pass—welded [m^3/kg] (UZ)	0.0	Triay, et al. (1997)	**
constant	MatrixKD_UCF_Cl[m3/kg]	Matrix K_d of Cl for Upper Crater Flat [m^3/kg] (UZ)	0.0	Triay, et al. (1997)	**

constant	MatrixKD_BFw_Cl[m3/kg]	Matrix K_d of Cl for Bullfrog—welded [m³/kg] (UZ)	0.0	Triay, et al. (1997)	**
constant	MatrixKD_UFZ_Cl[m3/kg]	Matrix K_d of Cl for Unsaturated Fracture Zone [m³/kg] (UZ)	0.0	Triay, et al. (1997)	**
constant	MatrixKD_TSw_Cm[m3/kg]	Matrix K_d of Cm for Topopah Springs—welded [m³/kg] (UZ)	0.0	No Data	**
constant	MatrixKD_CHnvCm[m3/kg]	Matrix K_d of Cm for Calico Hills—nonwelded vitric [m³/kg] (UZ)	0.0	No Data	**
constant	MatrixKD_CHnzCm[m3/kg]	Matrix K_d of Cm for Calico Hills—nonwelded zeolitic [m³/kg] (UZ)	0.0	No Data	**
constant	MatrixKD_PPw_Cm[m3/kg]	Matrix K_d of Cm for Prow Pass—welded [m³/kg] (UZ)	0.0	No Data	**
constant	MatrixKD_UCF_Cm[m3/kg]	Matrix K_d of Cm for Upper Crater Flat [m³/kg] (UZ)	0.0	No Data	**
constant	MatrixKD_BFw_Cm[m3/kg]	Matrix K_d of Cm for Bullfrog—welded [m³/kg] (UZ)	0.0	No Data	**
constant	MatrixKD_UFZ_Cm[m3/kg]	Matrix K_d of Cm for Unsaturated Fracture Zone [m³/kg] (UZ)	0.0	No Data	**
loguniform	MatrixKD_TSw_U[m3/kg]	Matrix K_d of U for Topopah Springs—welded [m³/kg] (UZ)	4.0e-8, 0.004	Triay, et al. (1997). A value 5 orders of magnitude smaller than the maximum value was used as the lower bound.	**
loguniform	MatrixKD_CHnvU[m3/kg]	Matrix K_d of U for Calico Hills—nonwelded vitric [m³/kg] (UZ)	3.0e-8, 0.003	Triay, et al. (1997). A value 5 orders of magnitude smaller than the maximum value was used as the lower bound.	**
loguniform	MatrixKD_CHnzU[m3/kg]	Matrix K_d of U for Calico Hills—nonwelded zeolitic [m³/kg] (UZ)	3.0e-7, 0.030	Triay, et al. (1997). A value 5 orders of magnitude smaller than the maximum value was used as the lower bound.	**
constant	MatrixKD_PPw_U[m3/kg]	Matrix K_d of U for Prow Pass—welded [m³/kg] (UZ)	0.0	Triay, et al. (1997)	**

NUREG–1668

loguniform	MatrixKD_UCF_U[m3/kg]	Matrix K_d of U for Upper Crater Flat [m³/kg] (UZ)	4.0e-8, 0.004	Triay, et al. (1997). A value 5 orders of magnitude smaller than the maximum value was used as the lower bound.	**
loguniform	MatrixKD_BFw_U[m3/kg]	Matrix K_d of U for Bullfrog—welded [m³/kg] (UZ)	4.0e-8, 0.004	Triay, et al. (1997). A value 5 orders of magnitude smaller than the maximum value was used as the lower bound.	**
loguniform	MatrixKD_UFZ_U[m3/kg]	Matrix K_d of U for Unsaturated Fracture Zone [m³/kg] (UZ)	4.0e-8, 0.004	Triay, et al. (1997). A value 5 orders of magnitude smaller than the maximum value was used as the lower bound.	**
loguniform	MatrixKD_TSw_Pu[m3/kg]	Matrix K_d of Pu for Topopah Springs—welded [m³/kg] (UZ)	0.004, 0.010	Triay, et al. (1996a)	**
loguniform	MatrixKD_CHnvPu[m3/kg]	Matrix K_d of Pu for Calico Hills—nonwelded vitric [m³/kg] (UZ)	0.020, 0.080	Triay, et al. (1996a)	**
loguniform	MatrixKD_CHnzPu[m3/kg]	Matrix K_d of Pu for Calico Hills—nonwelded zeolitic [m³/kg] (UZ)	0.090, 0.400	Triay, et al. (1996a)	**
loguniform	MatrixKD_PPw_Pu[m3/kg]	Matrix K_d of Pu for Prow Pass—welded [m³/kg] (UZ)	0.004, 0.010	Triay, et al. (1996a)	**
loguniform	MatrixKD_UCF_Pu[m3/kg]	Matrix K_d of Pu for Upper Crater Flat [m³/kg] (UZ)	0.004, 0.010	Triay, et al. (1996a)	**
loguniform	MatrixKD_BFw_Pu[m3/kg]	Matrix K_d of Pu for Bullfrog—welded [m³/kg] (UZ)	0.004, 0.010	Triay, et al. (1996a)	**
loguniform	MatrixKD_UFZ_Pu[m3/kg]	Matrix K_d of Pu for Unsaturated Fracture Zone [m³/kg] (UZ)	0.004, 0.010	Triay, et al. (1996a)	**
loguniform	MatrixKD_TSw_Th[m3/kg]	Matrix K_d of Th for Topopah Springs—welded [m³/kg] (UZ)	0.100, 2.0	Triay, et al. (1997)	**
loguniform	MatrixKD_CHnvTh[m3/kg]	Matrix K_d of Th for Calico Hills—nonwelded vitric [m³/kg] (UZ)	0.100, 1.0	Triay, et al. (1997)	**

loguniform	MatrixKD_CHnzTh[m3/kg]	Matrix K_d of Th for Calico Hills—nonwelded zeolitic [m³/kg] (UZ)	0.100, 1.0	Triay, et al. (1997)	**
loguniform	MatrixKD_PPw_Th[m3/kg]	Matrix K_d of Th for Prow Pass—welded [m³/kg] (UZ)	0.100, 2.0	Triay, et al. (1997)	**
loguniform	MatrixKD_UCF_Th[m3/kg]	Matrix K_d of Th for Upper Crater Flat [m³/kg] (UZ)	0.100, 2.0	Triay, et al. (1997)	**
loguniform	MatrixKD_BFw_Th[m3/kg]	Matrix K_d of Th for Bullfrog—welded [m³/kg] (UZ)	0.100, 2.0	Triay, et al. (1997)	**
loguniform	MatrixKD_UFZ_Th[m3/kg]	Matrix K_d of Th for Unsaturated Fracture Zone [m³/kg] (UZ)	0.100, 2.0	Triay, et al. (1997)	**
loguniform	MatrixKD_TSw_Ra[m3/kg]	Matrix K_d of Ra for Topopah Springs—welded [m³/kg] (UZ)	0.100, 0.500	Triay, et al. (1997)	**
loguniform	MatrixKD_CHnvRa[m3/kg]	Matrix K_d of Ra for Calico Hills—nonwelded vitric [m³/kg] (UZ)	0.050, 0.100	Triay, et al. (1997)	**
loguniform	MatrixKD_CHnzRa[m3/kg]	Matrix K_d of Ra for Calico Hills—nonwelded zeolitic [m³/kg] (UZ)	1.0, 5.0	Triay, et al. (1997)	**
loguniform	MatrixKD_PPw_Ra[m3/kg]	Matrix K_d of Ra for Prow Pass—welded [m³/kg] (UZ)	0.1, 0.5	Triay, et al. (1997)	**
loguniform	MatrixKD_UCF_Ra[m3/kg]	Matrix K_d of Ra for Upper Crater Flat [m³/kg] (UZ)	0.1, 0.5	Triay, et al. (1997)	**
loguniform	MatrixKD_BFw_Ra[m3/kg]	Matrix K_d of Ra for Bullfrog—welded [m³/kg] (UZ)	0.1, 0.5	Triay, et al. (1997)	**
loguniform	MatrixKD_UFZ_Ra[m3/kg]	Matrix K_d of Ra for Unsaturated Fracture Zone [m³/kg] (UZ)	0.1, 0.5	Triay, et al. (1997)	**
loguniform	MatrixKD_TSw_Pb[m3/kg]	Matrix K_d of Pb for Topopah Springs—welded [m³/kg] (UZ)	0.100, 0.500	Triay, et al. (1997)	**

loguniform	MatrixKD_CHnvPb[m3/kg]	Matrix K_d of Pb for Calico Hills—nonwelded vitric [m³/kg] (UZ)	0.100, 0.500	Triay, et al. (1997)	**
loguniform	MatrixKD_CHnzPb[m3/kg]	Matrix K_d of Pb for Calico Hills—nonwelded zeolitic [m³/kg] (UZ)	0.100, 0.500	Triay, et al. (1997)	**
loguniform	MatrixKD_PPw_Pb[m3/kg]	Matrix K_d of Pb for Prow Pass—welded [m³/kg] (UZ)	0.100, 0.500	Triay, et al. (1997)	**
loguniform	MatrixKD_UCF_Pb[m3/kg]	Matrix K_d of Pb for Upper Crater Flat [m³/kg] (UZ)	0.100, 0.500	Triay, et al. (1997)	**
loguniform	MatrixKD_BFw_Pb[m3/kg]	Matrix K_d of Pb for Bullfrog—welded [m³/kg] (UZ)	0.100, 0.500	Triay, et al. (1997)	**
loguniform	MatrixKD_UFZ_Pb[m3/kg]	Matrix K_d of Pb for Unsaturated Fracture Zone [m³/kg] (UZ)	0.100, 0.500	Triay, et al. (1997)	**
loguniform	MatrixKD_TSw_Cs[m3/kg]	Matrix K_d of Cs for Topopah Springs—welded [m³/kg] (UZ)	0.020, 1.000	Triay, et al. (1997)	**
loguniform	MatrixKD_CHnvCs[m3/kg]	Matrix K_d of Cs for Calico Hills—nonwelded vitric [m³/kg] (UZ)	0.010, 0.100	Triay, et al. (1997)	**
loguniform	MatrixKD_CHnzCs[m3/kg]	Matrix K_d of Cs for Calico Hills—nonwelded zeolitic [m³/kg] (UZ)	0.500, 5.000	Triay, et al. (1997)	**
loguniform	MatrixKD_PPw_Cs[m3/kg]	Matrix K_d of Cs for Prow Pass—welded [m³/kg] (UZ)	0.020, 1.000	Triay, et al. (1997)	**
loguniform	MatrixKD_UCF_Cs[m3/kg]	Matrix K_d of Cs for Upper Crater Flat [m³/kg] (UZ)	0.020, 1.000	Triay, et al. (1997)	**
loguniform	MatrixKD_BFw_Cs[m3/kg]	Matrix K_d of Cs for Bullfrog—welded [m³/kg] (UZ)	0.020, 1.000	Triay, et al. (1997)	**
loguniform	MatrixKD_UFZ_Cs[m3/kg]	Matrix K_d of Cs for Unsaturated Fracture Zone [m³/kg] (UZ)	0.020, 1.000	Triay, et al. (1997)	**

loguniform	MatrixKD_TSw_Ni[m3/kg]	Matrix K_d of Ni for Topopah Springs—welded [m³/kg] (UZ)	5.0e-6, 0.500	Triay, et al. (1997). A value 5 orders of magnitude smaller than the maximum value was used as the lower bound.	**
loguniform	MatrixKD_CHnvNi[m3/kg]	Matrix K_d of Ni for Calico Hills—nonwelded vitric [m³/kg] (UZ)	1.0e-6, 0.100	Triay, et al. (1997). A value 5 orders of magnitude smaller than the maximum value was used as the lower bound.	**
loguniform	MatrixKD_CHnzNi[m3/kg]	Matrix K_d of Ni for Calico Hills—nonwelded zeolitic [m³/kg] (UZ)	5.0e-6, 0.500	Triay, et al. (1997). A value 5 orders of magnitude smaller than the maximum value was used as the lower bound.	**
loguniform	MatrixKD_PPw_Ni[m3/kg]	Matrix K_d of Ni for Prow Pass—welded [m³/kg] (UZ)	5.0e-6, 0.500	Triay, et al. (1997). A value 5 orders of magnitude smaller than the maximum value was used as the lower bound.	**
loguniform	MatrixKD_UCF_Ni[m3/kg]	Matrix K_d of Ni for Upper Crater Flat [m³/kg] (UZ)	5.0e-6, 0.500	Triay, et al. (1997). A value 5 orders of magnitude smaller than the maximum value was used as the lower bound.	**
loguniform	MatrixKD_BFw_Ni[m3/kg]	Matrix K_d of Ni for Bullfrog - welded [m³/kg] (UZ)	5.0e-6, 0.500	Triay, et al. (1997). A value 5 orders of magnitude smaller than the maximum value was used as the lower bound.	**
loguniform	MatrixKD_UFZ_Ni[m3/kg]	Matrix K_d of Ni for Unsaturated Fracture Zone [m³/kg] (UZ)	5.0e-6, 0.500	Triay, et al. (1997). A value 5 orders of magnitude smaller than the maximum value was used as the lower bound.	**
constant	MatrixKD_TSw_C[m3/kg]	Matrix K_d of C for Topopah Springs—welded [m³/kg] (UZ)	0.0	Triay, et al. (1997)	**
constant	MatrixKD_CHnvC[m3/kg]	Matrix K_d of C for Calico Hills—nonwelded vitric [m³/kg] (UZ)	0.0	Triay, et al. (1997)	**
constant	MatrixKD_CHnzC[m3/kg]	Matrix K_d of C for Calico Hills—nonwelded zeolitic [m³/kg] (UZ)	0.0	Triay, et al. (1997)	**
constant	MatrixKD_PPw_C[m3/kg]	Matrix K_d of C for Prow Pass—welded [m³/kg] (UZ)	0.0	Triay, et al. (1997)	**
constant	MatrixKD_UCF_C[m3/kg]	Matrix K_d of C for Upper Crater Flat [m³/kg] (UZ)	0.0	Triay, et al. (1997)	**

NUREG-1668

constant	MatrixKD_BFw_C[m3/kg]	Matrix K_d of C for Bullfrog—welded [m³/kg] (UZ)	0.0	Triay, et al. (1997)	**
constant	MatrixKD_UFZ_C[m3/kg]	Matrix K_d of C for Unsaturated Fracture Zone [m³/kg] (UZ)	0.0	Triay, et al. (1997)	**
loguniform	MatrixKD_TSw_Se[m3/kg]	Matrix K_d of Se for Topopah Springs—welded [m³/kg] (UZ)	3.0e-7, 0.030	Triay, et al. (1997). A value 5 orders of magnitude smaller than the maximum value was used as the lower bound.	**
loguniform	MatrixKD_CHnvSe[m3/kg]	Matrix K_d of Se for Calico Hills—nonwelded vitric [m³/kg] (UZ)	2.0e-7, 0.020	Triay, et al. (1997). A value 5 orders of magnitude smaller than the maximum value was used as the lower bound.	**
loguniform	MatrixKD_CHnzSe[m3/kg]	Matrix K_d of Se for Calico Hills—nonwelded zeolitic [m³/kg] (UZ)	1.5e-7, 0.015	Triay, et al. (1997). A value 5 orders of magnitude smaller than the maximum value was used as the lower bound.	**
loguniform	MatrixKD_PPw_Se[m3/kg]	Matrix K_d of Se for Prow Pass—welded [m³/kg] (UZ)	3.0e-7, 0.030	Triay, et al. (1997). A value 5 orders of magnitude smaller than the maximum value was used as the lower bound.	**
loguniform	MatrixKD_UCF_Se[m3/kg]	Matrix K_d of Se for Upper Crater Flats [m³/kg] (UZ)	3.0e-7, 0.030	Triay, et al. (1997). A value 5 orders of magnitude smaller than the maximum value was used as the lower bound.	**
loguniform	MatrixKD_BFw_Se[m3/kg]	Matrix K_d of Se for Bullfrog [m³/kg] (UZ)	3.0e-7, 0.030	Triay, et al. (1997). A value 5 orders of magnitude smaller than the maximum value was used as the lower bound.	**
loguniform	MatrixKD_UFZ_Se[m3/kg]	Matrix K_d of Se for Unsaturated Fracture Zone [m³/kg] (UZ)	3.0e-7, 0.030	Triay, et al. (1997). A value 5 orders of magnitude smaller than the maximum value was used as the lower bound.	**
constant	MatrixKD_TSw_Nb[m3/kg]	Matrix K_d of Nb for Topopah Springs—welded [m³/kg] (UZ)	0.10	Triay, et al. (1997)	**
constant	MatrixKD_CHnvNb[m3/kg]	Matrix K_d of Nb for Calico Hills—nonwelded vitric [m³/kg] (UZ)	0.10	Triay, et al. (1997)	**
constant	MatrixKD_CHnzNb[m3/kg]	Matrix K_d of Nb for Calico Hills—nonwelded zeolitic [m³/kg] (UZ)	0.10	Triay, et al. (1997)	**

constant	MatrixKD_PPw_Nb[m3/kg]	Matrix K_d of Nb for Prow Pass—welded [m³/kg] (UZ)	0.10	Triay, et al. (1997)	**
constant	MatrixKD_UCF_Nb[m3/kg]	Matrix K_d of Nb for Upper Crater Flats [m³/kg] (UZ)	0.10	Triay, et al. (1997)	**
constant	MatrixKD_BFw_Nb[m3/kg]	Matrix K_d of Nb for Bullfrog—welded [m³/kg] (UZ)	0.10	Triay, et al. (1997)	**
constant	MatrixKD_UFZ_Nb[m3/kg]	Matrix K_d of Nb for Unsaturated Fracture Zone [m³/kg] (UZ)	0.10	Triay, et al. (1997)	**
constant	FractureRD_TSw_Am	Fracture retardation coefficient of Am for Topopah Springs—welded (UZ)	1.0	Conservative assumption of no retardation	**
constant	FractureRD_CHnvAm	Fracture retardation coefficient of Am for Calico Hills—nonwelded vitric (UZ)	1.0	Conservative assumption of no retardation	**
constant	FractureRD_CHnzAm	Fracture retardation coefficient of Am for Calico Hills—nonwelded zeolitic (UZ)	1.0	Conservative assumption of no retardation	**
constant	FractureRD_PPw_Am	Fracture retardation coefficient of Am for Prow Pass—welded (UZ)	1.0	Conservative assumption of no retardation	**
constant	FractureRD_UCF_Am	Fracture retardation coefficient of Am for Upper Crater Flat (UZ)	1.0	Conservative assumption of no retardation	**
constant	FractureRD_BFw_Am	Fracture retardation coefficient of Am for Bullfrog—welded (UZ)	1.0	Conservative assumption of no retardation	**
constant	FractureRD_UFZ_Am	Fracture retardation coefficient of Am for Unsaturated Fracture Zone (UZ)	1.0	Conservative assumption of no retardation	**
constant	FractureRD_TSw_Np	Fracture retardation coefficient of Np for Topopah Springs—welded (UZ)	1.0	Conservative assumption of no retardation	**
constant	FractureRD_CHnvNp	Fracture retardation coefficient of Np for Calico Hills—nonwelded vitric (UZ)	1.0	Conservative assumption of no retardation	**

constant	FractureRD_CHnzNp	Fracture retardation coefficient of Np for Calico Hills—nonwelded zeolitic (UZ)	1.0	Conservative assumption of no retardation	**
constant	FractureRD_PPw_Np	Fracture retardation coefficient of Np for Prow Pass—welded (UZ)	1.0	Conservative assumption of no retardation	**
constant	FractureRD_UCF_Np	Fracture retardation coefficient of Np for Upper Crater Flat (UZ)	1.0	Conservative assumption of no retardation	**
constant	FractureRD_BFw_Np	Fracture retardation coefficient of Np for Bullfrog—welded (UZ)	1.0	Conservative assumption of no retardation	**
constant	FractureRD_UFZ_Np	Fracture retardation coefficient of Np for Unsaturated Fracture Zone (UZ)	1.0	Conservative assumption of no retardation	**
constant	FractureRD_TSw_I	Fracture retardation coefficient of I for Topopah Springs—welded (UZ)	1.0	Conservative assumption of no retardation	**
constant	FractureRD_CHnvI	Fracture retardation coefficient of I for Calico Hills—nonwelded vitric (UZ)	1.0	Conservative assumption of no retardation	**
constant	FractureRD_CHnzI	Fracture retardation coefficient of I for Calico Hills—nonwelded zeolitic (UZ)	1.0	Conservative assumption of no retardation	**
constant	FractureRD_PPw_I	Fracture retardation coefficient of I for Prow Pass—welded (UZ)	1.0	Conservative assumption of no retardation	**
constant	FractureRD_UCF_I	Fracture retardation coefficient of I for Upper Crater Flat (UZ)	1.0	Conservative assumption of no retardation	**
constant	FractureRD_BFw_I	Fracture retardation coefficient of I for Bullfrog—welded (UZ)	1.0	Conservative assumption of no retardation	**
constant	FractureRD_UFZ_I	Fracture retardation coefficient of I for Unsaturated Fracture Zone (UZ)	1.0	Conservative assumption of no retardation	**
constant	FractureRD_TSw_Tc	Fracture retardation coefficient of Tc for Topopah Springs—welded (UZ)	1.0	Conservative assumption of no retardation	**

constant	FractureRD_CHnvTc	Fracture retardation coefficient of Tc for Calico Hills—nonwelded vitric (UZ)	1.0	Conservative assumption of no retardation	**
constant	FractureRD_CHnzTc	Fracture retardation coefficient of Tc for Calico Hills—nonwelded zeolitic (UZ)	1.0	Conservative assumption of no retardation	**
constant	FractureRD_PPw_Tc	Fracture retardation coefficient of Tc for Prow Pass—welded (UZ)	1.0	Conservative assumption of no retardation	**
constant	FractureRD_UCF_Tc	Fracture retardation coefficient of Tc for Upper Crater Flat (UZ)	1.0	Conservative assumption of no retardation	**
constant	FractureRD_BFw_Tc	Fracture retardation coefficient of Tc for Bullfrog—welded (UZ)	1.0	Conservative assumption of no retardation	**
constant	FractureRD_UFZ_Tc	Fracture retardation coefficient of Tc for Unsaturated Fracture Zone (UZ)	1.0	Conservative assumption of no retardation	**
constant	FractureRD_TSw_Cl	Fracture retardation coefficient of Cl for Topopah Springs—welded (UZ)	1.0	Conservative assumption of no retardation	**
constant	FractureRD_CHnvCl	Fracture retardation coefficient of Cl for Calico Hills—nonwelded vitric (UZ)	1.0	Conservative assumption of no retardation	**
constant	FractureRD_CHnzCl	Fracture retardation coefficient of Cl for Calico Hills—nonwelded zeolitic (UZ)	1.0	Conservative assumption of no retardation	**
constant	FractureRD_PPw_Cl	Fracture retardation coefficient of Cl for Prow Pass—welded (UZ)	1.0	Conservative assumption of no retardation	**
constant	FractureRD_UCF_Cl	Fracture retardation coefficient of Cl for Upper Crater Flat (UZ)	1.0	Conservative assumption of no retardation	**
constant	FractureRD_BFw_Cl	Fracture retardation coefficient of Cl for Bullfrog—welded (UZ)	1.0	Conservative assumption of no retardation	**
constant	FractureRD_UFZ_Cl	Fracture retardation coefficient of Cl for Unsaturated Fracture Zone (UZ)	1.0	Conservative assumption of no retardation	**

NUREG–1668

constant	FractureRD_TSw_Cm	Fracture retardation coefficient of Cm for Topopah Springs—welded (UZ)	1.0	Conservative assumption of no retardation	**
constant	FractureRD_CHnvCm	Fracture retardation coefficient of Cm for Calico Hills—nonwelded vitric (UZ)	1.0	Conservative assumption of no retardation	**
constant	FractureRD_CHnzCm	Fracture retardation coefficient of Cm for Calico Hills—nonwelded zeolitic (UZ)	1.0	Conservative assumption of no retardation	**
constant	FractureRD_PPw_Cm	Fracture retardation coefficient of Cm for Prow Pass—welded (UZ)	1.0	Conservative assumption of no retardation	**
constant	FractureRD_UCF_Cm	Fracture retardation coefficient of Cm for Upper Crater Flat (UZ)	1.0	Conservative assumption of no retardation	**
constant	FractureRD_BFw_Cm	Fracture retardation coefficient of Cm for Bullfrog—welded (UZ)	1.0	Conservative assumption of no retardation	**
constant	FractureRD_UFZ_Cm	Fracture retardation coefficient of Cm Unsaturated Fracture Zone (UZ)	1.0	Conservative assumption of no retardation	**
constant	FractureRD_TSw_U	Fracture retardation coefficient of U for Topopah Springs—welded (UZ)	1.0	Conservative assumption of no retardation	**
constant	FractureRD_CHnvU	Fracture retardation coefficient of U for Calico Hills—nonwelded vitric (UZ)	1.0	Conservative assumption of no retardation	**
constant	FractureRD_CHnzU	Fracture retardation coefficient of U for Calico Hills—nonwelded zeolitic (UZ)	1.0	Conservative assumption of no retardation	**
constant	FractureRD_PPw_U	Fracture retardation coefficient of U for Prow Pass—welded (UZ)	1.0	Conservative assumption of no retardation	**
constant	FractureRD_UCF_U	Fracture retardation coefficient of U for Upper Crater Flat (UZ)	1.0	Conservative assumption of no retardation	**
constant	FractureRD_BFw_U	Fracture retardation coefficient of U for Bullfrog—welded (UZ)	1.0	Conservative assumption of no retardation	**

constant	FractureRD_UFZ_U	Fracture retardation coefficient of U for Unsaturated Fracture Zone (UZ)	1.0	Conservative assumption of no retardation	**
constant	FractureRD_TSw_Pu	Fracture retardation coefficient of Pu for Topopah Springs—welded (UZ)	1.0	Conservative assumption of no retardation	**
constant	FractureRD_CHnvPu	Fracture retardation coefficient of Pu for Calico Hills—nonwelded vitric (UZ)	1.0	Conservative assumption of no retardation	**
constant	FractureRD_CHnzPu	Fracture retardation coefficient of Pu for Calico Hills—nonwelded zeolitic (UZ)	1.0	Conservative assumption of no retardation	**
constant	FractureRD_PPw_Pu	Fracture retardation coefficient of Pu for Prow Pass—welded (UZ)	1.0	Conservative assumption of no retardation	**
constant	FractureRD_UCF_Pu	Fracture retardation coefficient of Pu for Upper Crater Flat (UZ)	1.0	Conservative assumption of no retardation	**
constant	FractureRD_BFw_Pu	Fracture retardation coefficient of Pu for Bullfrog—welded (UZ)	1.0	Conservative assumption of no retardation	**
constant	FractureRD_UFZ_Pu	Fracture retardation coefficient of Pu for Unsaturated Fracture Zone (UZ)	1.0	Conservative assumption of no retardation	**
constant	FractureRD_TSw_Th	Fracture retardation coefficient of Th for Topopah Springs—welded (UZ)	1.0	Conservative assumption of no retardation	**
constant	FractureRD_CHnvTh	Fracture retardation coefficient of Th for Calico Hills—nonwelded vitric (UZ)	1.0	Conservative assumption of no retardation	**
constant	FractureRD_CHnzTh	Fracture retardation coefficient of Th for Calico Hills—nonwelded zeolitic (UZ)	1.0	Conservative assumption of no retardation	**
constant	FractureRD_PPw_Th	Fracture retardation coefficient of Th for Prow Pass—welded (UZ)	1.0	Conservative assumption of no retardation	**
constant	FractureRD_UCF_Th	Fracture retardation coefficient of Th for Upper Crater Flat (UZ)	1.0	Conservative assumption of no retardation	**

constant	FractureRD_BFw_Th	Fracture retardation coefficient of Th for Bullfrog—welded (UZ)	1.0	Conservative assumption of no retardation	**
constant	FractureRD_UFZ_Th	Fracture retardation coefficient of Th for Unsaturated Fracture Zone (UZ)	1.0	Conservative assumption of no retardation	**
constant	FractureRD_TSw_Ra	Fracture retardation coefficient of Ra for Topopah Springs—welded (UZ)	1.0	Conservative assumption of no retardation	**
constant	FractureRD_CHnvRa	Fracture retardation coefficient of Ra for Calico Hills—nonwelded vitric (UZ)	1.0	Conservative assumption of no retardation	**
constant	FractureRD_CHnzRa	Fracture retardation coefficient of Ra for Calico Hills—nonwelded zeolitic (UZ)	1.0	Conservative assumption of no retardation	**
constant	FractureRD_PPw_Ra	Fracture retardation coefficient of Ra for Prow Pass—welded (UZ)	1.0	Conservative assumption of no retardation	**
constant	FractureRD_UCF_Ra	Fracture retardation coefficient of Ra for Upper Crater Flat (UZ)	1.0	Conservative assumption of no retardation	**
constant	FractureRD_BFw_Ra	Fracture retardation coefficient of Ra for Bullfrog—welded (UZ)	1.0	Conservative assumption of no retardation	**
constant	FractureRD_UFZ_Ra	Fracture retardation coefficient of Ra for Unsaturated Fracture Zone (UZ)	1.0	Conservative assumption of no retardation	**
constant	FractureRD_TSw_Pb	Fracture retardation coefficient of Pb for Topopah Springs—welded (UZ)	1.0	Conservative assumption of no retardation	**
constant	FractureRD_CHnvPb	Fracture retardation coefficient of Pb for Calico Hills—nonwelded vitric (UZ)	1.0	Conservative assumption of no retardation	**
constant	FractureRD_CHnzPb	Fracture retardation coefficient of Pb for Calico Hills—nonwelded zeolitic (UZ)	1.0	Conservative assumption of no retardation	**
constant	FractureRD_PPw_Pb	Fracture retardation coefficient of Pb for Prow Pass—welded (UZ)	1.0	Conservative assumption of no retardation	**

constant	FractureRD_UCF_Pb	Fracture retardation coefficient of Pb for Upper Crater Flat (UZ)	1.0	Conservative assumption of no retardation	**
constant	FractureRD_BFw_Pb	Fracture retardation coefficient of Pb for Bullfrog—welded (UZ)	1.0	Conservative assumption of no retardation	**
constant	FractureRD_UFZ_Pb	Fracture retardation coefficient of Pb for Unsaturated Fracture Zone (UZ)	1.0	Conservative assumption of no retardation	**
constant	FractureRD_TSw_Cs	Fracture retardation coefficient of Cs for Topopah Springs—welded (UZ)	1.0	Conservative assumption of no retardation	**
constant	FractureRD_CHnvCs	Fracture retardation coefficient of Cs for Calico Hills—nonwelded vitric (UZ)	1.0	Conservative assumption of no retardation	**
constant	FractureRD_CHnzCs	Fracture retardation coefficient of Cs for Calico Hills—nonwelded zeolitic (UZ)	1.0	Conservative assumption of no retardation	**
constant	FractureRD_PPw_Cs	Fracture retardation coefficient of Cs for Prow Pass—welded (UZ)	1.0	Conservative assumption of no retardation	**
constant	FractureRD_UCF_Cs	Fracture retardation coefficient of Cs for Upper Crater Flat (UZ)	1.0	Conservative assumption of no retardation	**
constant	FractureRD_BFw_Cs	Fracture retardation coefficient of Cs for Bullfrog—welded (UZ)	1.0	Conservative assumption of no retardation	**
constant	FractureRD_UFZ_Cs	Fracture retardation coefficient of Cs for Unsaturated Fracture Zone (UZ)	1.0	Conservative assumption of no retardation	**
constant	FractureRD_TSw_Ni	Fracture retardation coefficient of Ni for Topopah Springs—welded (UZ)	1.0	Conservative assumption of no retardation	**
constant	FractureRD_CHnvNi	Fracture retardation coefficient of Ni for Calico Hills—nonwelded vitric (UZ)	1.0	Conservative assumption of no retardation	**
constant	FractureRD_CHnzNi	Fracture retardation coefficient of Ni for Calico Hills—nonwelded zeolitic (UZ)	1.0	Conservative assumption of no retardation	**

NUREG–1668

constant	FractureRD_PPw_Ni	Fracture retardation coefficient of Ni for Prow Pass—welded (UZ)	1.0	Conservative assumption of no retardation	**
constant	FractureRD_UCF_Ni	Fracture retardation coefficient of Ni for Upper Crater Flat (UZ)	1.0	Conservative assumption of no retardation	**
constant	FractureRD_BFw_Ni	Fracture retardation coefficient of Ni for Bullfrog—welded (UZ)	1.0	Conservative assumption of no retardation	**
constant	FractureRD_UFZ_Ni	Fracture retardation coefficient of Ni for Unsaturated Fracture Zone (UZ)	1.0	Conservative assumption of no retardation	**
constant	FractureRD_TSw_C	Fracture retardation coefficient of C for Topopah Springs—welded (UZ)	1.0	Conservative assumption of no retardation	**
constant	FractureRD_CHnvC	Fracture retardation coefficient of C for Calico Hills—nonwelded vitric (UZ)	1.0	Conservative assumption of no retardation	**
constant	FractureRD_CHnzC	Fracture retardation coefficient of C for Calico Hills—nonwelded zeolitic (UZ)	1.0	Conservative assumption of no retardation	**
constant	FractureRD_PPw_C	Fracture retardation coefficient of C for Prow Pass—welded (UZ)	1.0	Conservative assumption of no retardation	**
constant	FractureRD_UCF_C	Fracture retardation coefficient of C for Upper Crater Flat (UZ)	1.0	Conservative assumption of no retardation	**
constant	FractureRD_BFw_C	Fracture retardation coefficient of C for Bullfrog—welded (UZ)	1.0	Conservative assumption of no retardation	**
constant	FractureRD_UFZ_C	Fracture retardation coefficient of C for Unsaturated Fracture Zone (UZ)	1.0	Conservative assumption of no retardation	**
constant	FractureRD_TSw_Se	Fracture retardation coefficient of Se for Topopah Springs—welded (UZ)	1.0	Conservative assumption of no retardation	**
constant	FractureRD_CHnvSe	Fracture retardation coefficient of Se for Calico Hills—nonwelded vitric (UZ)	1.0	Conservative assumption of no retardation	**

Type	Variable name	Description	Value	Comment	
constant	FractureRD_CHnzSe	Fracture retardation coefficient of Se for Calico Hills—nonwelded zeolitic (UZ)	1.0	Conservative assumption of no retardation	**
constant	FractureRD_PPw_Se	Fracture retardation coefficient of Se for Prow Pass—welded (UZ)	1.0	Conservative assumption of no retardation	**
constant	FractureRD_UCF_Se	Fracture retardation coefficient of Se for Upper Crater Flat (UZ)	1.0	Conservative assumption of no retardation	**
constant	FractureRD_BFw_Se	Fracture retardation coefficient of Se for Bullfrog—welded (UZ)	1.0	Conservative assumption of no retardation	**
constant	FractureRD_UFZ_Sc	Fracture retardation coefficient of Se for Unsaturated Fracture Zone (UZ)	1.0	Conservative assumption of no retardation	**
constant	FractureRD_TSw_Nb	Fracture retardation coefficient of Nb for Topopah Springs—welded (UZ)	1.0	Conservative assumption of no retardation	**
constant	FractureRD_CHnvNb	Fracture retardation coefficient of Nb for Calico Hills—nonwelded vitric (UZ)	1.0	Conservative assumption of no retardation	**
constant	FractureRD_CHnzNb	Fracture retardation coefficient of Nb for Calico Hills—nonwelded zeolitic (UZ)	1.0	Conservative assumption of no retardation	**
constant	FractureRD_PPw_Nb	Fracture retardation coefficient of Nb for Prow Pass—welded (UZ)	1.0	Conservative assumption of no retardation	**
constant	FractureRD_UCF_Nb	Fracture retardation coefficient of Nb for Upper Crater Flat (UZ)	1.0	Conservative assumption of no retardation	**
constant	FractureRD_BFw_Nb	Fracture retardation coefficient of Nb for Bullfrog—welded (UZ)	1.0	Conservative assumption of no retardation	**
constant	FractureRD_UFZ_Nb	Fracture retardation coefficient of Nb for Unsaturated Fracture Zone (UZ)	1.0	Conservative assumption of no retardation	**
lognormal	MatrixPermeability_TSw_[m2]	Matrix permeability for Topopah Springs tuff—welded (UZ) (m^2)	0.2e-19, 0.2e-17	Numbers based on statistical average of subunits in report by Flint (1996). Distribution assumed as being typical for permeability values.	**

NUREG–1668

lognormal	MatrixPermeability_CHnv[m2]	Matrix permeability for Calico Hills—nonwelded vitric (UZ) (m²)	0.2e-14, 0.2e-12	Numbers based on statistical average of subunits in report by Flint (1996). Distribution assumed as being typical for permeability values. **
lognormal	MatrixPermeability_CHnz[m2]	Matrix permeability for Calico Hills—nonwelded zeolitic (UZ) (m²)	0.5e-18, 0.5e-16	Numbers based on statistical average of subunits in report by Flint (1996). Distribution assumed as being typical for permeability values. **
lognormal	MatrixPermeability_PPw_[n2]	Matrix permeability for Prow Pass—welded (UZ) (m²)	0.1e-17, 0.1e-15	Numbers based on statistical average of subunits in report by Flint (1996). Distribution assumed as being typical for permeability values. **
lognormal	MatrixPermeability_UCF_[m2]	Matrix permeability for Upper Crater Flat (UZ) (m²)	0.3e-18, 0.3e-16	Numbers based on statistical average of subunits in report by Flint (1996). Distribution assumed as being typical for permeability values. **
lognormal	MatrixPermeability_BFw_[m2]	Matrix permeability for Bullfrog—welded (UZ) (m²)	0.2e-19, 0.2e-17	Numbers based on statistical average of subunits in report by Flint (1996). Distribution assumed as being typical for permeability values. **
lognormal	MatrixPermeability_UFZ_[m2]	Matrix permeability for Unsaturated Fault Zone (UZ) (m²)	1.8e-18, 2.1e-16	Flint (1996). Current best estimate, but this should be further evaluated in the issue resolution status report (IRSR) update for USFIC. **
constant	MatrixPorosity_TSw_	Matrix porosity for Topopah Springs—welded (UZ) (m²)	0.12	Numbers based on statistical average of subunits in report by Flint (1996). Constant values assumed caused by small variability in data set **
constant	MatrixPorosity_CHnv	Matrix porosity for Calico Hills—nonwelded vitric (UZ)	0.33	Numbers based on statistical average of subunits in report by Flint (1996). Constant values assumed caused by small variability in data set **
constant	MatrixPorosity_CHnz	Matrix porosity for Calico Hills—nonwelded zeolitic (UZ)	0.32	Numbers based on statistical average of subunits in report by Flint (1996). Constant values assumed caused by small variability in data set **
constant	MatrixPorosity_PPw_	Matrix porosity for Prow Pass—welded (UZ)	0.28	Numbers based on statistical average of subunits in report by Flint (1996). Constant values assumed caused by small variability in data set **

constant	MatrixPorosity_UCF_	Matrix porosity for Upper Crater Flat (UZ)	0.28	Numbers based on statistical average of subunits in report by Flint (1996). Constant values assumed caused by small variability in data set	**
constant	MatrixPorosity_BFw_	Matrix porosity for Bullfrog—welded (UZ)	0.12	Numbers based on statistical average of subunits in report by Flint (1996). Constant values assumed caused by small variability in data set	**
constant	MatrixPorosity_UFZ_	Matrix porosity for Unsaturated Fault Zone (UZ)	0.12	Flint (1996). Current best estimate, but this should be further evaluated in the IRSR update for USFIC.	**
constant	MatrixBeta_TSw_	van Genuchten beta parameter for Topopah Springs—welded (matrix) (UZ)	1.5	Current best estimate by USFIC team. Also see Flint (1996).	**
constant	MatrixBeta_CH$_{uv}$	van Genuchten beta parameter for Calico Hills— nonwelded vitric (matrix) (UZ)	1.3	Current best estimate by USFIC team. Also see Flint (1996).	**
constant	MatrixBeta_CHnz	van Genuchten beta parameter for Calico Hills—nonwelded zeolitic (matrix) (UZ)	2.3	Current best estimate by USFIC team. Also see Flint (1996).	**
constant	MatrixBeta_PPw_	van Genuchten beta parameter for Prow Pass—welded (matrix) (UZ)	1.5	Current best estimate by USFIC team. Also see Flint (1996).	**
constant	MatrixBeta_UCF_	van Genuchten beta parameter for Upper Crater Flat (matrix) (UZ)	1.4	Current best estimate by USFIC team. Also see Flint (1996).	**
constant	MatrixBeta_BFw_	van Genuchten beta parameter for Bullfrog—welded (matrix) (UZ)	1.7	Current best estimate by USFIC team. Also see Flint (1996).	**
constant	MatrixBeta_UFZ_	van Genuchten beta parameter for Unsaturated Fault Zone (matrix) (UZ)	2.3	Current best estimate by USFIC team. Also see Flint (1996).	**
constant	MatrixGrainDensity_TSw_[kg/m3]	Matrix grain density for Topopah Springs—welded (UZ) (kg/m³)	2460.0	Best estimate by the USFIC team. Based on averaging of subunit data reported by Flint (1996). Constant values assumed caused by little variation in reported data	**

A-49

NUREG-1668

constant	MatrixGrainDensity_CHnv[kg/m3]	Matrix grain density for Calico Hills—nonwelded vitric (UZ) (kg/m^3)	2260.0	Best estimate by the USFIC team. Based on averaging of subunit data reported by Flint (1996). Constant values assumed caused by little variation in reported data	**
constant	MatrixGrainDensity_CHnz[kg/m3]	Matrix grain density for Calico Hills—nonwelded zeolitic (UZ) (kg/m^3)	2400.0	Best estimate by the USFIC team. Based on averaging of subunit data reported by Flint (1996). Constant values assumed caused by little variation in reported data	**
constant	MatrixGrainDensity_PPw_[kg/m3]	Matrix grain density for Prow Pass—welded (UZ) (kg/m^3)	2540.0	Best estimate by the USFIC team. Based on averaging of subunit data reported by Flint (1996). Constant values assumed caused by little variation in reported data	**
constant	MatrixGrainDensity_UCF_[kg/m3]	Matrix grain density for Upper Crater Flat (UZ) (kg/m^3)	2420.0	Best estimate by the USFIC team. Based on averaging of subunit data reported by Flint (1996). Constant values assumed caused by little variation in reported data	**
constant	MatrixGrainDensity_BFw_[kg/m3]	Matrix grain density for Bullfrog—welded (UZ) (kg/m^3)	2570.0	Best estimate by the USFIC team. Based on averaging of subunit data reported by Flint (1996). Constant values assumed caused by little variation in reported data	**
constant	MatrixGrainDensity_UFZ_[kg/m3]	Matrix grain density for Unsaturated Fault Zone (UZ) (kg/m^3)	2630.0	Best estimate by the USFIC team. Based on averaging of subunit data reported by Flint (1996). Constant values assumed caused by little variation in reported data	
lognormal	FracturePermeability_TSw_[m2]	Fracture permeability Topopah Springs—welded (m^2) (UZ)	8.0e-15, 8.0e-11	Current best estimate by the USFIC team based on Schenker, et al. (1995)	**
lognormal	FracturePermeability_CHnv[m2]	Calico Hills—nonwelded vitric fracture permeability (m^2) (UZ)	8.0e-15, 8.0e-11	Current best estimate by the USFIC team based on Schenker, et al. (1995)	**
lognormal	FracturePermeability_CHnz[m2]	Calico Hills—nonwelded zeolitic fracture permeability (m^2) (UZ)	6.0e-15, 6.0e-11	Current best estimate by the USFIC team based on Schenker, et al. (1995)	**
lognormal	FracturePermeability_PPw_[m2]	Prow Pass—welded fracture permeability (m^2) (UZ)	6.0e-15, 6.0e-11	Current best estimate by the USFIC team based on Klavetter and Peters (1986) and Nuclear Regulatory Commission (1995)	**

Distribution	Variable	Description	Value	Source	
lognormal	FracturePermeability_UCF_[m2]	Upper Crater Flat fracture permeability (m²) (UZ)	6.0e-15, 6.0e-11	Current best estimate by the USFIC team based on Klavetter and Peters (1986) and Nuclear Regulatory Commission (1995)	**
lognormal	FracturePermeability_BFw_[m2]	Bullfrog— welded fracture permeability (m²) (UZ)	3.0e-15, 3.0e-11	Current best estimate by the USFIC team based on Klavetter and Peters (1986) and Nuclear Regulatory Commission (1995)	**
lognormal	FracturePermeability_UFZ_[m2]	Unsaturated Fracture Zone fracture permeability (m²) (UZ)	1.0e-13, 1.0e-11	Current best estimate by the USFIC team based on Klavetter and Peters (1986) and Nuclear Regulatory Commission (1995)	**
loguniform	FracturePorosity_TSw_	Fracture porosity Topopah Springs—welded (UZ)	1.0e-3, 1.0e-2	Current best estimate by the USFIC team. Upper value based on Geldon, et al. (1997), citing Streltsova (1988), and lower values based on consideration of Geldon (1993)	**
loguniform	FracturePorosity_CHnv	Calico Hills - nonwelded vitric fracture porosity (UZ)	1.0e-3, 1.0e-2	Current best estimate by the USFIC team. Upper value based on Geldon, et al. (1997), citing Streltsova (1988), and lower values based on consideration of Geldon (1993)	**
loguniform	FracturePorosity_CHnz	Calico Hills—nonwelded zeolitic fracture porosity (UZ)	1.0e-3, 1.0e-2	Current best estimate by the USFIC team. Upper value based on Geldon, et al. (1997), citing Streltsova (1988), and lower values based on consideration of Geldon (1993)	**
loguniform	FracturePorosity_PPw_	Prow Pass—welded fracture porosity (UZ)	1.0e-3, 1.0e-2	Current best estimate by the USFIC team. Upper value based on Geldon, et al. (1997), citing Streltsova (1988), and lower values based on consideration of Geldon (1993)	**
loguniform	FracturePorosity_UCF_	Upper Crater Flat fracture porosity (UZ)	1.0e-3, 1.0e-2	Current best estimate by the USFIC team. Upper value based on Geldon, et al. (1997), citing Streltsova (1988), and lower values based on consideration of Geldon (1993)	**
loguniform	FracturePorosity_BFw_	Bullfrog—welded fracture porosity (UZ)	1.0e-3, 1.0e-2	Current best estimate by the USFIC team. Upper value based on Geldon, et al. (1997), citing Streltsova (1988), and lower values based on consideration of Geldon (1993)	**

loguniform	FracturePorosity_UFZ_	Unsaturated Fault Zone fracture porosity (UZ)	1.0e-3, 1.0e-2	Current best estimate by the USFIC team. Upper value based on Geldon, et al. (1997), citing Streltsova (1988), and lower values based on consideration of Geldon (1993)	**
constant	FractureBeta_TSw_	van Genuchten beta parameter for Topopah Springs—welded (UZ)	3.0	Best estimate by USFIC team based on Geldon (1993)	**
constant	FractureBeta_CHnv	van Genuchten beta parameter for Calico Hills—nonwelded vitric (UZ)	3.0	Best estimate by USFIC team based on Geldon (1993)	**
constant	FractureBeta_CHnz	van Genuchten beta parameter for Calico Hills—nonwelded zeolitic (UZ)	3.0	Best estimate by USFIC team based on Geldon (1993)	**
constant	FractureBeta_PPw_	van Genuchten beta parameter for Prow Pass—welded (UZ)	3.0	Best estimate by USFIC team based on Geldon (1993)	**
constant	FractureBeta_UCF_	van Genuchten beta parameter for Upper Crater Flat (UZ)	3.0	Best estimate by USFIC team based on Geldon (1993)	**
constant	FractureBeta_BFw_	van Genuchten beta parameter for Bullfrog—welded (UZ)	3.0	Best estimate by USFIC team based on Geldon (1993)	**
constant	FractureBeta_UFZ_	van Genuchten beta parameter for Unsaturated Fault Zone (UZ)	3.0	Best estimate by USFIC team based on Geldon (1993)	**
constant	InletArea__1SubArea[m2]	Inlet area (m²) - subarea 1 (UZ) (not used)	5.4e5	Center for Nuclear Waste Regulatory Analyses (CNWRA) staff best estimate	**
constant	InletArea__2SubArea[m2]	Inlet area (m²) - subarea 2 (UZ) (not used)	5.4e5	CNWRA staff best estimate	**
constant	InletArea__3SubArea[m2]	Inlet area (m²) - subarea 3 (UZ) (not used)	5.4e5	CNWRA staff best estimate	**
constant	InletArea__4SubArea[m2]	Inlet area (m²) - subarea 4 (UZ) (not used)	5.4e5	CNWRA staff best estimate	**
constant	InletArea__5SubArea[m2]	Inlet area (m²) - subarea 5 (UZ) (not used)	5.4e5	CNWRA staff best estimate	**
constant	InletArea__6SubArea[m2]	Inlet area (m²) - subarea 6 (UZ) (not used)	5.4e5	CNWRA staff best estimate	**

				CNWRA staff best estimate	**
constant	InletArea__7SubArea[m2}	Inlet area (m²) - subarea 7 (UZ) (not used)	5.4e5		**
constant	TSw_Thickness_1SubArea[m]	Topopah Springs—welded thickness Subarea1 (m) (UZ)	33.0	Derived directly from the CNWRA 3-D Geological Framework Model (Stirewalt and Henderson, 1995, pp. 3-5)	**
constant	CHnvThickness_1SubArea[m]	Subarea1 Calico Hills—nonwelded vitric thickness (m) (UZ)	0.0	Derived directly from the CNWRA 3-D Geological Framework Model (Stirewalt and Henderson, 1995, pp. 3-5). The identification of the vitric portions of the Calico Hills was developed from Fig. 1.1.3 in Bodvarsson and Bandurraga (1996).	**
constant	CHnzThickness_1SubArea[m]	Subarea1 Calico Hills—nonwelded zeolitic thickness (m) (UZ)	163.0	Derived directly from the CNWRA 3-D Geological Framework Model (Stirewalt and Henderson, 1995, pp. 3–5). The identification of the vitric portions of the Calico Hills was developed from Fig. 1.1.3 in Bodvarsson and Bandurraga (1996).	**
constant	PPw_Thickness_1SubArea[m]	Subarea1 Prow Pass—welded thickness (m) (UZ)	34.0	Derived directly from the CNWRA 3-D Geological Framework Model (Stirewalt and Henderson, 1995, pp. 3-5)	**
constant	UCF_Thickness_1SubArea[m]	Subarea1 Upper Crater Flat thickness (m) (UZ)	67.0	Derived directly from the CNWRA 3-D Geological Framework Model (Stirewalt and Henderson, 1995, pp. 3-5)	**
constant	BFw_Thickness_1SubArea[m}	Subarea1 Bullfrog—welded thickness (m) (UZ)	0.0	Derived directly from the CNWRA 3-D Geological Framework Model (Stirewalt and Henderson, 1995, pp. 3-5)	**
constant	UFZ_Thickness_1SubArea[m]	Subarea1 Unsaturated Fracture Zone thickness (m) (UZ)	0.0	Derived directly from the CNWRA 3-D Geological Framework Model (Stirewalt and Henderson, 1995, pp. 3-5)	**
constant	TSw_Thickness_2SubArea[m]	Subarea2 Topopah Springs—welded thickness (m) (UZ)	116.0	Derived directly from the CNWRA 3-D Geological Framework Model (Stirewalt and Henderson, 1995, pp. 3-5)	**

NUREG-1668

constant	CHnvThickness_2SubArea[m]	Subarea2 Calico Hills—nonwelded vitric thickness (m) (UZ)	0.0	Derived directly from the CNWRA 3-D Geological Framework Model (Stirewalt and Henderson, 1995, pp. 3–5). The identification of the vitric portions of the Calico Hills was developed from Fig. 1.1.3 in Bodvarsson and Bandurraga (1996).	**
constant	CHnzThickness_2SubArea[m]	Subarea2 Calico Hills—nonwelded zeolitic thickness (m) (UZ)	154.0	Derived directly from the CNWRA 3-D Geological Framework Model (Stirewalt and Henderson, 1995, pp. 3–5). The identification of the vitric portions of the Calico Hills was developed from Fig. 1.1.3 in Bodvarsson and Bandurraga (1996).	**
constant	PPw_Thickness_2SubArea[m]	Subarea2 Prow Pass—welded thickness (m) (UZ)	39.0	Derived directly from the CNWRA 3-D Geological Framework Model (Stirewalt and Henderson, 1995, pp. 3–5)	**
constant	UCF_Thickness_2SubArea[m]	Subarea2 Upper Crater Flat thickness (m) (UZ)	20.0	Derived directly from the CNWRA 3-D Geological Framework Model (Stirewalt and Henderson, 1995, pp. 3–5)	**
constant	BFw_Thickness_2SubArea[m]	Subarea2 Bullfrog—welded thickness (m) (UZ)	0.0	Derived directly from the CNWRA 3-D Geological Framework Model (Stirewalt and Henderson, 1995, pp. 3–5)	**
constant	UFZ_Thickness_2SubArea[m]	Subarea2 Unsaturated Fracture Zone thickness (m) (UZ)	0.0	Derived directly from the CNWRA 3-D Geological Framework Model (Stirewalt and Henderson, 1995, pp. 3–5)	**
constant	TSw_Thickness_3SubArea[m]	Subarea3 Topopah Springs—welded thickness (m) (UZ)	20.0	Derived directly from the CNWRA 3-D Geological Framework Model (Stirewalt and Henderson, 1995, pp. 3–5)	**
constant	CHnvThickness_3SubArea[m]	Subarea3 Calico Hills—nonwelded vitric thickness (m) (UZ)	0.0	Derived directly from the CNWRA 3-D Geological Framework Model (Stirewalt and Henderson, 1995, pp. 3–5). The identification of the vitric portions of the Calico Hills was developed from Fig. 1.1.3 in Bodvarsson and Bandurraga (1996).	**

constant	CHnzThickness_3SubArea[m]	Subarea3 Calico Hills—nonwelded zeolitic thickness (m) (UZ)	122.0	Derived directly from the CNWRA 3-D Geological Framework Model (Stirewalt and Henderson, 1995, pp. 3–5). The identification of the vitric portions of the Calico Hills was developed from Fig. 1.1.3 in Bodvarsson and Bandurraga (1996).	**
constant	PPw_Thickness_3SubArea[m]	Subarea3 Prow Pass—welded thickness (m) (UZ)	40.0	Derived directly from the CNWRA 3-D Geological Framework Model (Stirewalt and Henderson, 1995, pp. 3–5)	**
constant	UCF_Thickness_3SubArea[m]	Subarea3 Upper Crater Flat thickness (m) (UZ)	158.0	Derived directly from the CNWRA 3-D Geological Framework Model (Stirewalt and Henderson, 1995, pp. 3–5)	**
constant	BFw_Thickness_3SubArea[m]	Subarea3 Bullfrog—welded thickness (m) (UZ)	0.0	Derived directly from the CNWRA 3-D Geological Framework Model (Stirewalt and Henderson, 1995, pp. 3–5)	**
constant	UFZ_Thickness_3SubArea[m]	Subarea3 Unsaturated Fracture Zone thickness (m) (UZ)	0.0	Derived directly from the CNWRA 3-D Geological Framework Model (Stirewalt and Henderson, 1995, pp. 3–5)	**
constant	TSw_Thickness_4SubArea[m]	Subarea4 Topopah Springs—welded thickness (m) (UZ)	110.0	Derived directly from the CNWRA 3-D Geological Framework Model (Stirewalt and Henderson, 1995, pp. 3–5)	**
constant	CHnvThickness_4SubArea[m]	Subarea4 Calico Hills—nonwelded vitric thickness (m) (UZ)	0.0	Derived directly from the CNWRA 3-D Geological Framework Model (Stirewalt and Henderson, 1995, pp. 3–5). The identification of the vitric portions of the Calico Hills was developed from Fig. 1.1.3 in Bodvarsson and Bandurraga (1996).	**
constant	CHnzThickness_4SubArea[m]	Subarea4 Calico Hills—nonwelded zeolitic thickness (m) (UZ)	132.0	Derived directly from the CNWRA 3-D Geological Framework Model (Stirewalt and Henderson, 1995, pp. 3–5). The identification of the vitric portions of the Calico Hills was developed from Fig. 1.1.3 in Bodvarsson and Bandurraga (1996).	**

constant	PPw_Thickness_4SubArea[m]	Subarea4 Prow Pass—welded thickness (m) (UZ)	34.0	Derived directly from the CNWRA 3-D Geological Framework Model (Stirewalt and Henderson, 1995, pp. 3–5)	**
constant	UCF_Thickness_4SubArea[m]	Subarea4 Upper Crater Flat thickness (m) (UZ)	57.0	Derived directly from the CNWRA 3-D Geological Framework Model (Stirewalt and Henderson, 1995, pp. 3–5)	**
constant	BFw_Thickness_4SubArea[m]	Subarea4 Bullfrog—welded thickness (m) (UZ)	0.0	Derived directly from the CNWRA 3-D Geological Framework Model (Stirewalt and Henderson, 1995, pp. 3–5)	**
constant	UFZ_Thickness_4SubArea[m]	Subarea4 Unsaturated Fracture Zone thickness (m) (UZ)	0.0	Derived directly from the CNWRA 3-D Geological Framework Model (Stirewalt and Henderson, 1995, pp. 3–5)	**
constant	TSw_Thickness_5SubArea[m]	Subarea5 Topopah Springs—welded thickness (m) (UZ)	20.0	Derived directly from the CNWRA 3-D Geological Framework Model (Stirewalt and Henderson, 1995, pp. 3–5)	**
constant	CHnvThickness_5SubArea[m]	Subarea5 Calico Hills—nonwelded vitric thickness (m) (UZ)	113.0	Derived directly from the CNWRA 3-D Geological Framework Model (Stirewalt and Henderson, 1995, pp. 3–5). The identification of the vitric portions of the Calico Hills was developed from Fig. 1.1.3 in Bodvarsson and Bandurraga (1996).	**
constant	CHnzThickness_5SubArea[m]	Subarea5 Calico Hills—nonwelded zeolitic thickness (m) (UZ)	0.0	Derived directly from the CNWRA 3-D Geological Framework Model (Stirewalt and Henderson, 1995, pp. 3–5). The identification of the vitric portions of the Calico Hills was developed from Fig. 1.1.3 in Bodvarsson and Bandurraga (1996).	**
constant	PPw_Thickness_5SubArea[m]	Subarea5 Prow Pass—welded thickness (m) (UZ)	38.0	Derived directly from the CNWRA 3-D Geological Framework Model (Stirewalt and Henderson, 1995, pp. 3–5)	**
constant	UCF_Thickness_5SubArea[m]	Subarea5 Upper Crater Flat thickness (m) (UZ)	158.0	Derived directly from the CNWRA 3-D Geological Framework Model (Stirewalt and Henderson, 1995, pp. 3–5)	**

constant	BFw_Thickness_5SubArea[m]	Subarea5 Bullfrog—welded thickness (m) (UZ)	32.0	Derived directly from the CNWRA 3-D Geological Framework Model (Stirewalt and Henderson, 1995, pp. 3–5)	**
constant	UFZ_Thickness_5SubArea[m]	Subarea5 Unsaturated Fracture Zone thickness (m) (UZ)	0.0	Derived directly from the CNWRA 3-D Geological Framework Model (Stirewalt and Henderson, 1995, pp. 3–5)	**
constant	TSw_Thickness_6SubArea[m]	Subarea6 Topopah Springs—welded thickness (m) (UZ)	53.0	Derived directly from the CNWRA 3-D Geological Framework Model (Stirewalt and Henderson, 1995, pp. 3–5)	**
constant	CHnvThickness_6SubArea[m]	Subarea6 Calico Hills—nonwelded vitric thickness (m) (UZ)	125.0	Derived directly from the CNWRA 3-D Geological Framework Model (Stirewalt and Henderson, 1995, pp. 3–5). The identification of the vitric portions of the Calico Hills was developed from Fig. 1.1.3 in Bodvarsson and Bandurraga (1996).	**
constant	CHnzThickness_6SubArea[m]	Subarea6 Calico Hills—nonwelded zeolitic thickness (m) (UZ)	0.0	Derived directly from the CNWRA 3-D Geological Framework Model (Stirewalt and Henderson, 1995, pp. 3–5). The identification of the vitric portions of the Calico Hills was developed from Fig. 1.1.3 in Bodvarsson and Bandurraga (1996).	**
constant	PPw_Thickness_6SubArea[m]	Subarea6 Prow Pass—welded thickness (m) (UZ)	26.0	Derived directly from the CNWRA 3-D Geological Framework Model (Stirewalt and Henderson, 1995, pp. 3–5)	**
constant	UCF_Thickness_6SubArea[m]	Subarea6 Upper Crater Flat thickness (m) (UZ)	136.0	Derived directly from the CNWRA 3-D Geological Framework Model (Stirewalt and Henderson, 1995, pp. 3–5)	**
constant	BFw_Thickness_6SubArea[m]	Subarea6 Bullfrog—welded thickness (m) (UZ)	0.0	Derived directly from the CNWRA 3-D Geological Framework Model (Stirewalt and Henderson, 1995, pp. 3–5)	**
constant	UFZ_Thickness_6SubArea[m]	Subarea6 Unsaturated Fracture Zone thickness (m) (UZ)	0.0	Derived directly from the CNWRA 3-D Geological Framework Model (Stirewalt and Henderson, 1995, pp. 3–5)	**

NUREG-1668

constant	TSw_Thickness_7SubArea[m]	Subarea7 Topopah Springs—welded thickness (m) (UZ)	121.0	Derived directly from the CNWRA 3-D Geological Framework Model (Stirewalt and Henderson, 1995, pp. 3–5)	**
constant	CHnvThickness_7SubArea[m]	Subarea7 Calico Hills—nonwelded vitric thickness (m) (UZ)	0.0	Derived directly from the CNWRA 3-D Geological Framework Model (Stirewalt and Henderson, 1995, pp. 3–5). The identification of the vitric portions of the Calico Hills was developed from Fig. 1.1.3 in Bodvarsson and Bandurraga (1996).	**
constant	CHnzThickness_7SubArea[m]	Subarea7 Calico Hills—nonwelded zeolitic thickness (m) (UZ)	114.0	Derived directly from the CNWRA 3-D Geological Framework Model (Stirewalt and Henderson, 1995, pp. 3–5). The identification of the vitric portions of the Calico Hills was developed from Fig. 1.1.3 in Bodvarsson and Bandurraga (1996).	**
constant	PPw_Thickness_7SubArea[m]	Subarea7 Prow Pass—welded thickness (m) (UZ)	43.0	Derived directly from the CNWRA 3-D Geological Framework Model (Stirewalt and Henderson, 1995, pp. 3–5)	**
constant	UCF_Thickness_7SubArea[m]	Subarea7 Upper Crater Flat thickness (m) (UZ)	63.0	Derived directly from the CNWRA 3-D Geological Framework Model (Stirewalt and Henderson, 1995, pp. 3–5)	**
constant	BFw_Thickness_7SubArea[m]	Subarea7 Bullfrog—welded thickness (m) (UZ)	0.0	Derived directly from the CNWRA 3-D Geological Framework Model (Stirewalt and Henderson, 1995, pp. 3–5)	**
constant	UFZ_Thickness_7SubArea[m]	Subarea7 Unsaturated Fracture Zone thickness (m) (UZ)	0.0	Derived directly from the CNWRA 3-D Geological Framework Model (Stirewalt and Henderson, 1995, pp. 3–5)	**

Type	Variable	Description	Value	Comment	
constant	MixingZoneDispersionFraction	Dispersion fraction of the mixing zone (SZ)—longitudinal dispersion specified as a fraction of the path length	0.01	CNWRA best estimate. Assumed to be conservative	**
constant	DispersionFraction_STFF	Dispersion fraction of tuff (SZ)—longitudinal dispersion specified as a fraction of the path length	0.01	CNWRA best estimate. Assumed to be conservative	**
constant	DispersionFraction_SAV	Dispersion fraction of Amargosa Valley (SZ)—longitudinal dispersion specified as a fraction of the path length	0.1	de Marsily (1986) citing work of Lallemand-Barrès and Peaudcerf (1978)	**
constant	MinimumResidenceTime_STFF[yr]	Minimum residence time in tuff (SZ) (yr)	10.0	Value selected for code efficiency reasons	**
constant	MinimumResidenceTime_SAV[yr]	Minimum residence time in alluvium (SZ) (yr)	10.0	Value selected for code efficiency reason	**
constant	FractureRD_STFF_Am	Fracture retardation coefficient for Am (SZ)	1.0	Conservative assumption of no retardation	**
loguniform	AlluviumMatrixRD_SAV_Am	Alluvium retardation coefficient for Am (SZ)	1.0, 4.0e5	Wolfsberg, 1978; (Eu analog); Price, et al., 1992	**
constant	FractureRD_STFF_Np	Fracture retardation coefficient of tuff for Np (SZ)	1.0	Conservative assumption of no retardation	**
loguniform	AlluviumMatrixRD_SAV_Np	Alluvium retardation coefficient for Np (SZ)	1.0, 300.0	Triay, et al., 1996b. No data for alluvium, crushed tuff used as analog	**
constant	FractureRD_STFF_I	Fracture retardation coefficient of tuff for I (SZ)	1.0	Conservative assumption of no retardation	**
loguniform	AlluviumMatrixRD_SAV_I	Alluvium retardation coefficient for I (SZ)	1.0, 4.0	Daniels, 1981	**

constant	FractureRD_STFF_Tc	Fracture retardation coefficient of tuff for Tc (SZ)	1.0	Conservative assumption of no retardation	**
loguniform	AlluviumMatrixRD_SAV_Tc	Alluvium retardation coefficient for Tc (SZ)	1.0, 30.0	Meijer, 1990; crushed tuff analog	**
constant	FractureRD_STFF_Cl	Fracture retardation coefficient of tuff for Cl (SZ)	1.0	Conservative assumption of no retardation	**
constant	AlluviumMatrixRD_SAV_Cl	Alluvium retardation coefficient for Cl (SZ)	1.0	Conservative assumption of no retardation	**
constant	FractureRD_STFF_Cm	Fracture retardation coefficient of tuff for Cm (SZ)	1.0	Conservative assumption of no retardation	**
constant	AlluviumMatrixRD_SAV_Cm	Alluvium retardation coefficient for Cm (SZ)	1.0	No data and conservative assumption of no retardation	**
constant	FractureRD_STFF_U	Fracture retardation coefficient of tuff for U (SZ)	1.0	Conservative assumption of no retardation	**
loguniform	AlluviumMatrixRD_SAV_U	Alluvium retardation coefficient for U (SZ)	1.0, 300.0	Triay, et al., 1996a	**
constant	FractureRD_STFF_Pu	Fracture retardation coefficient of tuff for Pu (SZ)	1.0	Conservative assumption of no retardation	**
loguniform	AlluviumMatrixRD_SAV_Pu	Alluvium retardation coefficient for Pu (SZ)	9.0, 2.0e4	Meijer, 1990; Triay, 1996a	**
constant	FractureRD_STFF_Th	Fracture retardation coefficient of tuff for Th (SZ)	1.0	Conservative assumption of no retardation	**
constant	AlluviumMatrixRD_SAV_Th	Alluvium retardation coefficient for Th (SZ)	2.0e3, 3.0e4	Triay, et al., 1997	**
loguniform	FractureRD_STFF_Ra	Fracture retardation coefficient of tuff for Ra (SZ)	1.0	Conservative assumption of no retardation	**
constant	AlluviumMatrixRD_SAV_Ra	Alluvium retardation coefficient for Ra (SZ)	2.0e3, 8.0e3	Triay, et al., 1997; crushed tuff analog	**
loguniform	FractureRD_STFF_Pb	Fracture retardation coefficient of tuff for Pb (SZ)	1.0	Conservative assumption of no retardation	**

Distribution	Parameter	Description	Value	Comment	
loguniform	AlluviumMatrixRD_SAV_Pb	Alluvium retardation coefficient for Pb (SZ)	2.0e3, 8.0e3	Triay, et al., 1997; crushed tuff analog	**
constant	FractureRD_STFF_Cs	Fracture retardation coefficient of tuff for Cs (SZ)	1.0	Conservative assumption of no retardation	**
loguniform	AlluviumMatrixRD_SAV_Cs	Alluvium retardation coefficient for Cs (SZ)	9.0e4, 1.0e5	Wolfsberg, 1978	**
constant	FractureRD_STFF_Ni	Fracture retardation coefficient of tuff for Ni (SZ)	1.0	Conservative assumption of no retardation	**
loguniform	AlluviumMatrixRD_SAV_Ni	Alluvium retardation coefficient for Ni (SZ)	1.0, 8.0e3	Triay, et al., 1997; crushed tuff analog	**
constant	FractureRD_STFF_C	Fracture retardation coefficient of tuff for C (SZ)	1.0	Conservative assumption of no retardation	**
constant	AlluviumMatrixRD_SAV_C	Alluvium retardation coefficient for C (SZ)	1.0	Conservative assumption of no retardation	**
constant	FractureRD_STFF_Se	Fracture retardation coefficient of tuff for Se (SZ)	1.0	Conservative assumption of no retardation	**
loguniform	AlluviumMatrixRD_SAV_Se	Alluvium retardation coefficient for Se (SZ)	1.0, 500.0	Triay, et al., 1997; crushed devitrified tuff analog	**
constant	FractureRD_STFF_Nb	Fracture retardation coefficient of tuff for Nb (SZ)	1.0	Conservative assumption of no retardation	**
loguniform	AlluviumMatrixRD_SAV_Nb	Alluvium retardation coefficient for Nb (SZ)	2.0e3, 3.0e4	Triay, et al., 1997; crushed devitrified tuff analog	**
loguniform	FracturePorosity_STFF	Fracture porosity of saturated tuff	1.0e-3, 1.e-2	Best estimate by USFIC Team. For lower bound of 1e-3, see Geldon (1993). For upper bound of 1e-2, see Geldon (1997) citing Streltsova (1988).	**
uniform	AlluviumMatrixPorosity_SAV	Alluvium porosity of Amargosa Valley alluvium	1.0e-1, 1.5e-1	Best estimate by USFIC team. Also see Geldon (1993) and Oatfield and Czarnecki (1989).	**
constant	ImmobileRD_STFF_Am	Retardation factor for Am in matrix during matrix diffusion	1.8e4	Triay, et al., 1997	**

constant	ImmobileRD_STFF_Np	Retardation factor for Np in matrix during matrix diffusion	19.0	Triay, et al., 1997	**
constant	ImmobileRD_STFF_I	Retardation factor for I in matrix during matrix diffusion	1.0	Triay, et al., 1997	**
constant	ImmobileRD_STFF_Tc	Retardation factor for Tc in matrix during matrix diffusion	1.0	Triay, et al., 1997	**
constant	ImmobileRD_STFF_Cl	Retardation factor for Cl in matrix during matrix diffusion	1.0	Triay, et al., 1997	**
constant	ImmobileRD_STFF_Cm	Retardation factor for Cm in matrix during matrix diffusion	1.0	No Data	**
constant	ImmobileRD_STFF_U	Retardation factor for U in matrix during matrix diffusion	37.0	Triay, et al., 1997	**
constant	ImmobileRD_STFF_Pu	Retardation factor for Pu in matrix during matrix diffusion	1.8e3	Triay, et al., 1997	**
constant	ImmobileRD_STFF_Th	Retardation factor for Th in matrix during matrix diffusion	1.8e4	Triay, et al., 1997	**
constant	ImmobileRD_STFF_Ra	Retardation factor for Ra in matrix during matrix diffusion	5.4e3	Triay, et al., 1997	**
constant	ImmobileRD_STFF_Pb	Retardation factor for Pb in matrix during matrix diffusion	5.4e3	Triay, et al., 1997	**
constant	ImmobileRD_STFF_Cs	Retardation factor for Cs in matrix during matrix diffusion	9.0e3	Triay, et al., 1997	**
constant	ImmobileRD_STFF_Ni	Retardation factor for Ni in matrix during matrix diffusion	1.8e3	Triay, et al., 1997	**
constant	ImmobileRD_STFF_C	Retardation factor for C in matrix during matrix diffusion	1.0	Triay, et al., 1997	**
constant	ImmobileRD_STFF_Se	Retardation factor for Se in matrix during matrix diffusion	55.0	Triay, et al., 1997	**

constant	ImmobileRD_STFF_Nb	Retardation factor for Nb in matrix during matrix diffusion	1.8e4	Triay, et al., 1997	**
constant	ImmobilePorosity_STFF	Matrix porosity in tuff	0.01	CNWRA best estimate	**
constant	DiffusionRate_STFF	Diffusion rate in tuff. This parameter is set to a constant value of 0.0 for the base case in order to bypass matrix diffusion calculations.	0.0	CNWRA best estimate	**

NUREG–1668

constant	DistanceToCriticalGroup[km][should_be_5_or_20]	Distance to receptor group (km)	20.0	Farming is assumed to occur at a distance 20 km or greater at YM.	**
uniform	WellPumpingRateAtCriticalGroup5km[gal./day]	Well pumping rate for residential receptor group located for less than 20 km from YM (gal./day)	1.35e4, 2.64e5	Based on an assumed population of 100 to 880 (Fedors and Wittmeyer, 1998) and a range of per capita water use from 150 gpd for Tucson, AZ, to 300 gpd for Las Vegas, NV, water use (van der Leeden, et al., 1990; Table 5-16, P. 319).	**
uniform	WellPumpingRateAtCriticalGroup20km[gal./day]	Well pumping rate for farming receptor group located greater than 20 km from YM (gal./day)	4.5e6, 1.3e7	Equivalent number of quarter-section center-pivot irrigation plots under cultivation with alfalfa. From 1989 LANDSAT image (Wittmeyer, et al., 1996). Low value is based on actual irrigated area of 13 quarter-section plots. High value is based on evidence of as many as 27 quarter-section plots 126 acres per quarter-section center-pivot irrigation plot. Low value is based on consumptive water use for alfalfa in semiarid climates of 3.1 ft/yr near Los Angeles, CA (Linsley and Franzini, 1979, Table 14-2, P. 377). High value is based on 4.38 ft/yr in Mesa, AZ (van der Leeden, et al., 1990, Table 2-50, P. 99).	**
uniform	PlumeThickness5km[m]	Plume thickness at 5 km (m)	10.0, 100.0	—	**
uniform	AquiferThickness5km[m]	Aquifer thickness at 5 km (m)	300.0, 700.0	Luckey, et al. (1996), Table 2, P. 19. Computed by subtracting the altitude of the base of the lower volcanic aquifer from the altitude of the water level in the well.	**
uniform	MixingZoneThickness20km[m]	Mixing zone thickness at 20 km (m)	50.0, 200.0	Based on data on well screen depths obtained from the USGS GWSI database (Wittmeyer, et al., 1995), the distance from the water table to the bottom of the first screened section for wells in the Amargosa Farms area ranges from approximately 40 to 245 m.	**

Distribution	Parameter	Description	Value	Comment/Reference	
finite exponential	TimeOfNextFaultingEventinRegionOfInterest[yr]	Time of next faulting event (years from present)	100.0, 10000.0, 2.0e-5	Based on PSHA data, USGS (1996)	**
user distribution	ThresholdDisplacementforFaultDisruptionOfWP[m]	Threshold fault displacement for disruption (m). Data input order: number of fault displacement values to be provided followed by equiprobable displacement values	4, 0.1, 0.2, 0.3, 0.4	Estimated values because no direct analyses of WP stability in fault zones exist. The 0.2 m value is the mean value of the original distribution and is consistent with the proposed fault zone displacement.	**
uniform	XLocationOfFaultingEventInRegionOfInterest[m]	X location of center of faulting event in region of interest (m)	547400.0, 548600.0	Within repository	**
uniform	YLocationOfFaultingEventInRegionOfInterest[m]	Y location of center of faulting event in region of interest (m)	4076200.0, 4079040.0	Within repository	**
constant	ProbabilityForNWOrientationOfFaults	Probability for NW orientation of faults	0.05	Based on map of Scott and Bonk (1984)	**
uniform	RNtoDetermineFaultOrientation	A random number to determine fault orientation	0.0, 1.0	—	**
constant	NWFaultStrikeOrientationMeasuredfromNorthClockwise[degrees]	NW strike orientation measured from N clockwise (degrees)	-32.5	Based on map of Scott and Bonk (1984)	**
constant	NEFaultStrikeOrientationMeasuredfromNorthClockwise[degrees]	NE strike orientation measured from N clockwise (degrees)	10.0	Based on map of Scott and Bonk (1984)	**
constant	NWFaultTraceLength[m]	NW fault trace length (m)	4000.0	Based on maps of Scott and Bonk (1984) and Simonds, et al. (1995)	**
constant	NEFaultTraceLength[m]	NE fault trace length (m)	4000.0	Based on maps of Scott and Bonk (1984) and Simonds, et al. (1995)	**
beta	NWFaultZoneWidth[m]	NW fault zone width (m)	0.5, 275.0, 1.25, 15.0	Based on observational data from ESF, Ghosh, et al. (1997), pp. 2-5; Stirewalt, et al. (1995)	**

beta	NEFaultZoneWidth[m]	NE fault zone width (m)	0.5, 365.0, 1.25, 15.0	Based on observational data from ESF, Ghosh, et al. (1997), pp. 2-5; Stirewalt, et al. (1995)	**
constant	NWAmountOfLargestCredibleDisplacement[m]	NW largest credible displacement (m)	0.5	U.S. Geological Survey (1996), Chapters 4.3-4.13. Median value of measured paleoseismic ruptures	**
constant	NEAmountOfLargestCredibleDisplacement[m]	NE largest credible displacement (m)	0.5	U.S. Geological Survey (1996), Chapters 4.3-4.13. Median value of measured paleoseismic ruptures	**
constant	NWCumulativeDisplacementRate[mm/yr]	NW cumulative displacement rate (m/yr)	0.00005	Based on Electric Power Research Institute (1993)	**
constant	NECumulativeDisplacementRate[mm/yr]	NE cumulative displacement rate (m/yr)	0.00005	Based on Electric Power Research Institute (1993)	**

NUREG-1668

	Parameter	Description	Value	Comment	
finite exponential	TimeOfNextVolcanicEventin RegionOfInterest[yr]	Time of next volcanic event (yr)	100.0, 10000.0, 1.0e-7	Technical basis for probability of volcanic disruption of the proposed repository site is explained in the Connor, et al. (1997) issue resolution status report on probability of igneous events. Additional methodologies developed in Hill, et al. (1996), pp. 2-6 to 2-16, and Connor and Hill (1995), inclusive.	**
constant	XLocationInRegionOfInterest [m]	X-coordinate of the center of the dike	548000.0	Represents the x-coordinate of the center of the repository to ensure that the volcano will strike the repository when the volcanism scenario flag is turned on.	**
constant	YLocationInRegionOfInterest [m]	Y-coordinate of the center of the dike	4078000.0	Represents the y-coordinate of the center of the repository to ensure that the volcano will strike the repository when the volcanism scenario flag is turned on.	**
uniform	RNtoDetermineIfExtrusiveOr IntrusiveVolcanicEvent	Random numbers to determine volcanic event type	0.0, 1.0	This parameter was developed for previous versions of TPA but not used in TPA 3.1.4. If iflag = 1 for volcanic disruption, there is no reason to assume that a random fraction of igneous events is not volcanic but instead is wholly intrusive. This parameter currently is maintained as a place holder for future versions of TPA that will evaluate effects associated with intrusive igneous activity.	**
constant	FractionOfTimeVolcanicEven tIsExtrusive	Fraction of extrusive volcanic events	0.999	User-specified value to investigate simplistic effects of igneous events that are intrusive rather than extrusive. Results in failure of WPs based solely on area directly intersected by a dike that is constructed using subsequent uniform parameter distributions. This parameter was developed for previous versions of TPA and currently is maintained as a placeholder for future versions of TPA that will evaluate effects associated with intrusive igneous activity.	—
uniform	AngleOfVolcanicDikeMeasur edFromNorthClockwise[degr ees]	Dike angle (degree)	0.0, 15.0	The orientation of planar, vertical igneous intrusions (i.e., dikes) at shallow crustal levels is controlled by the 3-D distribution of crustal stress (Anderson, 1938; Delaney, et al., 1986). In addition, pre-existing crustal structures, such as faults, may serve as conduits for ascending magma at shallow crustal levels (Young, et al., 1994; Jolly and Sanderson, 1997). Although the current direction of maximum horizontal compressive stress at YM is oriented 28° from north, faults at the proposed repository site are oriented between 0° and 15° from north (Morris, et al., 1996, p. 277). This orientation serves as the most likely orientation direction for future igneous intrusions at the proposed repository site.	**

uniform	LengthOfVolcanicDike[m]	Volcanic dike length (m)	2000.0, 11000.0	Few data are available on dike lengths for the Yucca Mountain region (YMR). Connor, et al. (1997, pp. 4-29–4-34) summarize volcanic alignment data that are consistent with subsurface dike lengths between 2 km and 11 km for 5 Ma and younger YMR volcanoes. A detailed geologic map by Gartner and Delaney (1988) of the San Rafael volcanic field, Utah, provides length information for more than 200 basaltic dikes that were originally emplaced 0.5–1.5 km from the surface (Delaney and Gartner, 1997). The length range of 2–8 km accounts for the range observed by Gartner and Delaney (1988) and YMR vent alignment lengths (Connor, et al., 1997).	**
uniform	WidthOfVolcanicDike[m]	Volcanic dike width (m)	1.0, 10.0	Few data are available on dike widths for YMR. Staff have observed that dikes between 0.5 and 10 m currently are exposed at 3.8–12 Ma YMR basaltic volcanic and intrusive centers. A detailed geologic map by Gartner and Delaney (1988) of the San Rafael volcanic field, Utah, provides width information for more than 200 basaltic dikes that were originally emplaced 0.5–1.5 km from the surface (Delaney and Gartner, 1997, pp 1180-1185). The width range of 1–10 m accounts for the range observed by Delaney and Gartner (1988) and YMR dike widths.	**
uniform	DiameterOfVolcanicCone[m]	Cone diameter (m)	24.6, 77.9	Hill (1996, inclusive) provides the technical basis to conclude that subsurface conduits for the youngest YMR volcanoes likely widened from initial diameters of meters to tens of meters in diameter during late eruption stages. Lithologic and structural information constrains this widening to 49±7 m for an analog volcano site at Tolbachik, Russia. Insufficient deposits remain at YMR volcanoes to measure this parameter directly. The number of WPs disrupted, however, depends on the WP dimensions, WP spacing, and centerline spacing between emplacement drifts. Average repository waste loading values may significantly underestimate the number of WPs disrupted by a volcanic conduit. For example, using an average waste loading of 83 MTU/acre and a 9.76 MTU WP loading, a 50-m diameter conduit will only disrupt 4.1 WPs. Geometric relationships show that disruption of 1 to 10 WPs presents a reasonably conservative range for volcanic conduits that may vary between 1 and 50 m in diameter. This number of WPs corresponds to effective conduit diameters of 24.6–77.9 m using an average waste loading of 83 MTU/acre and a 9.76 MTU WP loading.	**

constant	DensityOfAirAtSTP[g/cm3]	Air density at standard temperature and pressure (g/cm³)	0.00129	Weast (1976), p. F-11	**
constant	ViscosityOfAirAtSTP[g/cm-s]	Air viscosity (g/cm/s)	0.00018	Weast (1976), pp. F-13 to F-16	**
constant	ConstantRelatingFallTimeToEddyDiffusivity[cm2/s5/2]	Constant relating eddy diffusion to particle fall time (cm²/s$^{5/2}$)	400.0	Suzuki (1983), p. 99	**
constant	MaximumParticleDiameterForParticleTransport[cm]	Maximum particle diameter for transport (cm)	10.0	Eruptions that are many orders of magnitude larger than the largest <10 Ma YMR basaltic eruption rarely transport particles >10 cm in diameter through convective dispersal (Pyle, 1989, pp. 7-11). Grain-size analyses at YMR analog deposits have average maximum particle diameters of <4 cm at distance 1–10 km from the vent (Hill and Connor, 1995, p. 149), which decreases to <1 cm at distances > 10 km from the vent (e.g., Hill, et al., 1997). Incompletely preserved YMR fall deposits exposed 0.5–3 km from the vent have maximum particle diameters <4 cm (Hill, et al., 1995, pp. 7-14 to 7-15).	**
constant	MinimumFuelParticulateSize[cm]	Minimum fuel particle diameter (cm)	0.0001	SF has initial grain-size average diameters on the order of 100s of μm (e.g., U.S. Department of Energy, 1988). Crush-impact studies (Nuclear Regulatory Commission, 1988) indicate an average diameter of crushed IILW on the order of 100 μm. The thermal, mechanical, and chemical loads of a basaltic volcanic eruption clearly exceed those imposed by a falling ceiling panel. Basaltic eruption processes will thus fragment HLW further than crush-impact processes, supporting the basic assumption of a 10-μm median diameter for HLW transported by a basaltic volcanic eruption. Minimum diameter for a log-triangular distribution is assumed to be 1 log unit below the median diameter, based on grain-size distributions in U.S. Department of Energy (1988) and Nuclear Regulatory Commission (1988).	**

constant	ModeFuelParticulateSize[cm]	Median fuel particle diameter (cm)	0.001	SF has initial grain-size average diameters on the order of 100s of μm (e.g., U.S. Department of Energy, 1988). Crush-impact studies (Nuclear Regulatory Commission, 1988) indicate an average diameter of crushed HLW on the order of 100 μm. The thermal, mechanical, and chemical loads of a basaltic volcanic eruption clearly exceed those imposed by a falling ceiling panel. Basaltic eruption processes will thus likely fragment HLW further than crush-impact processes, supporting the basic assumption of a 10-μm median diameter for HLW transported by a basaltic volcanic eruption.	**
constant	MaximumFuelParticulateSize[cm]	Maximum fuel particle diameter (cm)	0.01	SF has initial grain-size average diameters on the order of hundreds of microns (e.g., U.S. Department of Energy, 1988). Crush-impact studies (Nuclear Regulatory Commission, 1988) indicate an average diameter of crushed HLW on the order of 100 μm. The thermal, mechanical, and chemical loads of a basaltic volcanic eruption clearly exceed those imposed by a falling ceiling panel. Basaltic eruption processes will thus likely fragment HLW further than crush-impact processes, supporting the basic assumption of a 10-μm median diameter for HLW transported by a basaltic volcanic eruption. Maximum diameter for a log-triangular distribution is assumed to be 1 log unit above the median diameter, based on grain-size distributions in U.S. Department of Energy (1988) and Nuclear Regulatory Commission (1988).	**
constant	MinimumAshDensityForVariationWithSize[g/cm3]	Minimum ash density (g/cm³)	1.2	Basaltic tephra ranges in density from 2600 kg/m³ when non-vesiculated at 1100 °C, to about 1200 kg/m³, when vesiculated at 25 °C (e.g., Hill, et al., 1997, p. 10).	**
constant	MaximumAshDensityForVariationWithSize[g/cm3]	Maximum ash density (g/cm³)	2.0	Non-vesiculated YMR-type basalt has a density of around 2700 kg/m³ at 25 °C (e.g., Hill, et al., 1997, p. 10). The minimum amount of vesiculation (i.e., void spaces caused by volatile exsolution during depressurization) typically found in basaltic tephra is around 25% (e.g., Walker and Croasdale, 1972), which reduces the particle density to about 2,000 kg/m³.	**

constant	MinimumAshLogdiameterForDensityVariation	Minimum value of logarithm of ash diameter in cm	-2.0	The density of basaltic scoria is strongly controlled by average grain size. Basaltic scoria break along vesicle (i.e., bubble) walls during transport fragmentation, which results in a net loss of vesicle void-space and concomitant increase in bulk density. For example, Walker, et al. (1984) measured a 40-percent increase in scoria density as average grain size decreased from about 1 cm to about 1 mm. Jarzemba and LaPlante (1996, pp. 6–10) concluded that a range of 1 log unit in average grain size would effectively encompass the range in size-dependent density variations commonly observed in basaltic tephra-fall deposits (e.g., Hill, et al., 1997, p. 10).	**
constant	MaximumAshLogdiameterForDensityVariation	Maximum value of logarithm of ash diameter	-1.0	The density of basaltic scoria is strongly controlled by average grain size. Basaltic scoria break along vesicle (i.e., bubble) walls during transport fragmentation, which results in a net loss of vesicle void-space and concomitant increase in bulk density. For example, Walker, et al. (1984) measured a 40 percent increase in scoria density as average grain-size decreased from about 1 cm to about 1 mm. Jarzemba and LaPlante (1996, pp. 6–10) concluded that a range of 1 log unit in average grain size would effectively encompass the range in size-dependent density variations commonly observed in basaltic tephra-fall deposits (e.g., Hill, et al., 1997, p. 10).	**
constant	ParticleShapeParameter	Ash particle shape parameter	0.5	The shape of irregular particles with axial lengths a, b, and c can be described as [b+c]/2a, where a is greater than b or c (Suzuki, 1983, p. 100). Hill, et al. (1996, pp. 2-23) concluded shape parameters between 0.5 and 0.25 describe irregularly shaped basaltic tephra and that variations within this range do not affect transport properties in the ASHPLUME module.	**
constant	IncorporationRatio	SF incorporation ratio	0.3	Assumed value of 0.3 in Jarzemba and LaPlante (1996, pp. 9–12). For critical group locations 20 km from the proposed repository site, Hill and Trapp (1997, pp. 5–6) found that a uniform distribution of U[0.1, 1.0] for this parameter did not affect dose significantly for the default HLW grain-size characteristics used in the primary input file.	**

constant	WindDirection[degrees]	Wind direction relative to due east in the counter-clockwise direction (degrees)	−90.0	Setting wind direction to 90° clockwise from east (i.e., south) ensures that the wind blows to the receptor group during each run. This provides a conservative basis for evaluating the effects of volcanic eruptions on nuclear facilities (International Atomic Energy Agency, 1997, pp. 19–20). In addition, Hill and Trapp (1997, p. 6) calculated that ground-surface dose is within one order-of-magnitude of peak dose when the wind blows within a 50° sector centered about a main dispersion axis of 90°. Surface topography strongly controls wind direction and speed measured at surface weather stations to altitudes of at least 100 m (U.S. Department of Energy, 1997, pp. 3.63–3.71). Wind direction data are not available for altitudes of interest (2 km to about 6 km) above YM (U.S. Department of Energy, 1997, inclusive). The nearest wind direction data to YM and the receptor group location are from Desert Rock airstrip, located about 50 km SE of YM. For altitudes of about 2 and 4 km (700 and 500 mbar pressure level), wind direction above Desert Rock is within the 50° sector centered on south 18–15% of the time.	**
exponential	WindSpeed[cm/s]	Wind speed (cm/s)	0.00083	Wind speed data are not available for altitudes of interest (2 km to about 6 km) above YM (U.S. Department of Energy, 1988; U.S. Department of Energy, 1997, inclusive). The nearest wind speed data to YM and the receptor group location are from Desert Rock airstrip, located about 50 km SE of YM. For altitudes of about 2 km (700 mbar pressure level), wind speed above Desert Rock within the NNW–NNE sector averages 6 m/s (U.S. Department of Energy, 1997, pp. 3-63–3-71). For altitudes of about 4 km (500 mbar pressure level), wind speed above Desert Rock within the NNW–NNE sector averages 12 m/s (U.S. Department of Energy, 1997, pp. 3-63–3-71). Basaltic eruption column associated with YMR-type eruptions likely reached above-ground altitudes of at least 6 km (e.g., Jarzemba, 1997). A mean wind speed for 4-km altitudes thus presents a reasonable mean value for 2–6 km high eruption columns.	**
loguniform	VolcanicEventDuration[s]	Duration of volcanic event (s)	6.13e4, 7.24e6	Baca and Jarzemba (1997), pp. 4-3 to 4-5	**

NUREG-1668

					**
loguniform	VolcanicEventPower[W]	Event power (W)	2.57e9, 3.55e11	Baca and Jarzemba (1997), pp. 4-3 to 4-5	**
constant	VolcanicColumnConstantBeta	Shape parameter for the volcanic column	10.0	Hill and Trapp (1997, p. 10) concluded that this parameter was a relatively insensitive parameter and that setting this parameter to a constant value of 10 was sufficient for modeling YMR-type basaltic eruptions using ASHPLUME. This result was confirmed by sensitivity studies conducted by Hill, et al. (1997) on the 1995 Cerro Negro eruption deposits.	**
logtriangular	AshMeanParticleLogDiameter[d_in_cm]	Logarithm of the mean ash particle size in cm	0.01, 0.1, 1.0	Few data are available to characterize the mean particle diameter for an entire basaltic fall deposit. Hill, et al. (1997, pp. 4–5) calculated a mean diameter of 0.7 mm for the 1-mm and thicker fall deposits from the 1995 Cerro Negro eruption. Although this eruption was about an order of magnitude smaller than the smallest YMR eruptions, it is likely that overall grain-size characteristics remain fairly scale-independent for these types of basaltic eruptions (e.g., Walker and Croasdale, 1972).	**
constant	AshParticleSizeDistributionStandardDeviation	Ash particle size distribution standard deviation	1.0	Hill, et al. (1996, pp. 2-21-2-23) showed that this parameter does not demonstrably affect tephra-deposit thicknesses modeled using the ASHPLUME code. Hill and Trapp (1997) recommended setting this parameter to a constant value of 1.	**

constant	RelativeRateOfBlanketRemoval[1/yr]	Bulk removal rate from blanket (1/yr)	0.0001	Only traces of the tephra-fall deposit are preserved at the youngest YMR volcano, Lathrop Wells, which is around 100,000 years old. Recent work by Delgado-Granados, et al. (1998) at Xitle volcano, Mexico, has shown that substantial amounts of basaltic tephra-fall deposits are preserved from this 2,000-year-old eruption. The area around Xitle volcano receives around 500–700 mm of annual rainfall, in contrast to the roughly 100-mm annual rainfall in the YMR. Although many factors, such as topographic slope, degree of incision, and eolian sedimentation, affect tephra-fall erosion, areas of greater annual rainfall should experience higher erosion rates of nonconsolidated fall deposits than areas of lower annual rainfall. Thus, tephra-fall deposits in the YMR likely are significantly preserved for around 10,000 yrs.	**
constant	FractionOfPrecipitationLostToEvapotranspiration	Fraction of precipitation lost to evapotranspiration	0.68	Jarzemba and Manteufel (1997), p. 924. Based on the converse of the fraction of precipitation that can reach the deep soil	**
constant	FractionOfIrrigationLostToEvapotranspiration	Fraction of irrigation water lost to evapotranspiration	0.5	Jarzemba and Manteufel (1997), p. 924	**
constant	AnnualPrecipitation[m/yr]	Annual precipitation rate (m/yr)	0.085	Wilson, et al. (1994)	**
constant	AnnualIrrigation[m/yr]	Annual irrigation rate (m/yr)	1.52	Nuclear Regulatory Commission (1995) pp. 7-10	**
constant	FractionOfYearSoilIsSaturatedDueToPrecipitation	Fraction of year blanket saturated due to precipitation	0.0054	Jarzemba and Manteufel (1997), p. 924	**
constant	FractionOfYearSoilIsSaturatedDueToIrrigation	Fraction of year blanket saturated due to irrigation	0.2	LaPlante, et al. (1995), Table 2.1, pp. 2-6, based on average growing time of crops from Chambers and May (1994)	**
constant	AshBulkDensity[g/cm3]	Bulk density for volcanic ash (g/cm3)	1.4	Based on data in Fisher and Schmincke (1984), Table 5-6, p. 118 and Fig. 5-28, p. 120	**

Type	Name	Description	Value	Reference	
constant	AshVolumetricMoistureFractionAtSaturation	Volumetric fraction of moisture in volcanic ash at saturation	0.4	Calculated from bulk density of ash	**
constant	DepthOfTheRootingZone[m]	Depth of rooting zone	0.15	Napier, et al. (1988), p. 4.58	**
constant	KdOfUraniumInVolcanicAsh[cm3/g]	U distribution coefficient in volcanic ash (cm³/g)	35.0	Sheppard and Thibault (1990), Table 1, p. 472	**
constant	KdOfCuriumInVolcanicAsh[cm3/g]	Cm distribution coefficient in volcanic ash (cm³/g)	4000.0	Sheppard and Thibault (1990), Table 1, p. 472	**
constant	KdOfPlutoniumInVolcanicAsh[cm3/g]	Pu distribution coefficient in volcanic ash (cm³/g)	550.0	Sheppard and Thibault (1990), Table 1, p. 472	**
constant	KdOfAmericiumInVolcanicAsh[cm3/g]	Am distribution coefficient in volcanic ash (cm³/g)	1900.0	Sheppard and Thibault (1990), Table 1, p. 472	**
constant	KdOfThoriumInVolcanicAsh[cm3/g]	Th distribution coefficient in volcanic ash (cm³/g)	3200.0	Sheppard and Thibault (1990), Table 1, p. 472	**
constant	KdOfRadiumInVolcanicAsh[cm3/g]	Ra distribution coefficient in volcanic ash (cm³/g)	500.0	Sheppard and Thibault (1990), Table 1, p. 472	**
constant	KdOfLeadInVolcanicAsh[cm3/g]	Pb distribution coefficient in volcanic ash (cm³/g)	270.0	Sheppard and Thibault (1990), Table 1, p. 472	**
constant	KdOfProtactiniumInVolcanicAsh[cm3/g]	Pa distribution coefficient in volcanic ash (cm³/g)	550.0	Sheppard and Thibault (1990), Table 1, p. 472	**
constant	KdOfActiniumInVolcanicAsh[cm3/g]	Ac distribution coefficient in volcanic ash (cm³/g)	450.0	Sheppard and Thibault (1990), Table 1, p. 472	**

constant	KdOfNeptuniumInVolcanic Ash[cm3/g]	Np distribution coefficient in volcanic ash (cm³/g)	5.0	Sheppard and Thibault (1990), Table 1, p. 472	**
constant	KdOfSamariumInVolcanicA sh[cm3/g]	Sm distribution coefficient in volcanic ash (cm³/g)	245.0	Sheppard and Thibault (1990), Table 1, p. 472	**
constant	KdOfCesiumInVolcanicAsh[cm3/g]	Cs distribution coefficient in volcanic ash (cm³/g)	280.0	Sheppard and Thibault (1990), Table 1, p. 472	**
constant	KdOfIodineInVolcanicAsh[c m3/g]	I distribution coefficient in volcanic ash (cm³/g)	1.0	Sheppard and Thibault (1990), Table 1, p. 472	**
constant	KdOfTinInVolcanicAsh[cm3 /g]	Sn distribution coefficient in volcanic ash (cm³/g)	130.0	Sheppard and Thibault (1990), Table 1, p. 472	**
constant	KdOfSilverInVolcanicAsh[c m3/g]	Ag distribution coefficient in volcanic ash (cm³/g)	55.0	Sheppard and Thibault (1990), Table 1, p. 472	**
constant	KdOfPaladiumInVolcanicAs h[cm3/g]	Pd distribution coefficient in volcanic ash (cm³/g)	55.0	Sheppard and Thibault (1990), Table 1, p. 472	**
constant	KdOfTechnetiumInVolcanic Ash[cm3/g]	Tc distribution coefficient in volcanic ash (cm³/g)	0.1	Sheppard and Thibault (1990), Table 1, p. 472	**
constant	KdOfMolybdenumInVolcani cAsh[cm3/g]	Mo distribution coefficient in volcanic ash (cm³/g)	10.0	Sheppard and Thibault (1990), Table 1, p. 472	**
constant	KdOfNiobiumInVolcanicAsh [cm3/g]	Nb distribution coefficient in volcanic ash (cm³/g)	160.0	Sheppard and Thibault (1990), Table 1, p. 472	**
constant	KdOfZirconiumInVolcanicA sh[cm3/g]	Zr distribution coefficient in volcanic ash (cm³/g)	600.0	Sheppard and Thibault (1990), Table 1, p. 472	**

constant	KdOfStrontiumInVolcanicAsh[cm3/g]	Sr distribution coefficient in volcanic ash (cm³/g)	15.0	Sheppard and Thibault (1990), Table 1, p. 472	**
constant	KdOfSeleniumInVolcanicAsh[cm3/g]	Se distribution coefficient in volcanic ash (cm³/g)	150.0	Sheppard and Thibault (1990), Table 1, p. 472	**
constant	KdOfNickelInVolcanicAsh[cm3/g]	Ni distribution coefficient in volcanic ash (cm³/g)	400.0	Sheppard and Thibault (1990), Table 1, p. 472	**
constant	KdOfChlorineInVolcanicAsh[cm3/g]	Cl distribution coefficient in volcanic ash (cm³/g)	0.0	Sheppard and Thibault (1990), Table 1, p. 472	**
constant	KdOfCarbonInVolcanicAsh[cm3/g]	C distribution coefficient in volcanic ash (cm³/g)	5.0	Sheppard and Thibault (1990), Table 1, p. 472	**
constant	SolubilityOfUraniumInVolcanicAsh[moles/liter]	U solubility (mol/L)	4.5e-5	Wilson, et al. (1993), Table 9-2b, p. 9-7	**
constant	SolubilityOfCuriumInVolcanicAsh[moles/liter]	Cm solubility (mol/L)	1.0e-6	Kerrisk (1985), Table 9, p. 14	**
constant	SolubilityOfPlutoniumInVolcanicAsh[moles/liter]	Pu solubility (mol/L)	5.0e-6	Wilson, et al. (1993), Table 9-2b, p. 9-7; Kerrisk (1985), Table 9. p. 14	**
constant	SolubilityOfAmericiumInVolcanicAsh[moles/liter]	Am solubility (mol/L)	1.0e-6	Kerrisk (1985), Table 9, p. 14	**
constant	SolubilityOfThoriumInVolcanicAsh[moles/liter]	Th solubility (mol/L)	3.2e-9	Wilson, et al. (1993), Table 9-2b, p. 9-7	**
constant	SolubilityOfRadiumInVolcanicAsh[moles/liter]	Ra solubility (mol/L)	1.0e-7	Wilson, et al. (1993), Table 9-2b, p. 9-7	**
constant	SolubilityOfLeadInVolcanicAsh[moles/liter]	Pb solubility (mol/L)	3.2e-7	Wilson, et al. (1993), Table 9-2b, p. 9-7	**
constant	SolubilityOfProtactiniumInVolcanicAsh[moles/liter]	Pa solubility (mol/L)	3.2e-8	Wilson, et al. (1993), Table 9-2b, p. 9-7	**

constant	SolubilityOfActiniumInVolcanicAsh[moles/liter]	Ac solubility (mol/L)	1.0e-6	Wilson, et al. (1993), Table 9-2b, p. 9-7	**
constant	SolubilityOfNeptuniumInVolcanicAsh[moles/liter]	Np solubility (mol/L)	1.0e-4	Wilson, et al. (1993), Table 9-2b, p. 9-7	**
constant	SolubilityOfSamariumInVolcanicAsh[moles/liter]	Sm solubility (mol/L)	5.0e-6	Kerrisk (1985), Table 9, p. 14	**
constant	SolubilityOfCesiumInVolcanicAsh[moles/liter]	Cs solubility (mol/L)	1.0	Wilson, et al. (1993), Table 9-2b, p. 9-7; Kerrisk (1985), Table 9, p. 14	**
constant	SolubilityOfIodineInVolcanicAsh[moles/liter]	I solubility (mol/L)	1.0	Wilson, et al. (1993), p. 9-9; Kerrisk (1985), Table 9, p. 14	**
constant	SolubilityOfTinInVolcanicAsh[moles/liter]	Sn solubility (mol/L)	5.0e-8	Wilson, et al. (1993), Table 9-2b, p. 9-7; Kerrisk (1985), Table 9, p. 14	**
constant	SolubilityOfSilverInVolcanicAsh[moles/liter]	Ag solubility (mol/L)	1.0	Conservative estimation due to lack of data	**
constant	SolubilityOfPaladiumInVolcanicAsh[moles/liter]	Pd solubility (mol/L)	9.5e-4	Wilson, et al. (1993), Table 9-2b, p. 9-7; Kerrisk (1985), Table 9, p. 14	**
constant	SolubilityOfTechnetiumInVolcanicAsh[moles/liter]	Tc solubility (mol/L)	1.0	Wilson, et al. (1993), p. 9-9; Kerrisk (1985), Table, p. 14	**
constant	SolubilityOfMolybdenumInVolcanicAsh[moles/liter]	Mo solubility (mol/L)	1.0	Conservative estimation due to lack of data	**
constant	SolubilityOfNiobiumInVolcanicAsh[moles/liter]	Nb solubility (mol/L)	1.0e-8	Wilson, et al. (1993), Table 9-2b, p. 9-7	**
constant	SolubilityOfZirconiumInVolcanicAsh[moles/liter]	Zr solubility (mol/L)	3.2e-10	Wilson, et al. (1993), Table 9-2b, p. 9-7	**
constant	SolubilityOfStrontiumInVolcanicAsh[moles/liter]	Sr solubility (mol/L)	1.3e-4	Wilson, et al. (1993), Table 9-2b, p. 9-7	**
constant	SolubilityOfSeleniumInVolcanicAsh[moles/liter]	Se solubility (mol/L)	1.0	Wilson, et al. (1993), p. 9-9	**
constant	SolubilityOfNickelInVolcanicAsh[moles/liter]	Ni solubility (mol/L)	2.0e-3	Wilson, et al. (1993), Table 9-2b, p. 9-7	**

constant	SolubilityOfChlorineInVolcanicAsh[moles/liter]	Cl solubility (mol/L)	1.0	Wilson, et al. (1993), p. 9-9	**
constant	SolubilityOfCarbonInVolcanicAsh[moles/liter]	C solubility (mol/L)	1.0	Wilson, et al. (1993), p. 9-9; Kerrisk (1985), Table 9, p. 14	**

NUREG–1668

constant	DistanceCutoffForDoseConversionDualityInDCAGS[km]	Cutoff point between a farming and a residential group	19.99	Jarzemba and Weldy (1997), p. 8	**
loguniform	AirborneMassLoadForVolcanismDoseCalculation[g/m3]	Mass of soil in the air	1.0e-4, 1.0e-2	Best estimate based on data from Sehmel (1977) and Tegen and Fung (1994)	**
constant	OccupancyFactorForVolcanismDoseCalculation[-]	Fraction of time that receptor is exposed to airborne volcanic ash	0.24	Based on 40 hours per week	**
constant	DepthOfResuspendableLayer[cm]	Thickness of the upper soil layer that is available for resuspension (cm)	0.3	---	**

**   ***>>> correlated parameters <<<***

correlateinputs	SubareaWetFraction	ArealAverageMeanAnnualInfiltrationAtStart[mm/yr]	0.631	—	**
correlateinputs	SubareaWetFraction	MatrixPermeability_TSw_[m2]	-0.623	—	**
correlateinputs	FowFactor	ArealAverageMeanAnnualInfiltrationAtStart[mm/yr]	-0.224	—	**
correlateinputs	FowFactor	MatrixPermeability_TSw_[m2]	0.13	—	**
correlateinputs	FowFactor	SubareaWetFraction	-0.366	—	**

**

endoffile

APPENDIX A REFERENCES

Alvarez, M.G., and J.R. Galvele. 1984. The mechanism of pitting of high-purity iron in NaCl solutions. *Corrosion Science* 24: 27–8.

American Society for Testing and Materials. 1995. *Pressure Vessel Plates, Carbon Steel, for Moderate- and Lower-Temperature Service.* A516–0. West Conshohocken, PA: American Society for Testing and Materials.

Anderson, K. 1938. The dynamics of sheet intrusion. *Proceedings of the Royal Society of Edinburgh* 58: 242–251.

Andersson, K. 1988. *SKI Project-90: Chemical Data.* SKI TR 91:21. Stockholm, Sweden: Swedish Nuclear Power Inspectorate (SKI).

Baca, R.G., and M.S. Jarzemba, eds. 1997. *Detailed Review of Selected Aspects of Total System Performance Assessment — 1995.* San Antonio, TX: Center for Nuclear Waste Regulatory Analyses.

Belle, J., ed. 1961. *Uranium Dioxide: Properties and Nuclear Applications.* Washington, DC: U.S. Atomic Energy Commission.

Bockris, J., and A. Reddy, 1970. *Modern Electrochemistry.* New York: Plenum Press: 2.

Bodvarsson, G.S., and T.M. Bandurraga. 1996. *Development and Calibration of the Three-Dimensional Site-Scale Unsaturated Zone Model of Yucca Mountain, Nevada.* LBNL–9315. Berkeley, CA: Berkeley National Laboratory.

Brechtel, C.E., M. Lin, E. Martin, and D.S. Kessel. 1995. *Geotechnical Characterization of the North Ramp of the Exploratory Studies Facility.* SAND 95–0488/1 UC-814. Volume I: Data Summary. Albuquerque, NM: Sandia National Laboratories.

Calvo, E.J. 1979. *Study of the Electroreduction Reaction of Oxygen on Passive Metals in Different Aqueous Media* (in Spanish). Ph.D. dissertation. La Plata, Argentina: Universidad Nacional de La Plata.

Calvo, E.J., and D.J. Schriffrin. 1988. The electrochemical reduction of oxygen on passive iron in alkaline solutions. *Journal of Electroanalytical chemistry* 243: 171–185.

Chambers, D., and L. Mays. 1994. *The American Garden Guides: Vegetable Gardening.* New York: Pantheon Books.

Civilian Radioactive Waste Management System Management and Operating Contractor. 1994. *Initial Summary Report for Repository/Waste Package Advance Conceptual Design*. B00000000–01717–57–5–00015. Rev. 01. Las Vegas, NV: Civilian Radioactive Waste Management System.

Civilian Radioactive Waste Management System Management and Operating Contractor. 1996. *Mined Geologic Disposal System Advanced Conceptual Design Report*. B00000000–0717–5705–0027, Rev 00. Las Vegas, NV: Civilian Radioactive Waste Management System.

Connor, C.B., and B.E. Hill. 1995. Three nonhomogeneous Poisson models for the probability of basaltic volcanism: Application to the Yucca Mountain region. *Journal of Geophysical Research* 100: 10,107–10,125.

Connor, C.B., et al. 1997. Magnetic surveys help reassess volcanic hazards at Yucca Mountain. EOS, *Transactions of the American Geophysical Union* 74(7): 73–78.

Cragnolino, G.A., Dunn, D.S., P. Angell, Y.-M. Pang, and N. Sridhar. 1998. Factors influencing the performance of carbon steel overpacks in the proposed high-level nuclear waste repository. *Proceedings of the CORROSION 98 Conference*. Paper No. 147. Houston, TX: NACE International.

Daniels, W.R. 1981. *Laboratory and Field Studies Related to Radionuclide Migration Project*. LA–9192–PR. Albuquerque, NM: Los Alamos National Laboratory.

Delaney, P.T., and A.E. Gartner. 1997. Physical processes of shallow mafic dike emplacement near the San Rafael Swell, Utah. *Geological Society of America Bulletin* 109: 1177–1192.

Delaney, P.T., D.D. Pollard, J.I. Ziony, and E.H. McKee. 1986. Field relations between dikes and joints: Emplacement processes and paleostress analysis. *Journal of Geophysical Research* 91(B5): 4920–4938.

Delgado-Granados, H., et al. 1998. Geology of Xitle volcano (southern Mexico City): I. Stratigraphy and age. *Colima Volcano Sixth International Meeting*. Colima, Mexico: University of Colima: 114.

de Marsily, G. 1986. *Quantitative Hydrogeology*. Orlando, FL: Academic Press, Inc.

DeWispelare, A.R., L.T. Herren, M.P. Miklas, and R.T. Clemen. 1993. *Expert Elicitation of Future Climate in the Yucca Mountain Vicinity*. CNWRA 93–016. San Antonio, TX: Center for Nuclear Waste Regulatory Analyses.

Doering, T.W. 1995. *Dimensions of Waste Packages Barriers*. Civilian Radioactive Waste Management System Management and Operating Interoffice Correspondence IOC.LV.WP.TWD.5/95.182, May 26. Las Vegas, NV: Civilian Radioactive Waste Management System Management and Operations.

Einziger, R.E., L.E. Thomas, H.C. Buchanan, and R.B. Stout. 1992. Oxidation of spent fuel in air at 175 to 195 °C. *Journal of Nuclear Materials* 190: 53–60.

Electrical Power Research Institute. 1993. *Earthquakes and Tectonics Expert Judgement Elicitation Project.* Geomatrix Project 3055-13. EPRI TR-102000. San Francisco, CA: Geomatrix Consultants, Inc.

Fedors, R.W., and G.W. Wittmeyer. 1998. *Initial Assessment of Dilution Effects Induced by Water Well Pumping in the Amargosa Farms Area.* San Antonio, TX: Center for Nuclear Waste Regulatory Analyses.

Fisher, R.V., and H.-U. Schmincke. 1984. *Pyroclastic Rocks.* New York: Springer-Verlag.

Flint, L. 1996. *Matrix Properties of Hydrogeologic Units at Yucca Mountain, Nevada.* Draft. Denver, CO: U.S. Geological Survey.

Fuger, J. 1992. Thermodynamic properties of actinide aqueous species relevant to geochemical processes. *Radiochimica Acta* 58/59: 81–92.

Galvele, J.R. 1978. Present State of Understanding of the breakdown of passivity and repassivation. *Proceedings of the Fourth International Symposium on Passivity.* R.P. Frakenthal and J. Kruger, ed. Princeton, NJ: The Electrochemical Society, Inc.

Gartner, A.E., and P.T. Delaney. 1988. *Geologic Map Showing a Late Cenozoic Basaltic Intrusive Complex, Emery, Sevier, and Wayne Counties, Utah.* Miscellaneous Field Studies Map MF-2052. Reston, VA: U.S. Geological Survey.

Gauthier, J. 1993. *Expert Elicitation of the Solubility Distributions to be Used in TSPA#2 Calculations.* Draft Report. June 1.

Geldon, A.L. 1993. *Preliminary Hydrogeologic Assessment of Boreholes UE–25c#1, UE–25c#2, and UE–25c#3, Yucca Mountain, Nye County, Nevada.* Water Resources Investigations Report 92–4016. Denver, CO: U.S. Geological Survey.

Geldon, A.L., et al. 1997. *Results of Hydraulic and Conservative Tracer Tests in Miocene Tuffaceous Rocks at the C-Hole Complex, 1995 to 1997, Yucca Mountain, Nevada.* Milestone Report SP23PM3. Las Vegas, NV: U.S. Geological Survey.

Ghosh, A., R.D. Manteufel, and G.L. Stirewalt. 1997. *FAULTING Version 1.0—A Code for Simulation of Direct Fault Disruption: Technical Description and User's Guide.* CNWRA 97–002. San Antonio, TX: Center for Nuclear Waste Regulatory Analyses.

Guenther, R.J., et al. 1991. *Characterization of Spent Fuel Approved Testing Material–ATM–104.* PNL–5109-104. Richland, WA: Pacific Northwest Laboratories.

Heusler, K.E. 1976. Influence of temperature and pressure on the kinetics of electrode processes. *High Temperature, High Pressure Electrochemistry in Aqueous Systems.* D.deG. Jones and R.W. Staehle, eds. Houston, TX: National Association of Corrosion Engineers: 387–399.

Hill, B.E. 1996. *Constraints on the Potential Subsurface Area of Disruption Associated with Yucca Mountain Region Basaltic Volcanoes.* San Antonio, TX: Center for Nuclear Waste Regulatory Analyses.

Hill, B.E., and C.B. Connor. 1995. Field volcanism. *NRC High-Level Radioactive Waste Research at CNWRA, July-December 1994.* CNWRA 94–02S. San Antonio, TX: Center for Nuclear Waste Regulatory Analyses: 141–154.

Hill, B.E., and J.S. Trapp. 1997. *Sensitivity Analysis for Key Parameters in the VOLCANO and ASHPLUME Modules of the TPA 3.1 Code.* San Antonio, TX: Center for Nuclear Waste Regulatory Analyses.

Hill, B.E., C.B. Connor, and J.S. Trapp. 1996. Igneous activity. *NRC High-Level Radioactive Waste Program Annual Progress Report, Fiscal Year 1996.* CNWRA 96–01A. San Antonio, TX: Center for Nuclear Waste Regulatory Analyses: 2-1 to 2-32.

Hill, B.E., et al. 1997. *1995 Eruptions of Cerro Negro Volcano, Nicaragua, and Risk Assessment for Future Eruptions.* San Antonio, TX: Center for Nuclear Waste Regulatory Analyses.

Hoek, E., and E.T. Brown. 1982. *Underground Excavations in Rock.* London, England: Institution of Mining and Metallurgy.

Incropera, F.P., and D.P. DeWitt, 1995. *Fundamentals of Heat and Mass Transfer.* 3rd ed, New York: John Wiley & Sons.

International Atomic Energy Agency. 1997. *Volcanoes and Associated Topics in Relation to Nuclear Power Plant Siting.* IAEA–TECDOC–VOLCANO8.WPW. Vienna, Austria: International Atomic Energy Agency.

Jarzemba, M.S. 1997. Stochastic radionuclide distributions after a basaltic eruption for performance assessments of Yucca Mountain. *Nuclear Technology* 118(2): 132–141.

Jarzemba, M.S., and P.A. LaPlante. 1996. *Preliminary Calculations of Expected Dose from Extrusive Volcanic Events at Yucca Mountain.* San Antonio, TX: Center for Nuclear Waste Regulatory Analyses.

Jarzemba, M.S., and R.D. Manteufel. 1997. An analytically based model for the simultaneous leaching-chain decay of radionuclides from contaminated ground surface soil layers. *Health Physics* 73(6): 919–927.

Jarzemba, M.S., and J.R. Weldy. 1997. *A Summary of Information Relevant to Defining Critical Groups and Reference Biospheres.* San Antonio, TX: Center for Nuclear Waste Regulatory Analyses.

Jolly, R.J.H., and D.L. Sanderson. 1977. A Mohr circle construction for the opening of a fracture. *Journal of Structural Geology* 19: 887–892.

Kennedy, W.E., and D.L. Strenge. 1992. *Residual Radioactive Contamination from Decommissioning: Technical Basis for Translating Contamination Levels to Annual Total Effective Dose Equivalent.* NUREG/CR–5512. Vol. 1. Washington, DC: Nuclear Regulatory Commission.

Kerrisk, J.F. 1984. *Solubility Limits on Radionuclide Dissolution at a Yucca Mountain Repository.* LA–9995–MS. NNA.870519.0049. Los Alamos, NM: Los Alamos National Laboratory.

Kerrisk, J.F. 1985. *An Assessment of the Important Radionuclides in Nuclear Waste.* LA–10414–MS. Los Alamos, NM: Los Alamos National Laboratory.

Klavetter, E.A., and R.R. Peters. 1986. *Estimation of Hydrologic Properties of an Unsaturated Fractured Rock Mass.* SAND 84–2642. Albuquerque, NM: Sandia National Laboratories.

Lallemand-Barres, A., and P. Peaudcerf. 1978. Recherche des relations entre les valeurs mesurees de la dispersivite macroscopique d'un millieu aquifere, ses autres caracteristiques et les conditions de mesure. Etude bibilographique. *Bull. Bur. Rech. Geol. Min.* Ser. 2, Sec. III, 4-1978, 277–284.

LaPlante, P.A., S.J. Maheras, and M.S. Jarzemba. 1995. *Initial Analysis of Selected Site-Specific Dose Assessment Parameters and Exposure Pathways Applicable to a Groundwater Release Scenario at Yucca Mountain.* CNWRA 95–018. San Antonio, TX: Center for Nuclear Waste Regulatory Analyses.

Linsley, R.K., and J.R. Franzini. 1979. Water-Resources Engineering. New York: McGraw-Hill Book Company.

Lobnig, R.E., H.P. Schmidt, K. Hennesen, and H.J. Grabke. 1992. Diffusion of cations in Chromia layers grown on iron-base alloys. *Oxidation of Metals* 37: 81–93.

Luckey, R.R., et al. 1996. *Status of Understanding of the Saturated-Zone Ground-Water Flow System at Yucca Mountain, Nevada, as of 1995.* WRI–96–4077. Denver, CO: U.S. Geological Survey.

Manteufel, R.D. 1997. Effects on ventilation and backfill on a mined waste disposal facility. *Nuclear Engineering and Design.* 172: 205–219.

Manteufel, R.D., and N.E. Todreas. 1994. Effective thermal conductivity and edge conductance model for a spent fuel assembly. *Nuclear Technology* 105(3): 421–440.

Marsh, G. and K. Taylor. 1988. An assessment of carbon steel containers for radioactive waste disposal. *Corrosion Science* 28: 289–320.

Meijer, A. 1990. *Yucca Mountain Project Far-Field Sorption Studies and Data Needs.* LA–11671–MS. Los Alamos, NM: Los Alamos National Laboratory.

Mohanty S., et. al. 1996. *Engineered Barrier System Performance Assessment Code: EBSPAC Version 1.0 β, Technical Description and User's Manual.* CNWRA 96–001. San Antonio, TX: Center for Nuclear Waste Regulatory Analyses.

Mohanty, S., et al. 1997. *Engineered Barrier System Performance Assessment Code: EBSPAC Version 1.1 β, Technical Description and User's Manual.* CNWRA 97–006. San Antonio, TX: Center for Nuclear Waste Regulatory Analyses.

Morris, A.P., D.A. Ferrill, and D.B. Henderson. 1996. Slip tendency analysis and fault reactivation. *Geology* 24: 275–278.

Murphy, W.M., E.C. Pearcy, and D.A. Pickett. 1997. *Natural Analog Studies at Peña Blanca and Santorini.* In von Maravic, N., and J. Smellie, eds. Seventh EC Natural Analogue Working Group meeting. EUR 17851 EN European Commission: 105–112.

Napier, B.A., R.A. Peloquin, D.L. Strenge, and J.V. Ramsdell. 1988. *GENII: The Hanford Environmental Radiation Dosimetry Software System, Volumes 1, 2, and 3: Conceptual Representation, User's Manual, Code Maintenance Manual.* PNL–6584. Richland, WA: Pacific Northwest Laboratory.

Nitsche, H., et al. 1993. *Measured Solubilities and Speciations of Neptunium, Plutonium, and Americium in a Typical Groundwater (J-13) from the Yucca Mountain Region.* Milestone Report 3010–WBS 1.2.3.4.1.3.1. LA–12562–MS UC-802. Los Alamos, NM: Los Alamos National Laboratory.

Nuclear Regulatory Commission. 1988. *Nuclear Fuel Cycle Facility Accident Analysis Handbook.* NUREG–1320. Washington, DC: U.S. Nuclear Regulatory Commission.

Nuclear Regulatory Commission. 1995. *NRC Iterative Performance Assessment Phase 2: Development of Capabilities for Review of a Performance Assessment for a High-Level Waste Repository.* NUREG–1464. Washington, DC: Nuclear Regulatory Commission.

Nuclear Regulatory Commission. 1997. *Issue Resolution Status Report on Methods to Evaluate Climatic Change and Associated Effects at Yucca Mountain.* Washington, DC: Nuclear Regulatory Commission.

Oatfield, W.J., and J.B. Czarnecki. 1991. Hydrogeologic inferences from drillers' logs and from gravity and resistivity surveys in the Amargosa Desert, southern Nevada. *Journal of Hydrology* 124(1-2): 131–158.

Östhols, E., J. Bruno, and I. Grenthe. 1994. On the influence of carbonate on mineral dissolution: III. The solubility of microcrystalline ThO_2 in CO_2-H_2O) media. *Geochimica et Cosmochimica Acta* 58: 613–623.

Park, U. 1992. *Regulatory Overview and Recommendations on a Repository's Release of C-14.* San Diego, CA: Science Applications International Corporation.

Pei-Lin, T., M.D. Siegel, C.D. Updegraf, K.K. Wahi, and R.V. Guzowski. 1985. *Repository Site Data Report for Unsaturated Tuff, Yucca Mountain, Nevada.* NUREG/CR–4410. Washington, DC: Nuclear Regulatory Commission.

Popov, E.P. 1970. *Mechanics of Materials.* New York: Prentice-Hall.

Price, L., B.A. Zimmerman, N.E. Olague, S.H. Conrad, and C.T. Harlan. 1992. *Preliminary Performance Assessment of the Greater Confinement Disposal Facility of the Nevada Test.* SAND 91–0047. Albuquerque, NM: Sandia National Laboratories.

Pyle, D.M. 1989. The thickness, volume and grain size of tephra fall deposits. *Bulletin of Volcanology* 51: 1–15.

Rai, D., A.R. Felmy, D.A. Moore, and M.J. Mason. 1995. The solubility of Th(IV) and U(IV) hydrous oxides in concentrated $NaHCO_3$ and Na_2CO_3 solutions. in T. Murakami and R.C. Ewing, eds. *Scientific Basis for Nuclear Waste Management XVIII, Materials Research Society Symposium Proceedings, Volume 353.* Materials Research Society: Pittsburgh, PA: 1143–1150.

Schenker, A.R., D.C. Guerin, T.H. Robey, C.A. Rautman, and R.W. Bernard, 1995. *Stochastic Hydrogeologic Units and Hydrogeologic Properties Development for Total System Performance Assessments.* SAND 94–0244. Albuquerque, NM: Sandia National Laboratories.

Scott, R.B., and J. Bonk. 1984. *Preliminary Geologic Map (1:12,000 scale) of Yucca Mountain, Nye County, Nevada, with Geologic Cross Sections.* U.S. Geological Survey Open-File Report 84–494. Denver, CO: U.S. Geological Survey.

Scully, J.R., and H.P. Hack. 1984. Galvanic corrosion prediction using long- and short-term polarization curves. *CORROSION 84.* Paper No. 34. Houston, TX: National Association of Corrosion Engineers.

Sehmel, G.A. 1977. *Radionuclide Particle Resuspension Research Experiments on the Hanford Reservation.* BNWL–2081. Richland, WA: Pacific Northwest Laboratories.

Sheppard, M.I., and D.H. Thibault. 1990. Default soil solid/liquid partition coefficients, Kds, for four major soil types: A compendium. *Health Physics* 59: 471–482.

Siegel, M.D., et al. 1993. Preliminary characterization of materials for reactive transport model validation experiment. *Proceedings of the Fourth International Conference on High-Level Radioactive Waste Management.* La Grange Park, IL: American Nuclear Society: 348–358.

Simonds, W.F., et al. 1995. *Map of Fault Activity of the Yucca Mountain Area, Nye County, Nevada.* U.S. Geological Survey Miscellaneous Investigations Series Map, 1–2520. Denver, CO: U.S. Geological Survey.

Smith, H., and Baldwin, D. 1989. An investigation of thermal release of C-14 from PWR spent fuel cladding. American Nuclear Society *(FOCUS '89); Proceedings of the Topical Meeting on Nuclear Waste Isolation in the Unsaturated Zone,* September 17–21, 1989. Las Vegas, NV: 46–49.

Sridhar, N., G.A. Cragnolino, and D.S. Dunn. 1995. Experimental Investigations of Failure Processes of High-Level Radioactive Waste Container Materials. CNWRA 95-010. San Antonio, TX: Center for Nuclear Waste Regulatory Analyses.

Stirewalt, G.L., and D.B. Henderson. 1995. *A Preliminary Three-Dimensional Geological Framework Model for Yucca Mountain, Nevada, with Hydrologic Application: Report to Accompany 1995 Model Transfer to the Nuclear Regulatory Commission.* CNWRA 94-023 Rev. 1. San Antonio, TX: Center for Nuclear Waste Regulatory Analyses.

Stirewalt, G.L., S.M. McDuffie, R.D. Manteufel, and R.W. Janetzke. 1995. *Technical Specifications for a Fault Displacement Module.* Report to Nuclear Regulatory Commission. San Antonio, TX: Center for Nuclear Waste Regulatory Analyses.

Stockman, C. 1997. Discussion in the presentation of the near-field environment in the total system performance assessment: waste form degradation expert elicitation. *Workshop of significant Issues and Available Data: Waste Form Degradation and Radionuclide Mobilization Expert Elicitation (WFEE) Project San Francisco, CA.* Albuquerque, NM: Sandia National Laboratories.

Streltsova, T.D. 1988. *Well Testing in Heterogeneous Formations.* Exxon Monograph: 1-413. New York: John Wiley and Sons.

Suzuki, T. 1983. *A Theoretical Model for Dispersion of Tephra. Arc Volcanism: Physics and Tectonics.* Tokyo, Japan: Terra Scientific Publishing: 95-113.

Tegen, I., and I. Fung. 1994. Modeling of mineral dust in the atmosphere: sources, transport, and optical thickness. *Journal of Geophysical Research* 99(11): 22,897-22,914.

Triay, I.R., et al. 1996a. *Summary and Synthesis Report on Radionuclide Retardation for the Yucca Mountain Site Characterization Project.* Milestone 3784. Los Alamos, NM: Chemical Science and Technology Division, Los Alamos National Laboratory.

Triay, I.R., et al. 1996b. *Batch Sorption Results for Neptunium Transport Through Yucca Mountain Tuffs. Yucca Mountain Site Characterization Program, Milestone 3349.* LA-12961-MS/UC-814. Los Alamos, NM: Los Alamos National Laboratory.

Triay, I.R., et al. 1997. *Summary and Synthesis Report on Radionuclide Retardation for the Yucca Mountain Site Characterization Project. Yucca Mountain Site Characterization Program Milestone 3784M.* Draft. Los Alamos, NM: Chemical Science and Technology Division, Los Alamos National Laboratory.

TRW Environmental Safety Systems, Inc. 1995. *Total System Performance Assessment — 1995: An Evaluation of the Potential Yucca Mountain Repository.* B00000000-01717-2200-00136. Rev. 01. Las Vegas, NV: TRW Environmental Safety Systems, Inc.

Turnbull A., and M.K. Gardner. 1982. Electrochemical polarization studies of BS 4360 50D steel in 3.5% NaCl. *Corrosion Science* 22(7): 661-673.

U.S. Department of Energy. 1987. *Characteristics of Spent Fuel, High Level Waste, and Other Radioactive Wastes Which May Require Long-Term Isolation.* DOE/RW-0184. Washington, DC: U.S. Department of Energy, Office of Civilian Radioactive Waste Management.

U.S. Department of Energy. 1988. *Site Characterization Plan: Yucca Mountain Site, Nevada Research and Development Area, Nevada.* DOE/RW–0199. Washington, DC: U.S. Department of Energy, Office of Civilian Radioactive Waste Management.

U.S. Department of Energy. 1990. *Yucca Mountain Project Reference Information Database.* YMP/CC–0002. Version 04.002. Las Vegas, NV: U.S. Department of Energy.

U.S. Department of Energy. 1993. *Yucca Mountain Site Characterization Project Reference Information Base.* YMP/93–02. Rev. 3. Las Vegas, NV: U.S. Department of Energy.

U.S. Geological Survey. 1996. *Seismotectonic Framework and Characterization of Faulting at Yucca Mountain, Nevada.* J.W. Whitney, Report Coordinator. Denver, CO: U.S. Geological Survey.

U.S. Department of Energy. 1997. *Regional and Local Wind Patterns Near Yucca Mountain.* B00000000–01717–5705–00081. Washington, DC: U.S. Department of Energy, Office of Civilian Radioactive Waste Management.

van der Leeden, F., F.L. Troise, and D.K. Todd. 1990. *The Water Encyclopedia.* Chelsea, MI: Lewis Publishers.

Wanner, H., and I. Forest, eds. 1992. *Chemical Thermodynamics of Uranium.* New York: Elsevier.

Walker, G.P.L., and R. Croasdale. 1972. Characteristics of some basaltic pyroclastics. *Bulletin of Volcanology* 35: 303–317.

Walker, G.P.L., S. Self, and L. Wilson. 1984. Tarawera 1886, New Zealand - A basaltic plinian fissure eruption. *Journal of Volcanology and Geothermal Research* 21: 61–78.

Weast, R.C. 1976. *CRC Handbook of Chemistry and Physics.* Cleveland, OH: CRC Press.

Wilson, C. 1990. *Results from NNWSI Series 3 Spent Fuel Dissolution Tests.* PNL–7170. Richland, WA: Pacific Northwest Laboratories.

Wilson, M.L., et al. 1994. *Total-System Performance Assessment for Yucca Mountain-SNL Second Iteration (TSPA-93).* SAND 93–2675, Vols. 1 and 2. Albuquerque, NM: Sandia National Laboratories.

Wittmeyer, G.W., R. Klar, G. Rice, and W. Murphy. 1995. *The CNWRA Regional Hydrogeology Geographic Information System Database.* CNWRA 95–009. San Antonio, TX: Center for Nuclear Waste Regulatory Analyses.

Wittmeyer, G.W., M.P. Miklas, R.V. Klar, D. Williams, and D. Balin. 1996. *Use of Groundwater in the Arid and Semi-Arid Western United States: Implications for Yucca Mountain Area.* San Antonio, TX: Center for Nuclear Waste Regulatory Analyses.

Wolfsberg, K. 1978. Sorption-desorption studies of Nevada Test Site alluvium and leaching studies of nuclear test debris. LA–7216–MS. *Society for Metals*: April 1.

Wong, I.G., S.K. Pezzopane, C.M. Menges, and R.C. Quittmeyer. 1995. Probabilistic seismic hazard analysis of the exploratory studies facilities at Yucca Mountain. *Proceedings of the Methods of Seismic Hazards Evaluation Focus '95 September 18–20, 1995*. La Grandge, Park, IL: American Nuclear Society: 44–51.

Young, S.R., H.L. McKague, and R.W. Terhune. *Influence of Faults on Ascent of Mafic Magma by Dike Intrusion*. CNWRA 94–025. San Antonio, TX: Center for Nuclear Waste Regulatory Analyses.

APPENDIX B
EBSREL FLOW FACTOR DERIVATION

The factors F_{mult}, F_{wet}, and F_{ow} are used in the Total-system Performance Assessment (TPA) code to describe how waste packages (WPs) and the waste form interact with infiltrating water. The factors F_{mult} and F_{ow} could have been combined into a single factor for use in the TPA Version 3.1.4 code; however, these factors were kept separate because F_{ow} is strongly correlated with the parameters $<I>$ and $<K_{sat}>$, whereas F_{mult} is considered to be an independent random variable. F_{ow} is also correlated to F_{wet}, which is the fraction of WPs that could be intercepted by dripping water (less than 1). The specified initial infiltration rate at Yucca Mountain (YM), $<I>$, and the specified matrix saturated hydraulic conductivity for the TS$_W$ layer at YM, $<K_{sat}>$, are used in the derivation of these flow factor values as described below.

Derivation of Estimates for F_{wet} and F_{ow}

F_{wet} is the probability of a WP getting a dripping flux greater than zero; [i.e. P(q_{drip}) >0]. The TPA code requires the assumption that the number of wetted WPs is invariant, and does not consider changing dripping locations. Some data (Wilder, 1996) indicate that drip locations in tunnels move over time. Therefore, the number of wetted WPs could be considered to spread the dripping eventually over all WPs, producing a lower average flow rate to each.

Wilder (1996) proposed that the net infiltration to the repository level flows in either the rock matrix or fractures depending on the local values of infiltration and matrix hydraulic conductivity. Variabilities of the local values for I and K_{sat} within a realization are characterized by their coefficients of variation C_{vi} and C_{vk}. Both I and K_{sat} are assumed to be lognormally distributed within the subarea and correlated with each other.

The dripping model requires that the spatial scale of variability in K_{sat} and I be greater than the size of an individual WP. If not (for example, the flowing fractures are closely spaced), then variability from point to point would, in the limit, cause every WP to always be wetted. This is similar in some respects to the "weeps" model of TSPA93 (Wilson, et al., 1994) that assumes that in the limit of many small seeps, all WPs would be wetted.

If the scale of the dripping phenomenon is larger than the WP scale, then only the interaction of the variability of I and K_{sat} needs to be considered. Correlation between I and K_{sat} may be important if there are definite paths with high conductivity, so infiltration at the drift would be determined by the high conductivity of these paths. Although the present model allows correlation between K_{sat} and I, only the uncorrelated case is considered here.

Assumption about the distribution of model parameters

For the derivation of F_{ow} and F_{wet} the mean infiltration $<I>$ is uniformly distributed from 1 to 20 mm/yr with a coefficient of variation (C_{vi}) range from 0 to 2, and $<K_{sat}>$ is uniformly distributed from 0.1 to 20 mm/yr with a C_{vk} range from 0 to 2. The local parameters K_{sat} and I are described by the mean and standard deviation in most reported data on YM. However, this conceptualization assumes that these local parameters are lognormally distributed; it is, therefore, necessary to convert between the two distributions. As an example using

K_{sat}, given the arithmetic mean and coefficients of variation, the standard deviation in the log-transformed space can be described (Benjamin and Cornell, 1970) as follows:

$$\sigma = \sqrt{\ln(C_{vk}^2 + 1)} \qquad (B\text{-}1)$$

$$\mu = \ln\left(<K_{sat}> \, e^{\frac{-\sigma^2}{2}}\right) \qquad (B\text{-}2)$$

The F_{ow} and F_{wet} factors are generated from Monte Carlo runs, which sample the parameters $<I>$, $<K_{sat}>$, C_{vi}, and C_{vk}. Referring to Figure B-1, consider the repository subarea to be broken into n equal parts (subareas). For each Monte Carlo sample of $<I>$, $<K_{sat}>$, C_{vi}, and C_{vk}, a lognormal distribution is generated for the local values I and K_{sat} for each of the subareas. If ($I - K_{sat}$) is greater than zero, the number of wet areas N_{wet} is incremented, and the sum of the positive values of ($I - K_{sat}$) is collected into q_{wet}, which represents the potential dripping water. For the current Monte Carlo sample of $<I>$, $<K_{sat}>$, C_{vi}, and C_{vk}, the factor F_{wet} is the ratio N_{wet}/N, where N is the total number of WPs. The ratio of $q_{wet}/I/N_{wet}$ averaged over the entire area is F_{ow}. The fraction of wetted cells represents the potential fraction of wetted WPs.

The histograms of F_{wet} and F_{ow}, based on 6000 samples, with 1000 subareas per sample, are plotted in Figures B-2 and B-3. The covariances among I, K_{sat}, F_{wet}, and F_{ow} are presented in Table 3-2.

Determination of the Flow Diversion Factor F_{mult}

The flow diversion factor F_{mult} is defined as the fraction of potentially dripping water that will enter the WP and contribute to the release and transport of radionuclides. Only a fraction of the water intercepting the drifts is expected to come into direct contact with the WPs. Only a portion of the water dripping onto the WPs is expected to get inside where it can interact with the waste form. Factors considered in estimating F_{mult} include:

- Flow to WPs is reduced because of diversion of water around the drift by the capillary barrier. This is estimated as a uniform distribution between 0.8 and 1.0. Water flow into the drift will face a capillary barrier if the fractures are small enough. Water can easily move across the capillary break, and can be diverted in the fracture around the drift opening. If the drift were in a homogenous, unsaturated porous medium, there would be a precise relationship between infiltration rate and flow, with a clear diversion around the opening, and perching of water at the crest of the opening [e.g, Phillip and Watcher (1989); Ross (1990)]. However, most of the flow is expected to occur in fractures, and may be largely episodic in nature. Under these conditions, the flow may appear to be more saturated than unsaturated (Birkholzer, et al., 1997). Flow in large fractures may occur as unstable fingers or rivulets on one face or the other (Glass, 1993), which, for all intents and purposes, is not tractable by mathematical models. The relatively minor importance attached to this factor (0.8 to 1.0 range) reflects the large uncertainties in the phenomena.

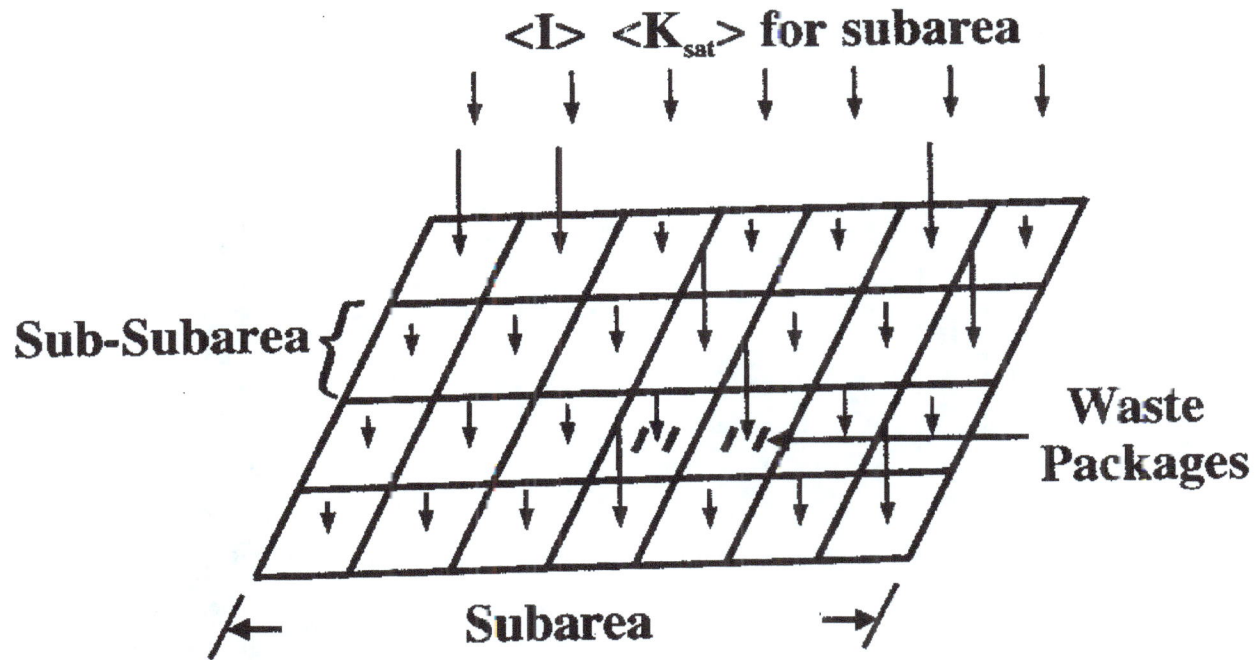

Figure B-1. Conceptual dripping model.

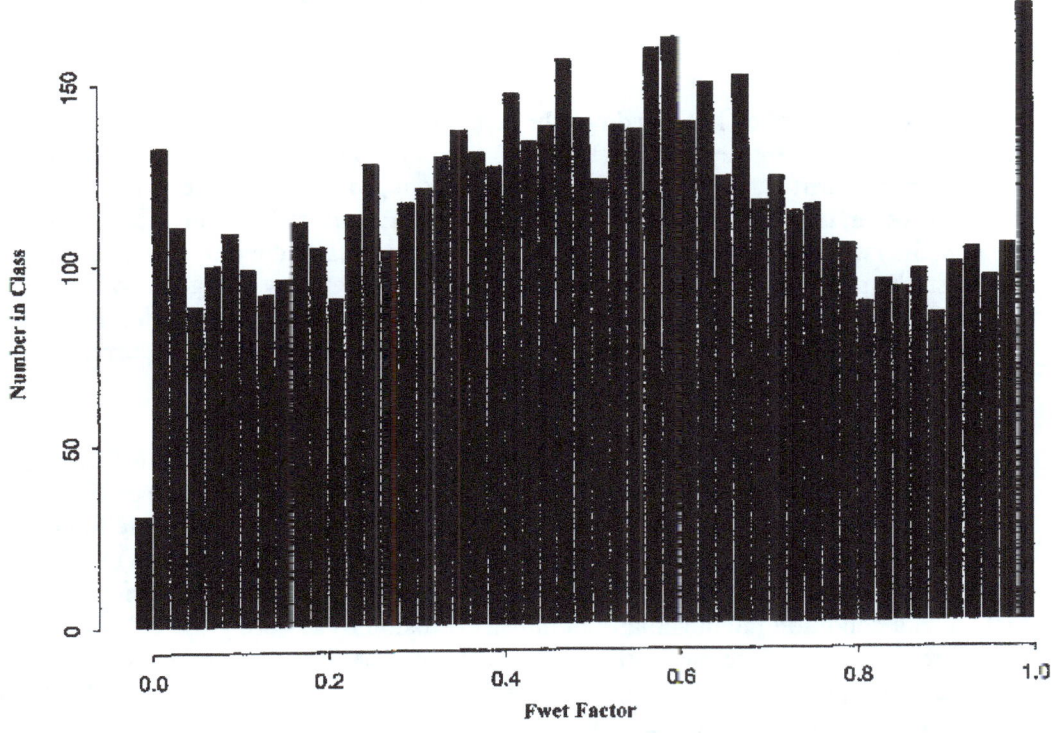

Figure B-2. F_wet histogram.

NUREG–1668

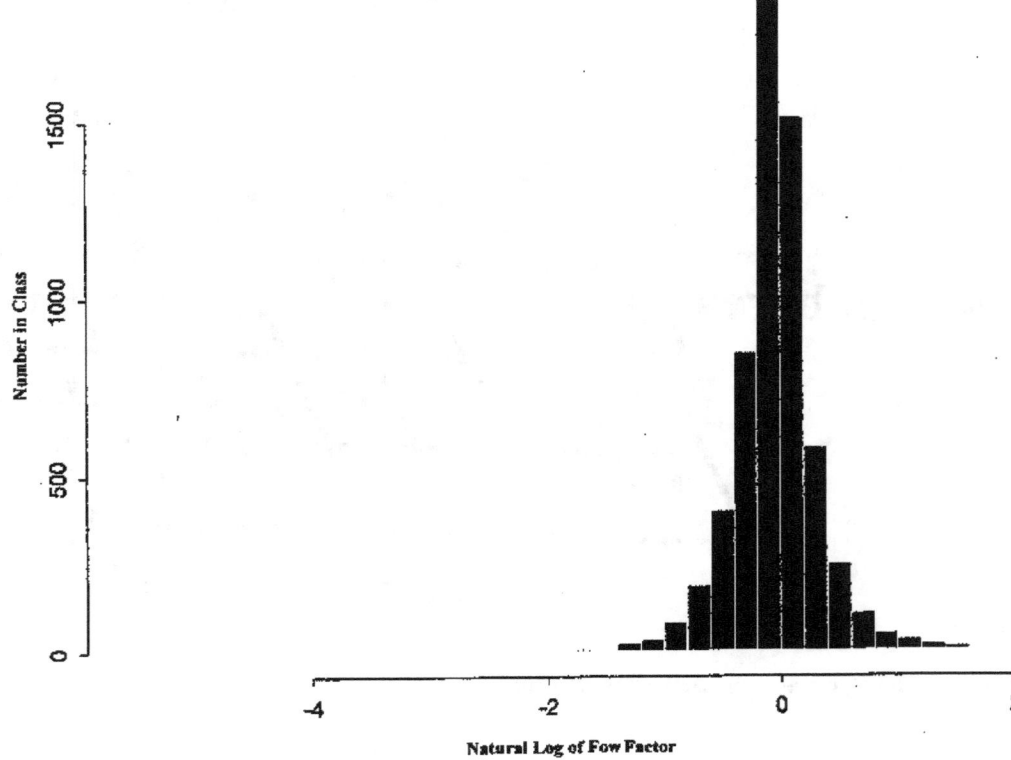

Figure B-3. F$_{ow}$ histogram.

- Water may be diverted by flowing down the drift wall. The fraction not diverted down the drift wall is estimated to be uniformly distributed from 0.5 to 0.8. Water flow managing to break the capillary barrier into the drift can drip from the ceiling or from protuberances along the drift. Much of the water, however, is likely to flow as a sheet along the drift walls rather than drip from the ceiling onto the containers. The closer to the crest of the tunnel, the greater the propensity to drip. Away from the crest, the slope of the tunnel walls diverts water to sheet flow along the walls. The placement of the canisters affect this factor. Some designs have the canisters aligned with the tunnel centerline, whereas other designs have the canisters offset from the centerline. The current design has the canister along the center of the placement drifts.

- Some fraction of dripping water will fall directly onto an open WP hole. This factor is assumed to be uniformly distributed with a range of 0.1 to 0.5. Water dripping from the ceiling would have to fall onto the WP in such a way that it could enter an open hole. If the drop were directly in the path of an open hole (e.g., the hole was caused by the corrosion from dripping water), then this condition would be fulfilled. However, the current corrosion model does not require dripping water, and, in fact, such a condition might actually diminish corrosion by lowering the chloride concentration. Considering that in the present model drip impingement has no direct bearing on the location of the corrosion holes, not all drips falling onto the container surface would fall directly into a hole. As more corrosion holes occur, this condition increases the likelihood of a drop falling into a corrosion hole. Otherwise, a large fraction of the water will run off the WP and not enter a hole. As time goes on, more pits form on a WP and the likelihood that water drips on a hole increases.

• Some fraction of drops may be poised to enter a corrosion hole, but may be unable to do so because of the presence of corrosion products. This factor is estimated as a uniformly distributed variable from 0.2 to 0.5. Products from the corrosion of the steel overpack and inner barrier may be flaky, porous, or gel-like. The density of the corrosion products will be considerably less than the steel itself, and, without a high rate of water flux, there will be no mechanism to dislodge them. If this is the case, then water dripping into the corrosion hole will have difficulty entering the canister and will simply flow off. Holes at the crest will have a higher probability that water will enter because there is a reduced propensity for water to flow off. Holes on the side will have higher probability that the water would flow off rather than enter.

Combining Factors into F_{mult}

The factor, F_{mult} shown in Figure B-4, was formulated by sampling each of the above factors and forming their product. The histogram of the resulting factor is approximately lognormal and is represented in the TPA input by a lognormal function ranging between 0.01 and 0.2. The fraction of WPs getting wetted and the average flow per WP are determined from the above procedure for fixed values of infiltration. Water infiltration at the repository horizon is determined by module NFENV and is time-varying because of changes in climate and thermal reflux. Under conditions of increased infiltration or reflux, it is expected both the wetting fraction and average flow to the WPs will increase. For purposes of computational expediency, the model currently assumes the factors F_{mult}, F_{ow}, and F_{wet} remain constant through the simulation period. Under conditions of increased infiltration or reflux, the wetting model should predict that the number of WPs wetted and the average flow per WP would increase. The current code, however, diverts the additional quantity of flow from increased infiltration or reflux to the initial number of wetted WPs, which partially compensates for this model shortcoming. Future versions of the code might allow for more realistic treatment of the transient nature of WP and fuel wetting. However, increasing the number of WPs wetted as a function of time in the middle of the calculation will make the code more complex and will demand significantly greater run times.

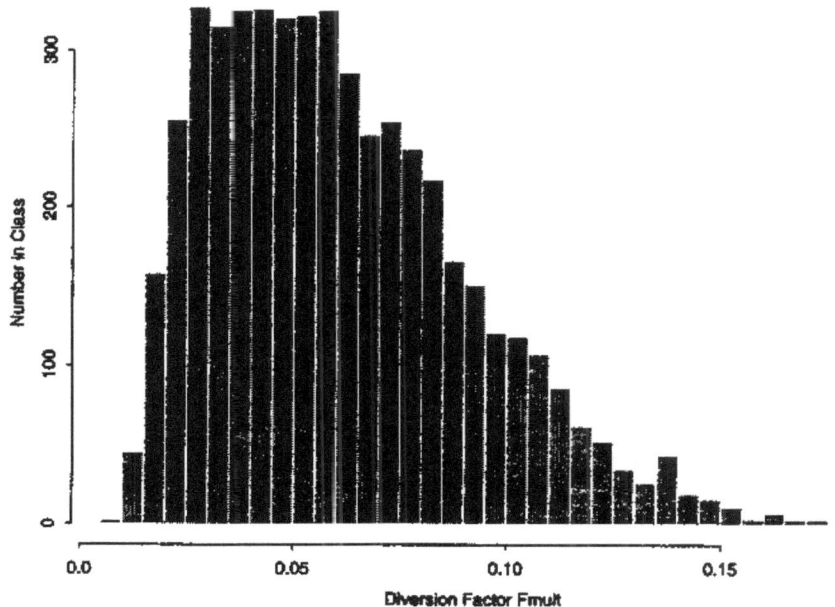

Figure B-4. F_{mult} histrogram

APPENDIX B REFERENCES

Benjamin, J.R., and C.A. Cornell. 1970. *Probability, Statistics, and Decision for Civil Engineers*. New York, NY: McGraw-Hill.

Birkholzer, J., G. Li, C. Tsang, T. Tsang. 1997. Drifts in unsaturated rock at Yucca Mountain. *FTAM Workshop*. December 15–16. Lawrence Berkeley Laboratories.

Glass, R.J. 1993. Modeling gravity-driven fingering in rough-walled fractures using modified percolation theory. *Proceeding of the 4th Annual International Conference of High-Level Radioactive Waste Management*. La Grange Park, IL: American Nuclear Society: 2042–2053.

Phillip, J.R. and R.T. Watcher. 1989. Unsaturated seepage and subterranean holes: conspectus and the exclusion problem for circular cylindrical cavities. *Water Resources Research* 25: 16–18.

Ross, B. 1990. Quasi-linear analysis of water flow in the unsaturated zone at Yucca Mountain, Nevada USA. *Memories of the 22nd Congress IAH, Vol XXII*, Lausanne, Switzerland: 166–173.

Wilder, D.G. 1996. *Near-field and altered zone environmental report, Volume II*. UCRL 124998. Lawrence Livermore National Laboratory.

Wilson, M.L., et al. 1994. *Total-System Performance Assessment for Yucca Mountain-SNL Second Iteration (TSPA-93)*. SAND 93-2675, Vols. 1 and 2. Albuquerque, NM: Sandia National Laboratories.9

NRC FORM 335
(2-89)
NRCM 1102,
3201, 3202

U.S. NUCLEAR REGULATORY COMMISSION

BIBLIOGRAPHIC DATA SHEET

(See instructions on the reverse)

1. REPORT NUMBER
(Assigned by NRC, Add Vol., Supp., Rev., and Addendum Numbers, if any.)

NUREG-1668
Vol. 1

2. TITLE AND SUBTITLE

NRC Sensitivity and Uncertainty Analyses for a Proposed Repository at Yucca Mountain, Nevada, Using TPA 3.1

Conceptual Models and Data

3. DATE REPORT PUBLISHED

MONTH	YEAR
February	2001

4. FIN OR GRANT NUMBER

5. AUTHOR(S)

S. Mohanty*, T. McCartin**

6. TYPE OF REPORT

Technical

7. PERIOD COVERED (Inclusive Dates)

8. PERFORMING ORGANIZATION - NAME AND ADDRESS (If NRC, provide Division, Office or Region, U.S. Nuclear Regulatory Commission, and mailing address; if contractor, provide name and mailing address.)

*Center for Nuclear Waste Regulatory Analyses, 6220 Culebra Road, San Antonio, TX 7828-0510

**Office of Nuclear Material Safety and Safeguards, U.S. Nuclear Regulatory Commission, Washington, DC 20555-0001

9. SPONSORING ORGANIZATION - NAME AND ADDRESS (If NRC, type "Same as above"; if contractor, provide NRC Division, Office or Region, U.S. Nuclear Regulatory Commission, and mailing address.)

Division of Waste Management
Office of Nuclear Materials Safety and Safeguards
U.S. Nuclear Regulatory Commission
Washington, DC 20555-0001

10. SUPPLEMENTARY NOTES

11. ABSTRACT (200 words or less)

The total-system performance assessment computer code (TPA Version 3.1.4) was developed to assist the U.S. Nuclear Regulatory Commission staff in evaluating the safety case for the proposed geologic repository at Yucca Mountain, Nevada. The TPA Version 3.1.4 code is designed to estimate total-system performance measures of annual individual dose or risk. The TPA Version 3.1.4 code consists of an executive driver, a set of consequence modules, and a library of utility modules. The executive driver controls the probabilistic sampling of input parameters, the calculational sequence and data transfers among consequence modules, and the generation of output files. The various output files are used in parameter sensitivity analyses, post-processing of time-dependent risk curves, and synthesis of statistical distributions (e.g., cumulative distribution functions and complementary cumulative distribution functions) for appropriate performance measures. Consequence models simulate physical processes and events such as unsaturated zone infiltration; evolution of the near-field thermal-hydrologic environment; failure of waste packages; dissolution and release of waste; transport of waste in the ground-water system; extraction of ground water; and consumption of ground water which, if contaminated, would expose future populations to radiation. In addition to considering climate change, the TPA Version 3.1.4 code is designed to calculate the effects of disruptive events such as faulting, seismicity, and volcanism. Utility modules ensure the consistency of algorithms and data sets used repeatedly by various consequence modules.

12. KEY WORDS/DESCRIPTORS (List words or phrases that will assist researchers in locating the report.)

Computer code
Dose
Ground water
Performance Assessment
Receptor group
Source term
Yucca Mountain
Disruptive Consequences
Geologic repository
High-level radioactive waste
Probabilistic risk assessment
Scenario
Waste package

13. AVAILABILITY STATEMENT

unlimited

14. SECURITY CLASSIFICATION

(This Page)

unclassified

(This Report)

unclassified

15. NUMBER OF PAGES

16. PRICE

NRC FORM 335 (2-89)

This form was electronically produced by Elite Federal Forms, Inc.

Federal Recycling Program